Advanced Information and Knowledge Processing

Series Editors

Professor Lakhmi Jain
Lakhmi.jain@unisa.edu.au
Professor Xindong Wu
xwu@cs.uvm.edu

Also in this series

Gregoris Mentzas, Dimitris Apostolou, Andreas Abecker and RonYoung
Knowledge Asset Management 1-85233-583-1

Michalis Vazirgiannis, Maria Halkidi and Dimitrios Gunopulos
Uncertainty Handling and Quality Assessment in Data Mining 1-85233-655-2

Asunción Gómez-Pérez, Mariano Fernández-López and Oscar Corcho
Ontological Engineering 1-85233-551-3

Arno Scharl (Ed.)
Environmental Online Communication 1-85233-783-4

Shichao Zhang, Chengqi Zhang and Xindong Wu
Knowledge Discovery in Multiple Databases 1-85233-703-6

Jason T.L. Wang, Mohammed J. Zaki, Hannu T.T. Toivonen and Dennis Shasha (Eds)
Data Mining in Bioinformatics 1-85233-671-4

C.C. Ko, Ben M. Chen and Jianping Chen
Creating Web-based Laboratories 1-85233-837-7

Manuel Graña, Richard Duro, Alicia d'Anjou and Paul P. Wang (Eds)
Information Processing with Evolutionary Algorithms 1-85233-886-0

Colin Fyfe
Hebbian Learning and Negative Feedback Networks 1-85233-883-0

Yun-Heh Chen-Burger and Dave Robertson
Automating Business Modelling 1-85233-835-0

Dirk Husmeier, Richard Dybowski and Stephen Roberts (Eds)
Probabilistic Modeling in Bioinformatics and Medical Informatics 1-85233-778-8

Ajith Abraham, Lakhmi Jain and Robert Goldberg (Eds)
Evolutionary Multiobjective Optimization 1-85233-787-7

K.C. Tan, E.F.Khor and T.H. Lee
Multiobjective Evolutionary Algorithms and Applications 1-85233-836-9

Nikhil R. Pal and Lakhmi Jain (Eds)
Advanced Techniques in Knowledge Discovery and Data Mining 1-85233-867-9

Amit Konar and Lakhmi Jain
Cognitive Engineering 1-85233-975-6

Miroslav Kárný (Ed.)
Optimized Bayesian Dynamic Advising 1-85233-928-4

Yannis Manolopoulos, Alexandros Nanopoulos, Apostolos N. Papadopoulos and Yannis Theodoridis
R-trees: Theory and Applications 1-85233-977-2

Sanghamitra Bandyopadhyay, Ujjwal Maulik, Lawrence B. Holder and Diane J. Cook (Eds)
Advanced Methods for Knowledge Discovery from Complex Data 1-85233-989-6

Marcus A. Maloof (Ed.)
Machine Learning and Data Mining for Computer Security 1-84628-029-X

Sifeng Liu and Yi Lin
Grey Information 1-85233-995-0

Vasile Palade, Cosmin Danut Bocaniala and Lakhmi Jain (Eds)
Computational Intelligence in Fault Diagnosis 1-84628-343-4

Mitra Basu and Tin Kam Ho (Eds)
Data Complexity in Pattern Recognition 1-84628-171-7

Samuel Pierre (Ed.)
E-learning Networked Environments and Architectures 1-84628-351-5

Arno Scharl and KlausTochtermann (Eds)
The Geospatial Web 1-84628-826-5

Ngoc Thanh Nguyen
Advanced Methods for Inconsistent Knowledge Management 1-84628-888-3

Francesco Camastra and Alessandro Vinciarelli
Machine Learning for Image, Video and Audio Analysis 978-1-84800-006-3

Amnon Meisels
Search by Constrained Agents
978-1-84800-039-1

Mikhail Prokopenko (Ed.)
Advances in Applied Self-organizing Systems
978-1-84628-981-1

András Kornai

Mathematical Linguistics

András Kornai
MetaCarta Inc.
350 Massachusetts Ave.
Cambridge, MA 02139
USA

ISBN: 978-1-84996-694-8 e-ISBN: 978-1-84628-986-6
DOI: 10.1007/978-1-84628-986-6

British Library Cataloguing in Publication Data
A catalogue record for this book is available from the British Library

Printed on acid-free paper

9 8 7 6 5 4 3 2 1

Springer Science+Business Media
Springer.com

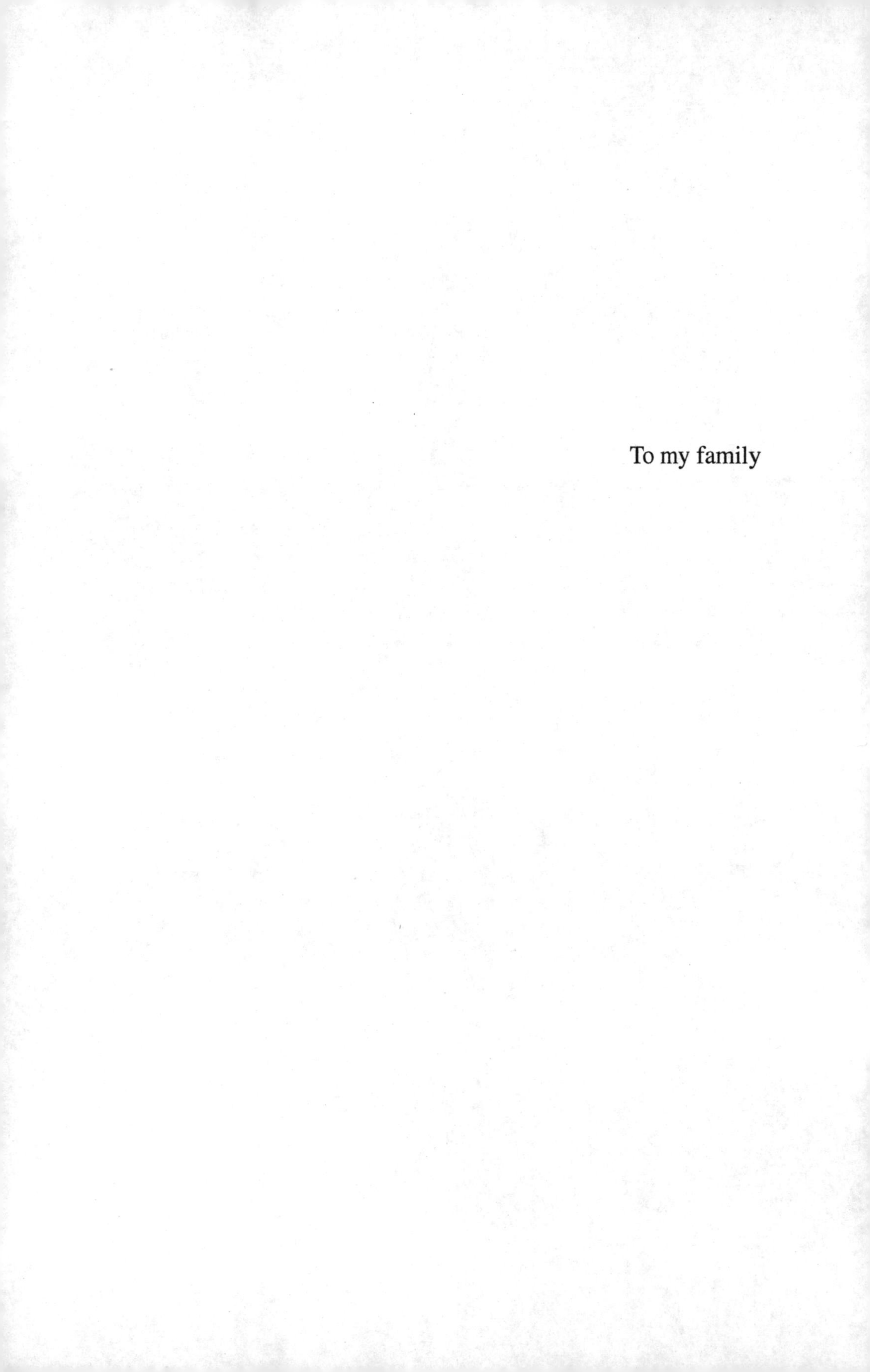

To my family

Preface

Mathematical linguistics is rooted both in Euclid's (circa 325–265 BCE) axiomatic method and in Pāṇini's (circa 520–460 BCE) method of grammatical description. To be sure, both Euclid and Pāṇini built upon a considerable body of knowledge amassed by their precursors, but the systematicity, thoroughness, and sheer scope of the *Elements* and the *Ashṭādhyāyī* would place them among the greatest landmarks of all intellectual history even if we disregarded the key methodological advance they made.

As we shall see, the two methods are fundamentally very similar: the axiomatic method starts with a set of statements assumed to be true and transfers truth from the axioms to other statements by means of a fixed set of logical rules, while the method of grammar is to start with a set of expressions assumed to be grammatical both in form and meaning and to transfer grammaticality to other expressions by means of a fixed set of grammatical rules.

Perhaps because our subject matter has attracted the efforts of some of the most powerful minds (of whom we single out A. A. Markov here) from antiquity to the present day, there is no single easily accessible introductory text in mathematical linguistics. Indeed, to the mathematician the whole field of linguistics may appear to be hopelessly mired in controversy, and neither the formidable body of empirical knowledge about languages nor the standards of linguistic argumentation offer an easy entry point.

Those with a more postmodern bent may even go as far as to doubt the existence of a solid core of mathematical knowledge, often pointing at the false theorems and incomplete or downright wrong proofs that slip through the peer review process at a perhaps alarming rate. Rather than attempting to drown such doubts in rivers of philosophical ink, the present volume will simply proceed *more geometrico* in exhibiting this solid core of knowledge. In Chapters 3–6, a mathematical overview of the traditional main branches of linguistics, phonology, morphology, syntax, and semantics, is presented.

Who should read this book?

The book is accessible to anyone with sufficient general mathematical maturity (graduate or advanced undergraduate). No prior knowledge of linguistics or languages is assumed on the part of the reader. The book offers a single entry point to the central methods and concepts of linguistics that are made largely inaccessible to the mathematician, computer scientist, or engineer by the surprisingly adversarial style of argumentation (see Section 1.2), the apparent lack of adequate definitions (see Section 1.3), and the proliferation of unmotivated notation and formalism (see Section 1.4) all too often encountered in research papers and monographs in the humanities. Those interested in linguistics can learn a great deal more about the subject here than what is covered in introductory courses just from reading through the book and consulting the references cited. Those who plan to approach linguistics through this book should be warned in advance that many branches of linguistics, in particular psycholinguistics, child language acquisition, and the study of language pathology, are largely ignored here – not because they are viewed as inferior to other branches but simply because they do not offer enough grist for the mathematician's mill. Much of what the linguistically naive reader may find interesting about language turns out to be more pertinent to cognitive science, the philosophy of language, and sociolinguistics, than to linguistics proper, and the Introduction gives these issues the shortest possible shrift, discussing them only to the extent necessary for disentangling mathematical linguistics from other concerns.

Conversely, issues that linguists sometimes view as peripheral to their enterprise will get more discussion here simply because they offer such a rich variety of mathematical techniques and problems that no book on mathematical linguistics that ignored them could be considered complete. After a brief review of information theory in Chapter 7, we will devote Chapters 8 and 9 to phonetics, speech recognition, the recognition of handwriting and machine print, and in general to issues of linguistic signal processing and pattern matching, including information extraction, information retrieval, and statistical natural language processing. Our treatment assumes a bit more mathematical maturity than the excellent textbooks by Jelinek (1997) and Manning and Schütze (1999) and intends to complement them. Kracht (2003) conveniently summarizes and extends much of the discrete (algebraic and combinatorial) work on mathematical linguistics. It is only because of the timely appearance of this excellent reference work that the first six chapters could be kept to a manageable size and we could devote more space to the continuous (analytic and probabilistic) aspects of the subject. In particular, expository simplicity would often dictate that we keep the underlying parameter space discrete, but in the later chapters we will be concentrating more on the case of continuous parameters, and discuss the issue of quantization losses explicitly.

In the early days of computers, there was a great deal of overlap between the concerns of mathematical linguistics and computer science, and a surprising amount of work that began in one field ended up in the other, sometimes explicitly as part of computational linguistics, but often as general theory with its roots in linguistics largely forgotten. In particular, the basic techniques of syntactic analysis are now firmly embedded in the computer science curriculum, and the student can already choose from a large variety of textbooks that cover parsing, automata, and formal language theory. Here we single out the classic monograph by Salomaa (1973), which shows the connection to formal syntax in a way readily accessible to the mathematically minded reader. We will selectively cover only those aspects of this field that address specifically linguistic concerns, and again our guiding principle will be mathematical content, as opposed to algorithmic detail. Readers interested in the algorithms should consult the many excellent natural language processing textbooks now available, of which we single out Jurafsky and Martin (2000, with a new edition planned in 2008).

How is the book organized?

To the extent feasible we follow the structure of the standard introductory courses to linguistics, but the emphasis will often be on points only covered in more advanced courses. The book contains many exercises. These are, for the most part, rather hard (over level 30 in the system of Knuth 1971) but extremely rewarding. Especially in the later chapters, the exercises are often based on classical and still widely cited theorems, so the solutions can usually be found on the web quite easily simply by consulting the references cited in the text. However, readers are strongly advised not to follow this route before spending at least a few days attacking the problem. Unsolved problems presented as exercises are marked by an asterisk, a symbol that we also use when presenting examples and counterexamples that native speakers would generally consider wrong (ungrammatical): *Scorsese is a great director* is a positive (grammatical) example while **Scorsese a great director is* is a negative (ungrammatical) example. Some exercises, marked by a dagger †, require the ability to manipulate sizeable data sets, but no in-depth knowledge of programming, data structures, or algorithms is presumed. Readers who write code effortlessly will find these exercises easy, as they rarely require more than a few simple scripts. Those who find such exercises problematic can omit them entirely. They may fail to gain direct appreciation of some empirical properties of language that drive much of the research in mathematical linguistics, but the research itself remains perfectly understandable even if the motivation is taken on faith. A few exercises are marked by a raised M – these are major research projects the reader is not expected to see to completion, but spending a few days on them is still valuable.

Because from time to time it will be necessary to give examples from languages that are unlikely to be familiar to the average undergraduate or graduate student of mathematics, we decided, somewhat arbitrarily, to split languages into two groups. *Major* languages are those that have a chapter in Comrie's (1990) *The World's Major Languages* – these will be familiar to most people and are left unspecified in the text. *Minor* languages usually require some documentation, both because language names are subject to a great deal of spelling variation and because different groups of people may use very different names for one and the same language. Minor languages are therefore identified here by their three-letter Ethnologue code (15th edition, 2005) given in square brackets [].

Each chapter ends with a section on further reading. We have endeavored to make the central ideas of linguistics accessible to those new to the field, but the discussion offered in the book is often skeletal, and readers are urged to probe further. Generally, we recommend those papers and books that presented the idea for the first time, not just to give proper credit but also because these often provide perspective and insight that later discussions take for granted. Readers who industriously follow the recommendations made here should do so for the benefit of learning the basic vocabulary of the field rather than in the belief that such reading will immediately place them at the forefront of research.

The best way to read this book is to start at the beginning and to progress linearly to the end, but the reader who is interested only in a particular area should not find it too hard to jump in at the start of any chapter. To facilitate skimming and alternative reading plans, a generous amount of forward and backward pointers are provided – in a hypertext edition these would be replaced by clickable links. The material is suitable for an aggressively paced one-semester course or a more leisurely paced two-semester course.

Acknowledgments

Many typos and stylistic infelicities were caught, and excellent references were suggested, by Márton Makrai (Budapest Institute of Technology), Dániel Margócsy (Harvard), Doug Merritt (San Jose), Reinhard Muskens (Tilburg), Almerindo Ojeda (University of California, Davis), Bálint Sass (Budapest), Madeleine Thompson (University of Toronto), and Gabriel Wyler (San Jose). The painstaking work of the Springer editors and proofreaders, Catherine Brett, Frank Ganz, Hal Henglein, and Jeffrey Taub, is gratefully acknowledged.

The comments of Tibor Beke (University of Massachusetts, Lowell), Michael Bukatin (MetaCarta), Anssi Yli-Jyrä (University of Helsinki), Péter Gács (Boston University), Marcus Kracht (UCLA), András Serény (CEU), Péter Siptár (Hungarian Academy of Sciences), Anna Szabolcsi (NYU), Péter Vámos (Budapest Institute of Technology), Károly Varasdi (Hungarian Academy of Sciences), and Dániel Varga (Budapest Institute of Technology) resulted in substantive improvements.

Writing this book would not have been possible without the generous support of MetaCarta Inc. (Cambridge, MA), the MOKK Media Research center at the Budapest Institute of Technology Department of Sociology, the Farkas Heller Foundation, and the Hungarian Telekom Foundation for Higher Education – their help and care is gratefully acknowledged.

Contents

Preface vii

1 Introduction 1
1.1 The subject matter 1
1.2 Cumulative knowledge 2
1.3 Definitions 3
1.4 Formalization 4
1.5 Foundations 6
1.6 Mesoscopy 6
1.7 Further reading 7

2 The elements 9
2.1 Generation 9
2.2 Axioms, rules, and constraints 13
2.3 String rewriting 17
2.4 Further reading 20

3 Phonology 23
3.1 Phonemes 24
3.2 Natural classes and distinctive features 28
3.3 Suprasegmentals and autosegments 33
3.4 Phonological computation 40
3.5 Further reading 49

4 Morphology 51
4.1 The prosodic hierarchy 53
4.1.1 Syllables 53
4.1.2 Moras 55
4.1.3 Feet and cola 56
4.1.4 Words and stress typology 56
4.2 Word formation 60

4.3 Optimality 67
4.4 Zipf's law 69
4.5 Further reading 75

5 Syntax 77
5.1 Combinatorical theories 78
5.1.1 Reducing vocabulary complexity 79
5.1.2 Categorial grammar 81
5.1.3 Phrase structure 84
5.2 Grammatical theories 88
5.2.1 Dependency 89
5.2.2 Linking 94
5.2.3 Valency 99
5.3 Semantics-driven theories 102
5.4 Weighted theories 111
5.4.1 Approximation 112
5.4.2 Zero density 116
5.4.3 Weighted rules 120
5.5 The regular domain 122
5.5.1 Weighted finite state automata 122
5.5.2 Hidden Markov models 129
5.6 External evidence 132
5.7 Further reading 137

6 Semantics 141
6.1 The explanatory burden of semantics 142
6.1.1 The Liar 142
6.1.2 Opacity 144
6.1.3 The Berry paradox 145
6.1.4 Desiderata 148
6.2 The standard theory 150
6.2.1 Montague grammar 151
6.2.2 Truth values and variable binding term operators 158
6.3 Grammatical semantics 167
6.3.1 The system of types 168
6.3.2 Combining signs 173
6.4 Further reading 177

7 Complexity 179
7.1 Information 179
7.2 Kolmogorov complexity 185
7.3 Learning 189
7.3.1 Minimum description length 190
7.3.2 Identification in the limit 194
7.3.3 Probable approximate correctness 198

7.4 Further reading ... 199

8 Linguistic pattern recognition ... 201
8.1 Quantization ... 202
8.2 Markov processes, hidden Markov models ... 205
8.3 High-level signal processing ... 209
8.4 Document classification ... 211
8.5 Further reading ... 217

9 Speech and handwriting ... 219
9.1 Low-level speech processing ... 220
9.2 Phonemes as hidden units ... 231
9.3 Handwriting and machine print ... 238
9.4 Further reading ... 245

10 Simplicity ... 247
10.1 Previous reading ... 250

References ... 251

Index ... 281

1

Introduction

1.1 The subject matter

What is *mathematical linguistics*? A classic book on the subject, (Jakobson 1961), contains papers on a variety of subjects, including a categorial grammar (Lambek 1961), formal syntax (Chomsky 1961, Hiż 1961), logical semantics (Quine 1961, Curry 1961), phonetics and phonology (Peterson and Harary 1961, Halle 1961), Markov models (Mandelbrot 1961b), handwriting (Chao 1961, Eden 1961), parsing (Oettinger 1961, Yngve 1961), glottochronology (Gleason 1961), and the philosophy of language (Putnam 1961), as well as a number of papers that are harder to fit into our current system of scientific subfields, perhaps because there is a void now where once there was cybernetics and systems theory (see Heims 1991).

A good way to understand how these seemingly so disparate fields cohere is to proceed by analogy to mathematical physics. Hamiltonians receive a great deal more mathematical attention than, say, the study of generalized incomplete Gamma functions, because of their relevance to mechanics, not because the subject is, from a purely mathematical perspective, necessarily more interesting. Many parts of mathematical physics find a natural home in the study of differential equations, but other parts fit much better in algebra, statistics, and elsewhere. As we shall see, the situation in mathematical linguistics is quite similar: many parts of the subject would fit nicely in algebra and logic, but there are many others for which methods belonging to other fields of mathematics are more appropriate. Ultimately the coherence of the field, such as it is, depends on the coherence of linguistics.

Because of the enormous impact that the works of Noam Chomsky and Richard Montague had on the postwar development of the discipline, there is a strong tendency, observable both in introductory texts such as Partee et al. (1990) and in research monographs such as Kracht (2003), to simply equate mathematical linguistics with formal syntax and semantics. Here we take a broader view, assigning syntax (Chapter 5) and semantics (Chapter 6) no greater scope than they would receive in any book that covers linguistics as a whole, and devoting a considerable amount of space to phonology (Chapter 2), morphology (Chapter 3), phonetics (Chapters 8 and 9), and other areas of traditional linguistics. In particular, we make sure that the reader will learn (in Chapter 7) the central mathematical ideas of information theory and algorithmic complexity that provide the foundations of much of the contemporary work in mathematical linguistics.

This does not mean, of course, that mathematical linguistics is a discipline entirely without boundaries. Since almost all social activity ultimately rests on linguistic communication, there is a great deal of temptation to reduce problems from other fields of inquiry to purely linguistic problems. Instead of understanding schizoid behavior, perhaps we should first ponder what the phrase *multiple personality* means. Mathematics already provides a reasonable notion of 'multiple', but what is 'personality', and how can there be more than one per person? Can a proper understanding of the suffixes *-al* and *-ity* be the key? This line of inquiry, predating the Schoolmen and going back at least to the *cheng ming* (rectification of names) doctrine of Confucius, has a clear and convincing rationale (*The Analects* 13.3, D.C. Lau transl.):

> When names are not correct, what is said will not sound reasonable; when what is said does not sound reasonable, affairs will not culminate in success; when affairs do not culminate in success, rites and music will not flourish; when rites and music do not flourish, punishments will not fit the crimes; when punishments do not fit the crimes, the common people will not know where to put hand and foot. Thus when the gentleman names something, the name is sure to be usable in speech, and when he says something this is sure to be practicable. The thing about the gentleman is that he is anything but casual where speech is concerned.

In reality, linguistics lacks the resolving power to serve as the ultimate arbiter of truth in the social sciences, just as physics lacks the resolving power to explain the accidents of biological evolution that made us human. By applying mathematical techniques we can at least gain some understanding of the limitations of the enterprise, and this is what this book sets out to do.

1.2 Cumulative knowledge

It is hard to find any aspect of linguistics that is entirely uncontroversial, and to the mathematician less steeped in the broad tradition of the humanities it may appear that linguistic controversies are often settled on purely rhetorical grounds. Thus it may seem advisable, and only fair, to give both sides the full opportunity to express their views and let the reader be the judge. But such a book would run to thousands of pages and would be of far more interest to historians of science than to those actually intending to learn mathematical linguistics. Therefore we will not necessarily accord equal space to both sides of such controversies; indeed often we will present a single view and will proceed without even attempting to discuss alternative ways of looking at the matter.

Since part of our goal is to orient the reader not familiar with linguistics, typically we will present the majority view in detail and describe the minority view only tersely. For example, Chapter 4 introduces the reader to morphology and will rely heavily on the notion of the morpheme – the excellent book by Anderson (1992) denying the utility, if not the very existence, of morphemes, will be relegated to footnotes. In some cases, when we feel that the minority view is the correct one, the emphasis will be inverted: for example, Chapter 6, dealing with semantics, is more informed by the 'surface compositional' than the 'logical form' view. In other cases, particularly in Chapter 5, dealing with syntax, we felt that such a bewildering variety of frameworks is available that the reader is better served by an impartial analysis that tries to bring out the common core than by in-depth formalization of any particular strand of research.

In general, our goal is to present linguistics as a cumulative body of knowledge. In order to find a consistent set of definitions that offer a rational reconstruction of the main ideas and techniques developed over the course of millennia, it will often be necessary to take sides in various controversies. There is no pretense here that mathematical formulation will necessarily endow a particular set of ideas with greater verity, and often the opposing view could be formalized just as well. This is particularly evident in those cases where theories diametrically opposed in their means actually share a common goal such as describing all and only the well-formed structures (e.g. syllables, words, or sentences) of languages. As a result, we will see discussions of many 'minority' theories, such as case grammar or generative semantics, which are generally believed to have less formal content than their 'majority' counterparts.

1.3 Definitions

For the mathematician, definitions are nearly synonymous with abbreviations: we say 'triangle' instead of describing the peculiar arrangement of points and lines that define it, 'polynomial' instead of going into a long discussion about terms, addition, monomials, multiplication, or the underlying ring of coefficients, and so forth. The only sanity check required is to exhibit an instance, typically an explicit set-theoretic construction, to demonstrate that the defined object indeed exists. Quite often, counterfactual objects such as the smallest group K not meeting some description, or objects whose existence is not known, such as the smallest nontrivial root of ζ not on the critical line, will play an important role in (indirect) proofs, and occasionally we find cases, such as *motivic cohomology*, where the whole conceptual apparatus is in doubt. In linguistics, there is rarely any serious doubt about the existence of the objects of inquiry. When we strive to define 'word', we give a mathematical formulation not so much to demonstrate that words exist, for we know perfectly well that we use words both in spoken and written language, but rather to handle the odd and unexpected cases. The reader is invited to construct a definition now and to write it down for comparison with the eventual definition that will emerge only after a rather complex discussion in Chapter 4.

In this respect, mathematical linguistics is very much like the empirical sciences, where formulating a definition involves at least three distinct steps: an *ostensive* definition based on positive and sometimes negative examples (vitriol is an acid, lye is not), followed by an *extensive* definition delineating the intended scope of the notion (every chemical that forms a salt with a base is an acid), and the *intensive* definition that exposes the underlying mechanism (in this case, covalent bonds) emerging rather late as a result of a long process of abstraction and analysis.

Throughout the book, the first significant instance of key notions will appear in *italics*, usually followed by ostensive examples and counterexamples in the next few paragraphs. (Italics will also be used for emphasis and for typesetting linguistic examples.) The empirical observables associated with these notions are always discussed, but textbook definitions of an extensive sort are rarely given. Rather, a mathematical notion that serves as a stand-in will be defined in a rigorous fashion: in the defining phrase, the same notion is given in **boldface**. Where an adequate mathematical formulation is lacking and we proceed by sheer analogy, the key terms will be *slanted* – such cases are best thought of as open problems in mathematical linguistics.

1.4 Formalization

In mathematical linguistics, as in any branch of applied mathematics, the issue of formalizing semiformally or informally stated theories comes up quite often. A prime example is the study of phrase structure, where Chomsky (1956) took the critical step of replacing the informally developed system of immediate constituent analysis (ICA, see Section 5.1) by the rigorously defined context-free grammar (CFG, see Section 2.3) formalism. Besides improving our understanding of natural language, a worthy goal in itself, the formalization opened the door to the modern theory of computer languages and their compilers. This is not to say that every advance in formalizing linguistic theory is likely to have a similarly spectacular payoff, but clearly the informal theory remains a treasure-house inasmuch as it captures important insights about natural language. While not entirely comparable to biological systems in age and depth, natural language embodies a significant amount of evolutionary optimization, and artificial communication systems can benefit from these developments only to the extent that the informal insights are captured by formal methods.

The quality of formalization depends both on the degree of faithfulness to the original ideas and on the mathematical elegance of the resulting system. Because the proper choice of formal apparatus is often a complex matter, linguists, even those as evidently mathematical-minded as Chomsky, rarely describe their models with full formal rigor, preferring to leave the job to the mathematicians, computer scientists, and engineers who wish to work with their theories. Choosing the right formalism for linguistic rules is often very hard. There is hardly any doubt that linguistic behavior is governed by rather abstract rules or constraints that go well beyond what systems limited to memorizing previously encountered examples could explain. Whether these rules have a stochastic aspect is far from settled: engineering applications are dominated by models that crucially rely on probabilities, while theoretical models, with the notable exception of the *variable rules* used in sociolinguistics (see Section 5.4.3), rarely include considerations relating to the frequency of various phenomena. The only way to shed light on such issues is to develop alternative formalizations and compare their mathematical properties.

The tension between faithfulness to the empirical details and the elegance of the formal system has long been familiar to linguists: Sapir (1921) already noted that "all grammars leak". One significant advantage that probabilistic methods have over purely symbolic techniques is that they come with their own built-in measure of leakiness (see Section 5.4). It is never a trivial matter to find the appropriate degree of idealization in pursuit of theoretical elegance, and all we can do here is to offer a couple of convenient stand-ins for the very real but still somewhat elusive notion of elegance.

The first stand-in, held in particularly high regard in linguistics, is *brevity*. The contemporary slogan of algorithmic complexity (see Section 7.2), that the best theory is the shortest theory, could have been invented by Pāṇini. The only concession most linguists are willing to make is that some of the complexity should be ascribed to principles of *universal grammar* (UG) rather than to the *parochial* rules specific to a given language, and since the universal component can be amortized over many languages, we should maximize its explanatory burden at the expense of the parochial component.

The second stand-in is *stability* in the sense that minor perturbations of the definition lead to essentially the same system. Stability has always been highly regarded in mathematics: for example, Birkhoff (1940) spent significant effort on establishing the value of lattices as legitimate objects of algebraic inquiry by investigating alternative definitions that ultimately lead to the same class of structures. There are many ways to formalize an idea, and when small changes in emphasis have a very significant impact on the formal properties of the resulting system, its mathematical value is in doubt. Conversely, when variants of formalisms as different as indexed grammars (Aho 1968), combinatory categorial grammar (Steedman 2001), head grammar (Pollard 1984), and tree adjoining grammar (Joshi 2003) define the same class of languages, the value of each is significantly enhanced.

One word of caution is in order: the fact that some idea is hard to formalize, or even seems so contradictory that a coherent mathematical formulation appears impossible, can be a reflection on the state of the art just as well as on the idea itself. Starting with Berkeley (1734), the intuitive notion of infinitesimals was subjected to all kinds of criticism, and it took over two centuries for mathematics to catch up and provide an adequate foundation in Robinson (1966). It is quite conceivable that equally intuitive notions, such as a *semantic theory of information*, which currently elude our mathematical grasp, will be put on firm foundations by later generations. In such cases, we content ourselves with explaining the idea informally, describing the main intuitions and pointing at possible avenues of formalization only programmatically.

1.5 Foundations

For the purposes of mathematical linguistics, the classical foundations of mathematics are quite satisfactory: all objects of interest are sets, typically finite or, rarely, denumerably infinite. This is not to say that nonclassical metamathematical tools such as Heyting algebras find no use in mathematical linguistics but simply to assert that the fundamental issues of this field are not foundational but definitional.

Given the finitistic nature of the subject matter, we will in general use the terms set, class, and collection interchangeably, drawing explicit cardinality distinctions only in the rare cases where we step out of the finite domain. Much of the classical linguistic literature of course predates Cantor, and even the modern literature typically conceives of infinity in the Gaussian manner of a potential, as opposed to actual, Cantorian infinity. Because of immediate empirical concerns, denumerable generalizations of finite objects such as ω-words and Büchi automata are rarely used,[1] and in fact even the trivial step of generalizing from a fixed constant to arbitrary n is often viewed with great suspicion.

Aside from the tradition of Indian logic, the study of languages had very little impact on the foundations of mathematics. Rather, mathematicians realized early on that natural language is a complex and in many ways unreliable construct and created their own simplified language of formulas and the mathematical techniques to investigate it. As we shall see, some of these techniques are general enough to cover essential facets of natural languages, while others scale much more poorly.

There is an interesting residue of foundational work in the Berry, Richard, Liar, and other paradoxes, which are often viewed as diagnostic of the vagueness, ambiguity, or even 'paradoxical nature' of natural language. Since the goal is to develop a mathematical theory of language, sooner or later we must define English in a formal system. Once this is done, the buck stops there, and questions like "what is the smallest integer not nameable in ten words?" need to be addressed anew.

We shall begin with the seemingly simpler issue of the first number not nameable in *one* word. Since it appears to be one hundred and one, a number already requiring *four* words to name, we should systematically investigate the number of words in number names. There are two main issues to consider: what is a word? (see Chapter 4); and what is a name? (see Chapter 6). Another formulation of the Berry paradox invokes the notion of syllables; these are also discussed in Chapter 4. Eventually we will deal with the paradoxes in Chapter 6, but our treatment concentrates on the linguistic, rather than the foundational, issues.

1.6 Mesoscopy

Physicists speak of mesoscopic systems when these contain, say, fifty atoms, too large to be given a microscopic quantum-mechanical description but too small for the classical macroscopic properties to dominate the behavior of the system. Linguistic

[1] For a contrary view, see Langendoen and Postal (1984).

systems are mesoscopic in the same broad sense: they have thousands of rules and axioms compared with the handful of axioms used in most branches of mathematics. Group theory explores the implications of five axioms, arithmetic and set theory get along with five and twelve axioms respectively (not counting members of axiom schemes separately), and the most complex axiom system in common use, that of geometry, has less than thirty axioms.

It comes as no surprise that with such a large number of axioms, linguistic systems are never pursued microscopically to yield implications in the same depth as group theory or even less well-developed branches of mathematics. What is perhaps more surprising is that we can get reasonable approximations of the behavior at the macroscopic level using the statistical techniques pioneered by A. A. Markov (see Chapters 7 and 8).

Statistical mechanics owes its success largely to the fact that in thermodynamics only a handful of phenomenological parameters are of interest, and these are relatively easy to link to averages of mechanical quantities. In mathematical linguistics the averages that matter (e.g. the percentage of words correctly recognized or correctly translated) are linked only very indirectly to the measurable parameters, of which there is such a bewildering variety that it requires special techniques to decide which ones to employ and which ones to leave unmodeled.

Macroscopic techniques, by their very nature, can yield only approximations for mesoscopic systems. Microscopic techniques, though in principle easy to extend to the mesoscopic domain, are in practice also prone to all kinds of bugs, ranging from plain errors of fact (which are hard to avoid once we deal with thousands of axioms) to more subtle, and often systematic, errors and omissions. Readers may at this point feel very uncomfortable with the idea that a given system is only 70%, 95%, or even 99.99% correct. After all, isn't a single contradiction or empirically false prediction enough to render a theory invalid? Since we need a whole book to develop the tools needed to address this question, the full answer will have to wait until Chapter 10.

What is clear from the outset is that natural languages offer an unparalleled variety of complex algebraic structures. The closest examples we can think of are in crystallographic topology, but the internal complexity of the groups studied there is a product of pure mathematics, while the internal complexity of the syntactic semigroups associated to natural languages is more attractive to the applied mathematician, as it is something found in vivo. Perhaps the most captivating aspect of mathematical linguistics is not just the existence of discrete mesoscopic structures but the fact that these come embedded, in ways we do not fully understand, in continuous signals (see Chapter 9).

1.7 Further reading

The first works that can, from a modern standpoint, be called mathematical linguistics are Markov's (1912) extension of the weak law of large numbers (see Theorem 8.2.2) and Thue's (1914) introduction of string manipulation (see Chapter 2), but pride of place must go to Pāṇini, whose inventions include not just grammatical rules but also a formal metalanguage to describe the rules and a set of principles governing their interaction. Although the *Ashṭādhyāyī* is available on the web in its entirety, the reader will be at a loss without the modern commentary literature starting with Böhtlingk (1887, reprinted 1964). For modern accounts of various aspects of the system see Staal (1962, 1967) Cardona (1965, 1969, 1970, 1976, 1988), and Kiparsky (1979, 1982a, 2002). Needless to say, Pāṇini did not work in isolation. Much like Euclid, he built on the inventions of his predecessors, but his work was so comprehensive that it effectively drove the earlier material out of circulation. While much of linguistics has aspired to formal rigor throughout the ages (for the Masoretic tradition, see Aronoff 1985, for medieval syntax see Covington 1984), the continuous line of development that culminates in contemporary formal grammar begins with Bloomfield's (1926) Postulates (see Section 3.1), with the most important milestones being Harris (1951) and Chomsky (1956, 1959).

Another important line of research, only briefly alluded to above, could be called mathematical antilinguistics, its goal being the elimination, rather than the explanation, of the peculiarities of natural language from the system. The early history of the subject is discussed in depth in Eco (1995); the modern mathematical developments begin with Frege's (1879) system of *Concept Writing* (Begriffsschrift), generally considered the founding paper of mathematical logic. There is no doubt that many great mathematicians from Leibniz to Russell were extremely critical of natural language, using it more for counterexamples and cautionary tales than as a part of objective reality worthy of formal study, but this critical attitude has all but disappeared with the work of Montague (1970a, 1970b, 1973). Contemporary developments in model-theoretic semantics or 'Montague grammar' are discussed in Chapter 6.

Major summaries of the state of the art in mathematical linguistics include Jakobson (1961), Levelt (1974), Manaster-Ramer (1987), and the subsequent Mathematics of Language (MOL) conference volumes. We will have many occasions to cite Kracht's (2003) indispensable monograph *The Mathematics of Language*.

The volumes above are generally more suitable for the researcher or advanced graduate student than for those approaching the subject as undergraduates. To some extent, the mathematical prerequisites can be learned from the ground up from classic introductory textbooks such as Gross (1972) or Salomaa (1973). Gruska (1997) offers a more modern and, from the theoretical computer science perspective, far more comprehensive introduction. The best elementary introduction to the logical prerequisites is Gamut (1991). The discrete side of the standard "mathematics for linguists" curriculum is conveniently summarized by Partee et al. (1990), and the statistical approach is clearly introduced by Manning and Schütze (1999). The standard introduction to pattern recognition is Duda et al. (2000). Variable rules were introduced in Cedergren and Sankoff (1974) and soon became the standard modeling method in sociolinguistics – we shall discuss them in Chapter 5.

2

The elements

A primary concern of mathematical linguistics is to effectively enumerate those sets of words, sentences, etc., that play some important linguistic role. Typically, this is done by means of *generating* the set in question, a definitional method that we introduce in Section 2.1 by means of examples and counterexamples that show the similarities and the differences between the standard mathematical use of the term 'generate' and the way it is employed in linguistics.

Because the techniques used in defining sets, functions, relations, etc., are not always directly useful for evaluating them at a given point, an equally important concern is to solve the membership problem for the sets, functions, relations, and other structures of interest. In Section 2.2 we therefore introduce a variety of *grammars* that can be used to, among other things, create *certificates* that a particular element is indeed a member of the set, gets mapped to a particular value, stands in a prescribed relation to other elements and so on, and compare generative systems to logical calculi.

Since *generative grammar* is most familiar to mathematicians and computer scientists as a set of rather loosely collected string-rewriting techniques, in Section 2.3 we give a brief overview of this domain. We put the emphasis on context-sensitive grammars both because they play an important role in phonology (see Chapter 3) and morphology (see Chapter 4) and because they provide an essential line of defense against undecidability in syntax (see Chapter 5).

2.1 Generation

To define a collection of objects, it is often expedient to begin with a fixed set of primitive elements E and a fixed collection of *rules* (we use this term in a broad sense that does not imply strict procedurality) R that describe permissible arrangements of the primitive elements as well as of more complex objects. If x, y, z are objects *satisfying* a (binary) rule $z = r(x, y)$, we say that z **directly generates** x and y (in this order) and use the notation $z \to_r xy$. The smallest collection of objects closed under direct generation by any $r \in R$ and containing all elements of E is called the set **generated** from E by R.

Very often the simplest or most natural definition yields a superset of the real objects of interest, which is therefore supplemented by some additional conditions to narrow it down. In textbooks on algebra, the symmetric group is invariably introduced before the alternating group, and the latter is presented simply as a subgroup of the former. In logic, closed formulas are typically introduced as a special class of well-formed formulas. In context-free grammars, the sentential forms produced by the grammar are kept only if they contain no nonterminals (see Section 2.3), and we will see many similar examples (e.g. in the handling of agreement; see Section 5.2.3).

Generative definitions need to be supported by some notion of *equality* among the defined objects. Typically, the notion of equality we wish to employ will abstract away from the derivational history of the object, but in some cases we will need a stronger definition of identity that defines two objects to be the same only if they were generated the same way. Of particular interest in this regard are derivational *strata*. A specific intermediary stage of a derivation (e.g. when a group or rules have been exhausted or when some well-formedness criterion is met) is often called a **stratum** and is endowed with theoretically significant properties, such as availability for interfacing with other modules of grammar. Theories that recognize strata are called *multistratal*, and those that do not are called *monostratal* – we shall see examples of both in Chapter 5.

In mathematical linguistics, the objects of interest are the collection of words in a language, the collection of sentences, the collection of meanings, etc. Even the most tame and obviously finite collections of this kind present great definitional difficulties. Consider, for example, the set of characters (graphemes) used in written English. Are uppercase and lowercase forms to be kept distinct? How about punctuation, digits, or Zapf dingbats? If there is a new character for the euro currency unit, as there is a special character for dollar and pound sterling, shall it be included on account of Ireland having already joined the euro zone or shall we wait until England follows suit? Before proceeding to words, meanings, and other more subtle objects of inquiry, we will therefore first refine the notion of a generative definition on some familiar mathematical objects.

Example 2.1.1 Wang tilings. Let C be a finite set of colors and S be a finite set of square tiles, each colored on the edges according to some function $e : S \to C^4$. We assume that for each coloring *type* we have an infinite supply of *tokens* colored with that pattern: these make up the set of primitive elements E. The goal is to tile the whole plane (or just the first quadrant) laying down the tiles so that their colors match at the edges. To express this restriction more precisely, we use a rule system R with four rules n, s, e, w as follows. Let $\mathbb{Z}$ be the set of integers, $'$ be the successor function "add one" and ‘ be its inverse "subtract one". For any $i, j \in \mathbb{Z}$, we say that the tile u whose bottom left corner is at (i, j) has a correct neighbor to the north if the third component of $e(u)$ is the same as the first component of $e(v)$ where v is the tile at (i, j'). Denoting the ith projection by π_i, we can write $\pi_3(e(u)) = \pi_1(e(v))$ for v at (i, j'). Similarly, the west rule requires $\pi_4(e(u)) = \pi_2(e(v))$ for v at (i', j), the east rule requires $\pi_2(e(u)) = \pi_4(e(v))$ for v at (i', j), and the south rule requires

$\pi_1(e(u)) = \pi_3(e(v))$ for v at (i, j'). We define **first-quadrant (plane) tilings** as functions from $\mathbb{N} \times \mathbb{N}$ ($\mathbb{Z} \times \mathbb{Z}$) to E that satisfy all four rules.

Discussion While the above may look very much like a generative definition, there are some crucial differences. First, the definition relies on a number of externally given objects, such as the natural numbers, the integers, the successor function, and Cartesian products. In contrast, the definitions we will encounter later, though they may require some minimal set-theoretical scaffolding, are almost always *noncounting*, both in the broad sense of being free of arithmetic aspects and in the narrower semigroup-theoretic sense (see Chapter 5.2.3).

Second, these rules are *well-formedness conditions* (WFCs, see Section 2.3) rather than procedural *rules of production*. In many cases, this is a distinction without a difference, since production rules can often be used to enforce WFCs. To turn the four WFCs n, s, e, w into rules of production requires some auxiliary definitions: we say that a new tile (i, j) is **north-adjacent** to a preexisting set of tiles T if $(i, j) \notin T$ but $(i, j') \in T$, and similarly for east-, west-, and south-adjacency (any combination of these relations may simultaneously obtain between a new tile and some suitably shaped T). A **(north, south, east, west)-addition** of a tile at (i, j) is an operation that is permitted between a set of tiles T and a (south, north, west, east)-adjacent tile (i, j) to form $T \cup (i, j)$ iff the n, s, w, e rules are satisfied, so there are 2^4 production rules.

It is a somewhat tedious but entirely trivial exercise to prove that these sixteen rules of production can be used to successively build all and only well-formed first quadrant (plane) tilings starting from a single tile placed at the origin. Obviously, for some tile inventories, the production rules can also yield partial tilings that can never be completed as well-formed first-quadrant (or plane) tilings, and in general we will often see reason to consider broader production processes, where ill-formed intermediate structures are integral to the final outcome.

This becomes particularly interesting in cases where the final result shows some regularity that is not shared by the intermediate structures. For example, in languages that avoid two adjacent vowels (a configuration known as *hiatus*), if a vowel-initial word would come after a vowel-final one, there may be several distinct processes that enforce this constraint; e.g., by deleting the last vowel of the first word or by inserting a consonant between them (as in *the very ideaR of it*). It has long been observed (see in particular Kisseberth 1970) that such processes can *conspire* to enforce WFCs, and an important generative model, optimality theory (see Section 4.3), takes this observation to be fundamental in the sense that surface regularities appear as the cause, rather than the effect, of production rules that conspire to maintain them. On the whole, there is great interest in *constraint-based* theories of grammar where the principal mechanism of capturing regularities is by stating them as WFCs.

Third, the four WFCs (as opposed to the 16 production rules) have no recursive aspect whatsoever. There is no notion of larger structures built via intermediate structures: we go from the atomic units (tiles) to the global structure (tiling of the first quadrant) in one leap. Linguistic objects, as we shall see, are generally organized in intermediate layers that are of interest in themselves: a typical example is provided by phonemes (sounds), which are organized in syllables, which in turn are organized in metrical feet, which may be organized in cola (superfeet) before reaching the word level (see Section 4.1). In contrast, Example 2.1.1 will lack recursive structure not only in the presentation chosen above but in any other presentation.

Theorem 2.1.1 (Berger 1966) It is recursively undecidable whether a given inventory of tiles E can yield a Wang tiling.

Example 2.1.2 Presentation of groups in terms of generators and relations. Let E be a set of generators $g_1, \cdots, g_k$ and their inverses $g_1^{-1}, \cdots, g_k^{-1}$, and let R contain both the formal product rule that forms a string of these from two such strings by means of concatenation and the usual rules of cancellation $g_i^{-1}g_i = g_i g_i^{-1} = \lambda$ as context-free string-rewriting rules. Formal products composed from the g_i and g_i^{-1} define the **free group** (or, if we omit inverses and cancellation, the **free monoid**) over k generators, with the usual conventions that the empty word is the multiplicative unit of the group (monoid) and that formal products containing canceling terms are equivalent to those with the canceling terms omitted. If a broader set of formal products is defined as canceling, representatives of this set are called **defining relations** for the group being presented, which is the factor of the free group by the cancellation kernel.

Discussion As is well-known, it is in general undecidable whether a formal product of generators and inverses is included in the kernel or not (Sims 1994) – we will discuss the relationship between combinatorial group theory and formal languages in Chapter 5. Note that it is a somewhat arbitrary technical decision whether we list the defining relations as part of our production rules or as part of the equality relation: we can keep one or the other (but not both) quite trivial without any loss of expressive power. In general, the *equality* clause of generative definitions can lead to just the same complications as the *rules* clause.

Example 2.1.3 Herbrand universes. As the reader will recall, a **first order language** (FOL) consists of logical symbols (variables, connectives, quantifiers, equal sign) plus some constants (e.g., distinguished elements of algebras), as well as function and relation symbols (each with finite arity). The primitive elements of the Herbrand universe are the object constants of the FOL under study (or an arbitrary constant if no object constant was available initially), and there are as many rules to describe permissible arrangements of elements as there are function/relation constants in the FOL under study: if f was such a constant of arity n, $f(x_1, \ldots, x_n)$ is in the Herbrand universe provided the x_i were.

Discussion Herbrand universes are used in building purely formula-based models which are in some sense canonical among the many models that first order theories have. It should come as no surprise that logic offers many *par excellence* examples of generative definition – after all, the techniques developed for formalizing mathematical statements grew out of the larger effort to render statements of all sorts formally. However, the definition of an FOL abstracts away from several important properties of natural language. In FOLs, functions and relations of arbitrary arity are permitted, while in natural language the largest number of arguments one needs to consider is five (see Section 5.2). Also, in many important cases (see Chapter 3), the freedom to utilize an infinite set of constants or variables is not required.

As a matter of fact, it is often tempting to replace natural languages, the true object of inquiry, by some well-regimented semiformal or fully formal construct used in mathematics. Certainly, there is nothing wrong with a bit of idealization, especially with ignoring factors best classified as noise. But a discussion about the English word *triangle* cannot rely too much on the geometrical object by this name since this would create problems where there aren't any; for example, it is evident that a hunter *circling* around a clearing does not require that her path keep the exact same distance from the center at all times. To say that this amounts to fuzzy definitions or sloppy language use is to put the cart before the horse: the fact to be explained is not how a cleaned-up language *could be* used for communication but how real language *is* used.

Exercise 2.1 The Fibonacci numbers are defined by $f_0 = 0, f_1 = 1, f_{n+1} = f_n + f_{n-1}$. Is this a generative definition? Why?

2.2 Axioms, rules, and constraints

There is an unbroken tradition of argumentation running from the Greek sophists to the Oxford Union, and the axiomatic method has its historic roots in the efforts to regulate the permissible methods of debate. As in many other fields of human activity, ranging from ritual to game playing, regulation will lay bare some essential features of the activity and thereby make it more enjoyable for those who choose to participate. Since it is the general experience that almost all statements are debatable, to manage argumentation one first needs to postulate a small set of primitive statements on which the parties agree – those who will not agree are simply excluded from the debate. As there is remarkable agreement about the validity of certain kinds of inference, the stage is set for a fully formal, even automatic, method of verifying whether a given argument indeed leads to the desired conclusion from the agreed upon premises.

There is an equally venerable tradition of protecting the full meaning and exact form of sacred texts, both to make sure that mispronunciations and other errors that may creep in over the centuries do not render them ineffectual and that misinterpretations do not confuse those whose task is to utter them on the right occasion. Even if we ignore the phonetic issues related to 'proper' pronunciation (see Chapter 8), writing down the texts is far from sufficient for the broader goals of preservation. With any material of great antiquity, we rarely have a single fully preserved and widely accepted version – rather, we have several imperfect variants and fragments. What is needed is not just a frozen description of some texts, say the Vedas, but also a grammar that defines what constitutes a proper Vedic text. The philological ability to determine the age of a section and undo subsequent modifications is especially important because the words of earlier sages are typically accorded greater weight.

In defining the language of a text, a period, or a speech community, we can propagate *grammaticality* the same way we propagate truth in an axiomatic system, by choosing an initial set of grammatical expressions and defining some permissible combinatorical operations that are guaranteed to preserve grammaticality. Quite often, such operations are conceptualized as being composed of a purely combinatorical step (typically concatenation) followed by some tidying up; e.g., adding a third-person suffix to the verb when it follows a third-person subject: compare *I see* to *He sees*. In logic, we mark the operators overtly by affixing them to the sequence of the operands – prefix (Polish), interfix (standard), and postfix (reverse Polish) notations are all in wide use – and tend not to put a great deal of emphasis on tidying up (omission of parentheses is typical). In linguistics, there is generally only one operation considered, concatenation, so no overt marking is necessary, but the tidying up is viewed as central to the enterprise of obtaining all and only the attested forms.

The same goal of characterizing all and only the grammatical forms can be accomplished by more indirect means. Rather than starting from a set of fully grammatical forms, we can begin with some more abstract inventory, such as the set of words W, elements of which need not in and of themselves be grammatical, and rather than propagating grammaticality from the parts to the whole, we perform some computation along the way to keep score.

Example 2.2.1 Balanced parentheses. We have two atomic expressions, the left and the right paren, and we assign the values $+1$ to '(' and -1 to ')'. We can successively add new paren symbols on the right as long as the score (overall sum of $+1$ and -1 values) does not dip below zero: the well-formed (balanced) expressions are simply those where this WFC is met and the overall score is zero.

Discussion The example is atypical for two reasons: first because linguistic theories are noncounting (they do not rely on the full power of arithmetic) and second because it is generally not necessary for a WFC to be met at every stage of the derivation. Instead of computing the score in $\mathbb{Z}$, a better choice is some finite structure G with well-understood rules of combination, and instead of assigning a single value to each atomic expression, it gives us much-needed flexibility to make the assignment disjunctive (taking any one of a set of values). Thus we have a mapping $c : W \to 2^G$ and consider grammatical only those sequences of words for which the rules of combination yield a desirable result. Demonstrating that the assigned elements of G indeed combine in the desired manner constitutes a **certificate** of membership according to the grammar defined by c.

Example 2.2.2 Categorial grammar. If G behaves like a free group except that formal inverses of generators do not cancel from both sides ($g \cdot g^{-1} = e$ is assumed but $g^{-1} \cdot g = e$ is not) and we consider only those word sequences $w_1.w_2 \ldots w_n$ for which there is at least one h_i in each $c(w_i)$ such that $h_1 \cdot \ldots \cdot h_n = g_0$ (i.e. the group-theoretical product of the h_i yields a distinguished generator g_0), we obtain a version of *bidirectional categorial grammar* (Bar-Hillel 1953, Lambek 1958). If we take G as the free Abelian group, we obtain *unidirectional categorial grammar* (Ajdukiewitz 1935). These notions will be developed further in Chapter 5.2.

Example 2.2.3 Unification grammar. By choosing G to be the set of rooted directed acyclic node-labeled graphs, where the labels are first order variables and constants, and considering only those word sequences for which the assigned graphs will unify, we obtain a class of *unification grammars*.

Example 2.2.4 Link grammar. By choosing G to satisfy a generalized version of the (horizontal) tiling rules of Example 2.1.1, we obtain the *link grammars* of Sleator and Temperley (1993).

We will investigate a variety of such systems in detail in Chapters 5 and 6, but here we concentrate on the major differences between truth and grammaticality. First, note that systems such as those above are naturally set up to define not only one distinguished set of strings but its cosets as well. For example, in a categorial grammar, we may inquire not only about those strings of words for which multiplication of the associated categories yields the distinguished generator but also about those for which the yield contains another generator or any specific word of G. This corresponds to the fact that e.g. *the house of the seven gables* is grammatical but only as a noun phrase and not as a sentence, while *the house had seven gables* is a grammatical sentence but not a grammatical noun phrase. It could be tempting to treat the cosets in analogy with n-valued logics, but this does not work well since the various stringsets defined by a grammar may overlap (and will in fact irreducibly overlap in every case where a primitive element is assigned more than one disjunct by c), while truth values are always uniquely assigned in n-valued logic.

Second, the various calculi for propagating truth values by specific rules of inference can be supported by an appropriately constructed theory of model structures. In logic, a model will be unique only in degenerate cases: as soon as there is an infinite model, by the Löwenheim-Skolem theorems we have at least as many non-isomorphic models as there are cardinalities. In grammar, the opposite holds: as soon as we fix the period, dialect, style, and possibly other parameters determining grammaticality, the model is essentially unique.

The fact that up to isomorphism there is only one model structure M gives rise to two notions peculiar to mathematical linguistics: *overgeneration* and *undergeneration*. If there is some string $w_1.w_2 \ldots w_n \notin M$ that appears in the yield of c, we say that c **overgenerates** (with respect to M), and if there is a $w_1.w_2 \ldots w_n \in M$ that does not appear in the yield of c, we say that c **undergenerates**. It is quite possible, indeed typical, for working grammars to have both kinds of errors at the same time. We will develop quantitative methods to compare the errors of different grammars in Section 5.4, and note here that neither undergeneration nor overgeneration is a definitive diagnostic of some fatal problem with the system. In many cases, overgeneration is benign in the sense that the usefulness of a system that e.g. translates English sentences to French is not at all impaired by the fact that it is also capable of translating an input that lies outside the confines of fully grammatical English. In other cases, the aim of the system may be to shed light only on a particular range of phenomena, say on the system of intransitive verbs, to the exclusion of transitive, ditransitive, etc., verbs. In the tradition of Montague grammar (see Section 6.2), such systems are explicitly called *fragments*. Constraint-based theories, which view the task of characterizing all and only the well-formed structures as one of (rank-prioritized) intersection of WFCs (see Section 4.2) can have the same under- and overgeneration problems as rule-based systems, as long as they have too many (too few) constraints.

In spite of these major differences, the practice of logic and that of grammar have a great deal in common. First, both require a systematic ability to analyze sentences in component parts so that generalizations involving only some part can be stated and the ability to construct new sentences from ones already seen. Chapter 5 will discuss such *syntactic* abilities in detail. We note here that the practice of logic is largely *normative* in the sense that constructions outside those explicitly permitted by its syntax are declared ill-formed, while the practice of linguistics is largely *descriptive* in the sense that it takes the range of existing constructions as given and strives to adjust the grammar so as to match this range.

Second, both logic and grammar are largely driven by an overall consideration of economy. As the reader will have no doubt noticed, having a separate WFC for the northern, southern, eastern, and western edges of a tile in Example 2.1.1 is quite unnecessary: any two orthogonal directions would suffice to narrow down the range of well-formed tilings. Similarly, in context-free grammars, we often find it sufficient to deal only with rules that yield only two elements on the right-hand side (Chomsky normal form), and there has to be some strong reason for departing from the simplest binary branching structure (see Chapter 5).

From the perspective of linguistics, logical calculi are generation devices, with the important caveat that in logic the rules of deduction are typically viewed as possibly having more than one premiss, while in linguistics such rules would generally be viewed as having only one premiss, namely the conjunction of the logically distinct premisses, and axiom systems would be viewed as containing a single starting point (the conjunction of the axioms). The deduction of theorems from the axiom by brute force enumeration of all proofs is what linguists would call **free generation**. The use of a single conjunct premiss instead of multiple premisses may look like a distinction without a difference, but it has the effect of making generative systems *invertible:* for each such system with rules $r_1, \dots, r_k$ we can construct an inverted system with rules $r_1^{-1}, \dots, r_k^{-1}$ that is now an **accepting**, rather than generating, device. This is very useful in all those cases where we are interested in characterizing both production (synthesis, generation) and perception (analysis, parsing) processes because the simplest hypothesis is that these are governed by the same set of abstract rules.

Clearly, definition by generation differs from deduction by a strict algorithmic procedure only in that the choice of the next algorithmic step is generally viewed as being completely determined by the current step, while in generation the next step is freely drawn from the set of generative rules. The all-important boundary between recursive and recursively enumerable (r.e.) is drawn the same way by certificates (derivation structures), but the systems of interest congregate on different sides of this boundary. In logic, proving the negation of a statement requires the same kind of certificate (a proof object rooted in the axioms and terminating in the desired conclusion) as proving the statement itself – the difficulty is that most calculi are r.e. but not recursive (decidable). In grammar, proving the ungrammaticality of a form requires an apparatus very different from proving its grammaticality: for the latter purpose an ordinary derivation suffices, while for the former we typically need to exhaustively survey all forms of similar and lesser complexity, which can be difficult, even though most grammars are not only r.c. but in fact recursive.

2.3 String rewriting

Given a set of atomic symbols Σ called the **alphabet**, the simplest imaginable operation is that of **concatenation**, whereby a complex symbol xy is formed from x and y by writing them in succession. Applying this operation recursively, we obtain **strings** of arbitrary *length*. Whenever such a distinction is necessary, the operation will be denoted by . (dot). The result of the dot operation is viewed as having no internal punctuation: $u.v = uv$ both for atomic symbols and for more complex strings, corresponding to the fact that concatenation is by definition associative. To forestall confusion, we mention here that in later chapters the . will also be used in *glosses* to connect a word stem to the complex of morphosyntactic (inflectional) features the word form carries: for example *geese* = *goose.PL* (the plural form of *goose* is *geese*) or Hungarian *házammal* = *house.POSS1SG.INS* 'with my house', where *POSS1SG* refers to the suffix that signifies possession by a first-person singular entity and *INS* refers to the instrumental case ending roughly analogous to English *with*. (The reader should be forewarned that translation across languages rarely proceeds as smoothly on a morpheme by morpheme basis as the example may suggest: in many cases morphologically expressed concepts of the source language have no exact equivalent in the language used for glossing.)

Of special interest is the **empty string** λ, which serves as a two-sided multiplicative unit of concatenation: $\lambda.u = u.\lambda = u$. The whole set of strings generated from Σ by concatenation is denoted by Σ^+ (**λ-free Kleene closure**) or, if the empty string is included, by Σ^* (**Kleene closure**). If $u.v = w$, we say that u (v) is a **left (right) factor** of w. If we define the **length** $l(x)$ of a string x as the number of symbols in x, counted with multiplicity (the empty word has length 0), l is a homomorphism from Σ^* to the additive semigroup of nonnegative integers. In particular, the semigroup of nonnegative integers (with ordinary addition) is isomorphic to the Kleene closure of a one-symbol alphabet (with concatenation): the latter may be called integers in **base one** notation.

Subsets of Σ^* are called **stringsets**, **formal languages**, or just **languages**. In addition to the standard Boolean operations, we can define the **concatenation** of strings and languages U and V as $UV = \{uv|u \in U, v \in V\}$, suppressing the distinction between a string and a one-member language, writing xU instead of $\{x\}U$, etc. The (λ-free) Kleene closure of strings and languages is defined analogously to the closure of alphabets. For a string w and a language U, we say $u \in L$ is a **prefix** of w if u is a left factor of w and no smaller left factor of w is in U.

Finite languages have the same distinguished status among all stringsets that the natural numbers $\mathbb{N}$ have among all numbers: they are, after all, all that can be directly listed without relying on any additional interpretative mechanism. And as in arithmetic, where the simplest natural superset of the integers includes not only finite decimal fractions but some infinite ones as well, the simplest natural superset of the finite languages is best defined by closure under operations (both Boolean and string operations) and will contain some infinite languages as well. We call **regular** all finite languages, and all languages that can be obtained from these by repeated application of union, intersection, complementation, concatenation, and Kleene closure. The classic Kleene theorem guarantees that regular languages have the same distinguished status among languages that the rationals in $\mathbb{Q}$ have among numbers in $\mathbb{R}$.

Exercise 2.2 Let F be the language of Fibonacci numbers written in base one. Is F finitely generated?

The generative grammars defining stringsets typically use an alphabet V that is a proper superset of Σ that contains the symbols of interest. Elements of $N = V \setminus \Sigma$ are called **nonterminal symbols** or just **nonterminals** to distinguish them from elements of Σ (called **terminal symbols** or **terminals**). Nonterminals play only a transient role in generating the objects of real interest, inasmuch as the yield of a grammar is explicitly restricted to terminal strings – the name nonterminal comes from the notion that a string containing them corresponds to a stage of the derivation that has not (yet) terminated. In **context-free grammars** (CFGs), we use a **start symbol** $S \in N$ and **productions** or **rewrite rules** of the form $A \rightarrow v$, where $A \in N$ and $v \in V^*$.

Example 2.3.1 A CFG for base ten integers. We use nonterminals `SIGN` and `DIGIT` and posit the rules $S \rightarrow$ `SIGN DIGIT`; $S \rightarrow$ `DIGIT`; `DIGIT` $\rightarrow$ `DIGIT DIGIT`; `DIGIT` $\rightarrow$ 0; `DIGIT` $\rightarrow$ 1; ... `DIGIT` $\rightarrow$ 9; `SIGN` $\rightarrow$ +; `SIGN` $\rightarrow$ –. (The nonterminals are treated here as atomic symbols rather than strings of Latin letters. We use whitespace to indicate token boundaries rather than the Algol convention of enclosing each token in $\langle\rangle$.) At the first step of the derivation, we can only choose the first or the second rule (since no other rule rewrites S) and we obtain the string `SIGN DIGIT` or `DIGIT`. Taking the first option and using the last rule to rewrite `SIGN`, we obtain `-DIGIT`, and using the third rule n times, we get `-DIGIT`$^{n+1}$. By eliminating the nonterminals, we obtain a sequence of $n+1$ decimal digits preceded by the minus sign.

Discussion Needless to say, base ten integers are easy to define by simpler methods (see Section 6.1), and the CFG used above is overkill also in the sense that strings with three or more digits will have more than one derivation. **Context-free languages** (languages generated by a CFG) are a proper superset of regular languages. For example, consider the CFG with nonterminal S, terminals a, b, and rewrite rules $S \rightarrow aSa$; $S \rightarrow bSb$; $S \rightarrow a$; $S \rightarrow b$; $S \rightarrow \lambda$. It is easily seen that this grammar defines the language of *palindromes* over $\{a, b\}$, which contains exactly those strings that are their own reversal (mirror image).

Exercise 2.3 Given a CFG G generating some CFL L not containing the empty string, create another CFG G' generating the same language such that every production has the form $A \rightarrow bC$, where A and C are nonterminals (members of N) and b is a terminal (member of Σ).

Continuing with the comparison to numbers, CFLs play the same role among languages that algebraic numbers play among the reals. To appreciate this, one needs to generalize from the view of languages as stringsets to a view of languages as mappings (from strings to weights in a semiring or similar structure). We take up this matter in Chapter 5.4.

Exercise 2.4 Prove that the language of palindromes is not regular.

If a context-free rewrite rule $A \rightarrow v$ is applied to a string iAj and l is a right factor of i (r is a left factor of j), we say that the rule is applied **in the left context** l (**in the right context** r). **Context-sensitive** rewrite rules are defined as triples (p, l, r), where p is a context-free production as above, and l and r are (possibly empty) strings defining the left and the right contexts of the rule in question. In keeping with the structuralist morphology and phonology of the 1950s, the # symbol is often used as an *edge marker* signifying the beginning or end of a string.

Traditionally, p is called the **structural change**, the context, written l_r, is called the **structural description**, and the triple is written as the structural change separated from the structural description by / (assumed to be outside the alphabet P). For example, a rule that deletes a leading zero from an unsigned decimal could be written $0 \rightarrow \lambda/\#\_$, and the more general rule that deletes it irrespective of the presence of a sign could be written $0 \rightarrow \lambda/\{\#, +, -\}\_$. Note that the right context of these rules is empty (it does not matter what digit, if any, follows the leading 0), while the left context 'edge of string' needs to be explicitly marked by the # symbol for the rules to operate correctly.

When interpreted as WFCs, the context statements simply act as filters on derivations: an otherwise legitimate rewriting step $iAj \rightarrow ivj$ is blocked (deemed ill-formed) unless l is a right factor of i and r is a left factor of j. This notion of context-sensitivity adds nothing to the generative power of CFGs: the resulting system is still capable only of generating CFLs.

Theorem 2.3.1 (McCawley 1968) Context-free grammars with context checking generate only context-free languages.

However, if context-sensitivity is part of the generation process, we can obtain context-sensitive languages (CSLs) that are not CFLs. If λ-rules (rewriting nonterminals as the empty string in some context) are permitted, every r.e. language can be generated. If such rules are disallowed, we obtain the CSL family proper (the case when CSLs contain the empty string has to be treated separately).

Theorem 2.3.2 (Jones 1966) The **context-sensitive (CS)** family of languages that can be generated by λ-free context-sensitive productions is the same as the family of languages that can be generated by using only length-increasing productions (i.e. productions of the form $u \rightarrow v$, where $l(v) \geq l(u)$ holds) and the same as the family of languages computable by **linear bounded automata** (LBAs).

LBA are one-tape Turing machines (TMs) that accept on the empty tape, with the additional restriction that at all stages of the computation, the reading head must remain on the portion of the tape that was used to store the string whose membership is to be decided.

These results are traditionally summarized in the *Chomsky hierarchy*: assigning regular languages to Type 3, CFLs to Type 2, CSLs to Type 1, and r.e. languages to Type 0, Chomsky (1956) demonstrated that each type is properly contained in the next lower one. These proofs, together with examples of context-free but not regular, context-sensitive but not context-free, and recursive but not context-sensitive languages, are omitted here, as they are discussed in many excellent textbooks of formal language theory such as Salomaa (1973) or Harrison (1978). To get a better feel for CSLs, we note the following results:

Theorem 2.3.3 (Karp 1972) The membership problem for CSLs is PSPACE-complete.

Theorem 2.3.4 (Szelepcsényi 1987, Immerman 1988) The complement of a CSL is a CSL.

Exercise 2.5 Construct three CSGs that generate the language F of Fibonacci numbers in base one, the language F_2 of Fibonacci numbers in base two, and the language F_{10} of Fibonacci numbers in base ten. Solve the membership problem for 117467.

Exercise 2.6 Call a set of natural numbers k-regular if their base k representations are a regular language over the alphabet of k digits. It is easy to see that a 1-regular language is 2-regular (3-regular) and that the converse is not true. Prove that a set that is both 2-regular and 3-regular is also 1-regular.

2.4 Further reading

Given that induction is as old as mathematics itself (the key idea going back at least to Euclid's proof that there are infinitely many primes) and that recursion can be traced back at least to Fibonacci's (1202) *Liber Abaci*, it is somewhat surprising that the closely related notion of generation is far more recent: the first systematic use is in von Dyck (1882) for free groups. See Chandler and Magnus (1982 Ch. I.7) for some fascinating speculation why the notion did not arise earlier within group theory. The kernel membership problem is known as the *word problem* in this setting (Dehn 1912). The use of freely generated pure formula models in logic was pioneered by Herbrand (1930); Wang tilings were introduced by Wang (1960). Theorem 2.1.1 was proven by Berger (1966), who demonstrated the undecidability by encoding the halting problem in tiles. For a discussion, see Gruska (1997 Sec. 6.4.3). The notion that linguistic structures are noncounting goes back at least to Chomsky (1965:55).

From Pāṇini to the *neogrammarians* of the 19th century, linguists were generally eager to set up the system so as to cover related styles, dialects, and historical stages of the same language by minor variants of the same theory. In our terms this would mean that e.g. British English and American English or Old English and Modern English would come out as models of a single 'abstract English'. This is one point where current practice (starting with de Saussure) differs markedly from the traditional approach. Since grammars are intended as abstract theories of the native speaker's competence, they cannot rely on data that are not observable by the ordinary language learner. In particular, they are restricted to a single temporal slice, called the *synchronic* view by de Saussure, as opposed to a view encompassing different historical stages (called the *diachronic* view). Since the lack of cross-dialectal or historical data is never an impediment in the process of children acquiring their native language (children are capable of constructing their internal grammar without access to such data), by today's standards it would raise serious methodological problems for the grammarian to rely on facts outside the normal range of input available to children. (De Saussure actually arrived at the synchrony/diachrony distinction based on somewhat different considerations.) The neogrammarians amassed a great deal of knowledge about *sound change*, the historical process whereby words change their pronunciation over the centuries, but some of their main tenets, in particular the exceptionlessness of sound change laws, have been found not to hold universally (see in particular Wang 1969, Wang and Cheng 1977, Labov 1981, 1994).

Abstract string manipulation begins with Thue (1914, reprinted in Nagell 1977), who came to the notion from combinatorial group theory. For Thue, rewriting is symmetrical: if AXB can be rewritten as AYB the latter can also be rewritten as the former. This is how Harris (1957) defined transformations. The direct precursors of the modern generative grammars and transformations that were introduced by Chomsky (1956, 1959) are semi-Thue systems, where rewriting need not necessarily work in both directions. The basic facts about regular languages, finite automata, and Kleene's theorem are covered in most textbooks about formal language theory or the foundations of computer science, see e.g. Salomaa (1973) or Gruska (1997). We will develop the connection between these notions and semigroup theory along the lines of Eilenberg (1974) in Chapter 5. Context-free grammars and languages are also well covered in computer science textbooks such as Gruska (1997), for more details on context-sensitivity, see Section 10 of Salomaa (1973). Theorem 2.3.1 was discovered in (McCawley 1968), for a rigorous proof see Peters and Ritchie (1973), and for a modern discussion, see Oehrle (2000).

Some generalizations of the basic finite state notions that are of particular interest to phonologists, namely regular relations, and finite k-automata, will be discussed in Chapter 3. Other generalizations, which are also relevant to syntax, involve weighted (probabilistic) languages, automata, and transducers – these are covered in Sections 5.4 and 5.5. Conspiracies were first pointed out by Kisseberth (1970) – we return to this matter in Section 4.3. The founding papers on categorial grammars are Ajdukiewicz (1935) and Lambek (1958). Unification grammars are discussed in Shieber (1986, 1992).

3

Phonology

The fundamental unit of linguistics is the *sign*, which, as a first approximation, can be defined as a conventional pairing of sound and meaning. By *conventional* we mean both that signs are handed down from generation to generation with little modification and that the pairings are almost entirely arbitrary, just as in bridge, where there is no particular reason for a bid of two clubs in response to one no trump to be construed as an inquiry about the partner's major suits. One of the earliest debates in linguistics, dramatized in Plato's *Cratylus*, concerns the arbitrariness of signs. One school maintained that for every idea there is a true sound that expresses it best, something that makes a great deal of sense for *onomatopoeic* words (describing e.g. the calls of various animals) but is hard to generalize outside this limited domain. Ultimately the other school prevailed (see Lyons 1968 Sec. 1.2 for a discussion) at least as far as the word-level pairing of sound and meaning is concerned.

It is desirable to build up the theory of sounds without reference to the theory of meanings both because the set of atomic units of sound promises to be considerably simpler than the set of atomic units of meanings and because sounds as linguistic units appear to possess clear physical correlates (acoustic waveforms; see Chapter 8), while meanings, for the most part, appear to lack any direct physical embodiment. There is at least one standard system of communication, Morse code, that gets by with only two units, dot (short beep) and dash (long beep) or possibly three, (if we count pause/silence as a separate unit; see Ex. 7.7). To be sure, Morse code is parasitic on written language, which has a considerably larger alphabet, but the enormous success of the alphabetic mode of writing itself indicates clearly that it is possible to analyze speech sounds into a few dozen atomic units, while efforts to do the same with meaning (such as Wilkins 1668) could never claim similar success.

There is no need to postulate the existence of some alphabetic system for transcribing sounds, let alone a meaning decomposition of some given kind. In Section 3.1 we will start with easily observable entities called *utterances*, which are defined as maximal pause-free stretches of speech, and describe the concatenative building blocks of sound structure called *phonemes*. For each natural language L these will act as a convenient set of atomic symbols P_L that can be manipulated by context-sensitive string-rewriting techniques, giving us what is called the *segmental phonology* of the language. This is not to say that the set of words W_L, viewed as a formal language over P_L, will be context-sensitive (Type 1) in the sense of formal language theory. On the contrary, we have good reasons to believe that W is in fact regular (Type 3).

To go beyond segments, in Section 3.2 we introduce some subatomic components called *distinctive features* and the formal linguistic mechanisms required to handle them. To a limited extent, distinctive features pertaining to tone and stress are already useful in describing the *suprasegmental phonology* of languages. To get a full understanding of suprasegmentals in Section 3.3 we introduce *multitiered* data structures more complex than strings, composed of *autosegments.* Two generalizations of regular languages motivated by phonological considerations, regular transducers and regular k-languages, are introduced in Section 3.4. The notions of prosodic hierarchy and optimality, being equally relevant for phonology and morphology, are deferred to Chapter 4.

3.1 Phonemes

We are investigating the very complex *interpretation relation* that obtains between certain structured kinds of sounds and certain structured kinds of meanings; our eventual goal is to define it in a generative fashion. At the very least, we must have some notion of identity that tells us whether two signs sound the same and/or mean the same. The key idea is that we actually have access to more information, namely, whether two utterances are *partially similar* in form and/or meaning. To use Bloomfield's original examples:

> A needy stranger at the door says *I'm hungry.* A child who has eaten and merely wants to put off going to bed says *I'm hungry*. Linguistics considers only those vocal features which are alike in the two utterances ... Similarly, *Put the book away* and *The book is interesting* are partly alike *(the book).*

That the same utterance can carry different meanings at different times is a fact we shall not explore until we introduce *disambiguation* in Chapter 6 – the only burden we now place on the theory of meanings is that it be capable of (i) distinguishing meaningful from meaningless and (ii) determining whether the meanings of two utterances share some aspect. Our expectations of the observational theory of sound are similarly modest: we assume we are capable of (i′) distinguishing pauses from speech and (ii′) determining whether the sounds of two utterances share some aspect.

We should emphasize at the outset that the theory developed on this basis does not rely on our ability to exercise these capabilities to the extreme. We have not formally defined what constitutes a pause or silence, though it is evident that observationally such phenomena correspond to very low acoustic energy when integrated over a period of noticeable duration, say 20 milliseconds. But it is not necessary to be able to decide whether a 19.2 millisecond stretch that contains exactly 1.001 times the physiological minimum of audible sound energy constitutes a pause or not. If this stretch is indeed a pause we can always produce another instance, one that will have a significantly larger duration, say 2000 milliseconds, and containing only one-tenth of the previous energy. This will show quite unambiguously that we had two utterances in the first place. If it was not a pause, but rather a functional part of sound formation such as a stop closure, the new 'utterances' with the artificially interposed pause will be deemed ill-formed by native speakers of the language. Similarly, we need not worry a great deal whether *Colorless green ideas sleep furiously* is meaningful, or what it exactly means. The techniques described here are robust enough to perform well on the basis of ordinary data without requiring us to make ad hoc decisions in the edge cases. The reason for this robustness comes from the fact that when viewed as a probabilistic ensemble, the edge cases have very little weight (see Chapter 8 for further discussion).

The domain of the interpretation relation I is the set of *forms* F, and the codomain is the set of *meanings* M, so we have $I \subset F \times M$. In addition, we have two *overlap* relations, $O_F \subset F \times F$ and $O_M \subset M \times M$, that determine partial similarity of form and meaning respectively. O_F is traditionally divided into *segmental* and *suprasegmental* overlaps. We will discuss mostly segmental overlap here and defer suprasegmentals such as tone and stress to Section 3.3 and Section 4.1, respectively. Since speech happens in time, we can define two forms α and β as *segmentally overlapping* if their temporal supports as intervals on the real line can be made to overlap, as in the *the book* example above. In the segmental domain at least, we therefore have a better notion than mere overlap: we have a partial ordering defined by the usual notion of interval containment. In addition to O_F, we will therefore use sub- and superset relations (denoted by $\subset_F, \supset_F$) as well as intersection, union, and complementation operations in the expected fashion, and we have

$$\alpha \cap_F \beta \neq \emptyset \Rightarrow \alpha O_F \beta \tag{3.1}$$

In the domain of I, we find obviously complex forms such as a full epic poem and some that are atomic in the sense that

$$\forall x \subset_F \alpha : x \notin dom(I) \tag{3.2}$$

These are called *minimum forms*. A form that can stand alone as an utterance is a *free form*; the rest (e.g. forms like *ity* or *al* as in *electricity, electrical*), which cannot normally appear between pauses, are called *bound forms*.

Typically, utterances are full phrases or sentences, but when circumstances are right, e.g. because a preceding question sets up the appropriate context, forms much smaller than sentences can stand alone as complete utterances. Bloomfield (1926) defines a *word* as a minimum free form. For example, *electrical* is a word because it is a free form (can appear e.g. as answer to the question *What kind of engine is in this car?*) and it cannot be decomposed further into free forms (*electric* would be free but *al* is bound). We will have reason to revise this definition in Chapter 4, but for now we can provisionally adopt it here because in defining phonemes it is sufficient to restrict ourselves to free forms.

For the rest of this section, we will only consider the set of words $W \subset F$, and we are in the happy position of being able to ignore the meanings of words entirely.

We may know that forms such as *city* and *velocity* have nothing in common as far as their meanings are concerned and that we cannot reasonably analyze the latter as containing the former, but we also know that the two rhyme, and as far as their forms are concerned *velocity* = *velo.city*. Similarly, *velo* and *kilo* share the form *lo* so we can isolate *ve*, *ki*, and *lo* as more elementary forms.

In general, if pOq, we have a nonempty u such that $p = aub$, $q = cud$. a, b, c, d, u will be called **word fragments** obtained from comparing p and q, and we say p **is a subword of** q, denoted $p \prec q$, if $a = b = \lambda$. We denote by $\tilde{W}$ the smallest set containing W and closed under the operation of taking fragments – $\tilde{W}$ contains all and only those fragments that can be obtained from W in finitely many steps.

By successively comparing forms and fragments, we can rapidly extract a set of short fragments P that is sufficiently large for each $w \in \tilde{W}$ to be a concatenation of elements of P and sufficiently small that no two elements of it overlap. A **phonemic alphabet** P is therefore defined by (i) $\tilde{W} \subset P^*$ and (ii) $\forall p, q \in P : pO_F q \Rightarrow p = q$. To forestall confusion, we emphasize here that P consists of mental rather than physical units, as should be evident from the fact that the method of obtaining them relies on human oracles rather than on some physical definition of (partial) similarity. The issue of relating these mental units to physical observables will be taken up in Chapters 8 and 9.

We emphasize here that the procedure for finding P does not depend on the existence of an alphabetic writing system. All it requires is an informant (oracle) who can render judgments about partial similarity, and in practice this person can just as well be illiterate. Although the number of unmapped languages is shrinking, to this day the procedure is routinely carried out whenever a new language is encountered. In some sense (to be made more precise in Section 7.3), informant judgments provide more information than is available to the language learner: the linguist's *discovery procedure* is driven both by the positive (grammatical) and negative (ungrammatical) data, while it is generally assumed that infants learning the language only have positive data at their disposal, an assumption made all the more plausible by the wealth of language acquisition research indicating that children ignore explicit corrections offered by adults.

For an arbitrary set W endowed with an arbitrary overlap relation O_F, there is no guarantee that a phonemic alphabet exists; for example, if W is the set of intervals $[0, 2^{-n}]$ with overlap defined in the standard manner, $P = \{[2^{-(i+1)}, 2^{-i}] | i \geq 0\}$ will enjoy (ii) but not (i). In actual word inventories W and their extensions $\tilde{W}$, we never see the phenomenon of an infinite descending chain of words or fragments $w_1, w_2, \ldots$ such that each w_{i+1} is a proper part of w_i, nor can we find a large number (say $> 2^8$) words or fragments such that no two of them overlap. We call such statements of contingent facts about the real world *postulates* to distinguish them from ordinary axioms, which are not generally viewed as subject to falsification.

Postulate 3.1.1 Foundation. Any sequence of words and word fragments $w_1, w_2, \ldots$ such that each $w_{i+1} \prec w_i, w_{i+1} \neq w_i$, terminates after a finite number of steps.

Postulate 3.1.2 Dependence. Any set of words or word fragments $w_1, w_2, \ldots w_m$ contains two different but overlapping words or fragments for any $m > 2^8$.

From these two postulates both the existence and uniqueness of phonetic alphabets follow. Foundation guarantees that every $w \in W$ contains at least one atom under $\prec$, and dependence guarantees that the set P of atoms is finite. Since different atoms cannot overlap, all that remains to be seen is that every word of $\tilde{W}$ is indeed expressible as a concatenation of atoms. Suppose indirectly that q is a word or fragment that could not be expressed this way: either q itself is atomic or we can find a fragment q_1 in it that is not expressible. Repeating the same procedure for q_1, we obtain $q_2, \ldots, q_n$. Because of Postulate 3.2.1, the procedure terminates in an atomic q_n. But by the definition of P, q_n is a member of it, a contradiction that proves the indirect hypothesis false.

Discussion Nearly every communication system that we know of is built on a finite inventory of discrete symbols. There is no law of nature that would forbid a language to use measure predicates such as *tall* that take different vowel lengths in proportion to the tallness of the object described. In such a hypothetical language, we could say *It was taaaaaaall* to express the fact that something was seven times as tall as some standard of comparison, and *It was taaall* to express that it was only three times as tall. The closest thing we find to this is in Arabic/Persian calligraphy, where joining elements are sometimes sized in accordance with the importance of a word, or in Web2.0-style tag clouds, where font size grows with frequency. Yet even though analog signals like these are always available, we find that in actual languages they are used only to convey a discrete set of possible values (see Chapter 9), and no communication system (including calligraphic text and tag clouds) makes their use obligatory.

Postulates 3.1.1 and 3.1.2 go some way toward explaining why discretization of continuous signals must take place. We can speculate that foundation is necessitated by limitations of perception (it is hard to see how a chain could descend below every perceptual threshold), and dependence is caused by limitations of memory (it is hard to see how an infinite number of totally disjoint atomic units could be kept in mind). No matter how valid these explanations turn out to be, the postulates have a clear value in helping us to distinguish linguistic systems from nonlinguistic ones. For example, the dance of bees, where the direction and size of figure-8 movements is directly related to the direction and distance from the hive to where food can be collected (von Frisch 1967), must be deemed nonlinguistic, while the genetic code, where information about the composition of proteins is conveyed by DNA/RNA strings, can at least provisionally be accepted as linguistic.

Following the tradition of Chomsky (1965), memory limitations are often grouped together with mispronunciations, lapses, hesitations, coughing, and other minor errors as **performance** factors, while more abstract and structural properties are treated as *competence* factors. Although few doubt that some form of the competence vs. performance distinction is valuable, at least as a means of keeping the noise out of the data, there has been a great deal of debate about where the line between the two should be drawn. Given the orthodox view that limitations of memory and perception are matters of performance, it is surprising that such a deeply structural property as the existence of phonetic alphabets can be derived from postulates rooted in these limitations.

3.2 Natural classes and distinctive features

Isolating the atomic segmental units is a significant step toward characterizing the phonological system of a language. Using the phonemic alphabet P, we can write every word as a string $w \in P^*$, and by adding just one extra symbol # to denote the pause between words, we can write all utterances as strings over $P \cup \{\#\}$. Since in actual *connected* speech pauses between words need not be manifest, we need an interpretative convention that # can be *phonetically realized* either as silence or as the empty string (zero realization). Silence, of course, is distinctly audible and has positive duration (usually 20 milliseconds or longer), while λ cannot be heard and has zero duration.

In fact, similar interpretative conventions are required throughout the alphabet, e.g. to take care of the fact that in English word-initial *t* is *aspirated* (released with a puff of air similar in effect to *h* but much shorter), while in many other positions *t* is *unaspirated* (released without an audible puff of air): compare *ton* to *stun*. The task of relating the abstract units of the alphabet to their audible manifestations is a complex one, and we defer the details to Chapter 9. We note here that the interpretation process is by no means trivial, and there are many unassailable cases, such as aspirated vs. unaspirated *t* and silenceful vs. empty #, where we permit two or more alternative realizations for the same segment. (Here and in what follows we reserve the term **segment** for alphabetic units; i.e. strings of length one.)

Since λ can be one of the alternatives, an interesting technical possibility is to permit cases where it is the only choice: i.e. to declare elements of a phonemic alphabet that never get realized. The use of such *abstract* or **diacritic** elements *(anubandha)* is already pivotal in Pāṇini's system and remains characteristic of phonology to this day. This is our first example of the linguistic distinction between *underlying* (abstract) and *surface* (concrete) forms – we will see many others later.

Because in most cases alternative realizations of a symbol are governed by the symbols in its immediate neighborhood, the mathematical tool of choice for dealing with most of segmental phonology is string rewriting by means of context-sensitive rules. Here a word of caution is in order: from the fact that context-sensitive rules are used it does not follow that the generated stringset over P, or over a larger alphabet Q that includes abstract elements as well, will be context-sensitive. We defer this issue to Section 3.4, and for now emphasize only the convenience of context-sensitive rules, which offer an easy and well-understood mechanism to express the phonological regularities or *sound laws* that have been discovered over the centuries.

Example 3.2.1 Final devoicing in Russian. The nominative form of Russian nouns can be predicted from their dative forms by removing the dative suffix *u* and inspecting the final consonant: if it is *b* or *p*, the final consonant of the nominative form will be *p*. This could be expressed in a phonological rule of *final b devoicing*: $b \rightarrow p/\_\#$.

When it is evident that the change is caused by some piece of the environment where the rule applies, we speak of the piece *triggering* the change; here the trigger is the final #.

Remarkably, we find that a similar rule links d to t, g to k, and in fact any voiced obstruent to its voiceless counterpart. The phenomenon that the structural description and/or the structural change in rules extends to some disjunction of segments is extremely pervasive. Those sets of segments that frequently appear together in rules (either as triggers or as undergoers) are called *natural classes*; for example, the class $\{p, t, k\}$ of *unvoiced stops* and the class $\{b, d, g\}$ of *voiced stops* are both natural, while the class $\{p, t, d\}$ is not. Phonologists would be truly astonished to find a language where some rule or regularity affects *p*, *t*, and *d* but no other segment.

The linguist has no control over the phonemic alphabet of a language: P is computed as the result of a specific (oracle-based, but otherwise deterministic) algorithm. Since the set $N \subset 2^P$ of natural classes is also externally given by the phonological patterning of the language, over the millennia a great deal of effort has been devoted to the problem of properly characterizing it, both in order to shed some light on the structure of P and to help simplify the statement of rules.

So far, we have treated P as an unordered set of alphabetic symbols. In the Ashṭādhyāyī, Pāṇini arranges elements of P in a linear sequence (the *śivasūtras*) with some abstract (phonetically unrealized) symbols *(anubandha)* interspersed. Simplifying his treatment somewhat (for a fuller discussion, see Staal 1962), natural classes *(pratyāhāra)* are defined in his 1.1.71 as those subintervals of the *śivasūtras* that end in some *anubandha*. If there are k symbols in P, in principle there could be as many as 2^k natural classes. However, the Pāṇinian method will generate at most $k(k+1)/2$ subintervals (or even fewer, if diacritics are used more sparingly), which is in accordance with the following postulate.

Postulate 3.2.1 In any language, the number of natural classes is small.

We do not exactly spell out what 'small' means here. Certainly it has to be polynomial, rather than exponential, in the size of P. The European tradition reserves names for many important natural classes such as the *apicals*, *aspirates*, *bilabials*, *consonants*, *continuants*, *dentals*, *fricatives*, *glides*, *labiodentals*, *linguals*, *liquids*, *nasals*, *obstruents*, *sibilants*, *stops*, *spirants*, *unaspirates*, *velars*, *vowels*, etc. – all told, there could be a few hundred, but certainly not a few thousand, such classes. As these names suggest, the reason why a certain class of sounds is natural can often be found in sharing some aspects of production (e.g. all sounds crucially involving a constriction at the lips are *labials*, and all sounds involving turbulent airflow are *fricatives*), but often the justification is far more complex and indirect. In some cases, the matter of whether a particular class is natural is heavily debated. For a particularly hard chestnut, the *ruki* class; see Section 9.2, Collinge's (1985) discussion of Pedersen's law I, and the references cited therein.

For the mathematician, the first question to ask about the set of natural classes N is neither its size nor its exact membership but rather its algebraic structure: under what operations is N closed? To the extent that Pāṇini is right, the structure is not fully Boolean: the complement of an interval typically will not be expressible as a single interval, but the intersection of two intervals *(pratyāhāra)* will again be an interval. We state this as the following postulate.

Postulate 3.2.2 In any language, the set of natural classes is closed under intersection.

This postulate makes N a meet semilattice, and it is clear that the structure is not closed under complementation since single segments are natural classes but their complements are not. The standard way of weakening the Boolean structure is to consider meet semilattices of linear subspaces. We embed P in a hypercube so that natural classes correspond to hyperplanes parallel to the axes. The basis vectors that give rise to the hypercube are called **distinctive features** and are generally assumed to be binary; a typical example is the *voiced/unvoiced* distinction that is defined by the presence/absence of periodic vocal fold movements. It is debatable whether the field underlying this vector space construct should be $\mathbb{R}$ or GF(2). We take the second option and use GF(2), but we will have reason to return to the notion of real-valued features in Chapters 8 and 9. Thus, we define a **feature assignment** as an injective mapping C from the set Q of segments into the linear space GF(2,n).

This is a special case of a general situation familiar from universal algebra: if A_i are algebras of the same signature and $A = \prod A_i$ is their direct product, we say that a subalgebra B of A is a **subdirect product** of the A_i if all its projections on the components A_i are surjective. A classic theorem of Birkhoff asserts that every algebra can be represented as a subdirect product of subdirectly irreducible algebras. Here the algebras are simply finite sets, and as the only subdirectly irreducible sets have one or two members (and one-member sets obviously cannot contribute to a product), we obtain distinctive feature representations (also called **feature decompositions**) for any set for free.

Since any set, not just phonological segments, could be defined as vectors (also called *bundles*) of features, to give feature decomposition some content that is specific to phonology we must go a step further and link natural classes to this decomposition. This is achieved by defining as **natural classes** those sets of segments that can be expressed by fewer features than their individual members (see Halle 1964:328). To further simplify the use of natural classes, we assume a theory of *markedness* (Chomsky and Halle 1968 Ch. IX) that supplies those features that are predictable from the values already given (see Section 7.3). For example, high vowels will be written as $\begin{bmatrix} +\text{syll} \\ +\text{high} \end{bmatrix}$, requiring only two features, because the other features that define this class, such as [−low] or [+voice], are predictable values already given.

In addition to using *pratyāhāra*, Pāṇini employs a variety of other devices, most notably the concept of 'homogeneity' *(sāvarṇya)*, as a means of cross-classification (see Cardona 1965). This device enables him to treat quality distinctions in vowels separately from length, nasality, and tone distinctions, as well as to treat place of articulation distinctions in consonants separately from nasality, voicing, and aspiration contrasts. Another subsidiary concept, that of *antara* 'nearness', is required to handle the details of mappings between natural classes. Since Pāṇinian rules always map classes onto classes, the image of a segment under a rule is decided by P1.1.50

sthāne 'ntaratamaḥ 'in replacement, the nearest'. The modern equivalent of P1.1.50 is the convention that features unchanged by a rule need not be explicitly mentioned, so that the Russian final devoicing rule that we began with may simply be stated as [+obstruent] → [−voice] / _#.

For very much the same empirical reasons that forced Pāṇini to introduce additional devices like *sāvarṇya*, the contemporary theory of features also relaxes the requirement of full orthogonality. One place where the standard (Chomsky and Halle 1968) theory of distinctive features shows some signs of strain is the treatment of vowel height. Phonologists and phoneticians are in broad agreement that vowels come in three varieties, *high*, *mid*, and *low*, which form an interval structure: we often have reason to group high and mid vowels together or to group mid and low vowels together, but we never see a reason to group high and low vowels together to the exclusion of mid vowels. The solution adopted in the standard theory is to use two binary features, [± high] and [± low], and to declare the conjunction [+high, +low] ill-formed.

Similar issues arise in many other corners of the system; e.g. in the treatment of *place of articulation* features. Depending on where the major constriction that determines the type of a consonant occurs, we distinguish several places of articulation, such as *bilabial*, *labiodental*, *dental*, *alveolar*, *postalveolar*, *retroflex*, *palatar*, *velar*, *pharyngeal*, *epiglottal*, and *glottal*, moving back from the lips to the glottis inside the vocal tract. No single language has phonemes at every point of articulation, but many show five-, or six-way contrasts. For example, Korean distinguishes bilabial, dental, alveolar, velar, and glottal, and the difference is noted in the basic letter shape (□, ∨, ㅅ, ㄱ, and ○, respectively). Generally, there is more than one consonant per point of articulation; for example, English has alveolars n, t, d, s, z, l. Consonants sharing the same place of articulation are said to be *homorganic* and they form a natural class (as can be seen e.g. from rules of nasal assimilation that replace e.g. *input* by *imput*).

Since the major classes (labial, coronal, dorsal, radical, laryngeal) show a five-way contrast, the natural way to deal with the situation would be the use of one GF(5)-valued feature rather than three (or more) underutilized GF(2) values, but for reasons to be discussed presently this is not a very attractive solution. What the system really needs to express is the fact that some features tend to occur together in rules to the exclusion of others, a situation somewhat akin to that observed among the segments. The first idea that leaps to mind would be to utilize the same solution, using features of features *(metafeatures)* to express natural classes of features. The Cartesian product operation that is used in the feature decomposition (subdirect product form) of P is associative, and therefore it makes no difference whether we perform the feature decomposition twice in a metafeature setup, or just once at the segment level. Also, the inherent ordering of places of articulation (for consonants) or height (for vowels) is very hard to cenvey by features, be they 2-valued or n-valued, without recourse to arithmetic notions, something we would very much like to avoid as it would make the system overly expressive.

The solution now widely accepted in phonology (Clements 1985, McCarthy 1988) is to arrange the features in a tree structure, using intermediate **class nodes**

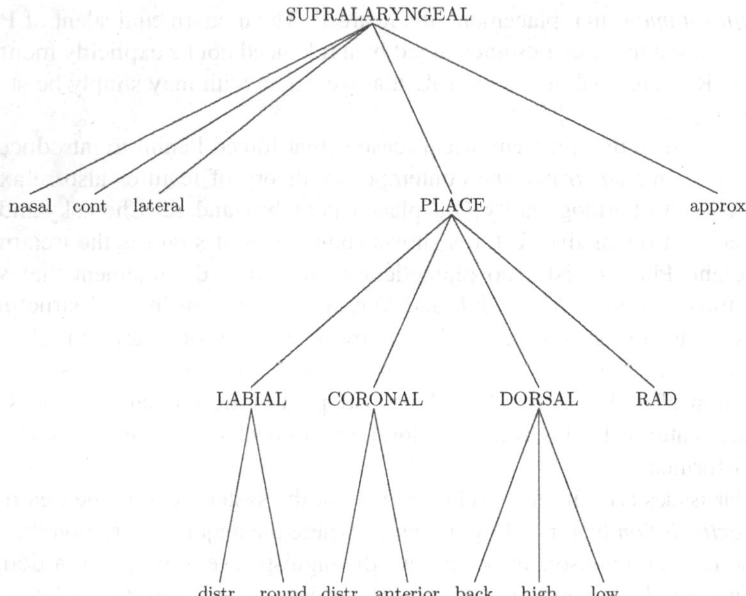

Fig. 3.1. Feature geometry tree. Rules that required the special principle of *sāvarṇya* can be stated using the *supralaryngeal class node*

to express the grouping together of some features to the exclusion of others (see Fig. 3.1). This solution, now permanently (mis)named **feature geometry**, is in fact a generalization of both the pratyāhāra and the standard feature decomposition methods. The linear intervals of the Pāṇinian model are replaced by generalized (lattice-theoretic) intervals in the subsumption lattice of the tree, and the Cartesian product appearing in the feature decomposition corresponds to the special case where the feature geometry tree is a star (one distinguished root node, all other nodes being leaves).

Discussion The segmental inventories P developed in Section 3.1 are clearly different from language to language. As far as natural classes and feature decomposition are concerned, many phonologists look for a single universal inventory of features arranged in a universally fixed geometry such as the one depicted in Fig. 3.1. Since the cross-linguistic identity of features such as [nasal] is anchored in their phonetic (acoustic and articulatory) properties rather than in some combinatorial subtleties of their intralanguage phonological patterning, this search can lead to a single object, unique up to isomorphism, that will, much like Mendeleyev's periodic table, encode a large number of regularities in a compact format.

Among other useful distinctions, Chomsky and Halle (1968) introduce the notion of *formal* vs. *substantive universals*. Using this terminology, meet semilattices are a formal, and a unique feature geometry tree such as the one in Fig. 3.1 would be a substantive, universal. To the extent that phonological research succeeds in identifying a unique feature geometry, every framework, such as semilattices, that permits a variety of geometries overgenerates. That said, any theory is interesting to people other than its immediate developers only to the extent that it can be generalized to problems other than the one it was originally intended to solve. Phonology, construed broadly as an abstract theory of linguistic form, applies not only to speech but to other forms of communication (handwritten, printed, signed, etc.) as well. In fact, phonemes, distinctive features, and feature geometry are widely used in the study of sign language (see e.g. Sandler 1989, Liddell and Johnson 1989); where substantive notions like nasality may lose their grip, the formal theory remains valuable. (However, as the abstract theory is rooted in the study of sound, we will keep on talking about 'utterances', 'phonemes', 'syllables', etc., rather than using 'gestures', 'graphemes', or other narrow terms.)

Exercise 3.2 What are the phonemes in the genetic code? How would you define feature decomposition and feature geometry there?

3.3 Suprasegmentals and autosegments

In Section 3.1 we noted that words can be partially alike even when they do not share any segments. For example, *blackbird* and *whitefish* share the property that they have a single stressed syllable, a property that was used by Bloomfield (1926) to distinguish them from multiword phrases such as *black bird* or *white fish*, which will often be pronounced without an intervening pause but never without both syllables stressed. In addition to stress, there are other *suprasegmentals*, such as *tone*, that appear to be capable of holding constant over multisegment stretches of speech, typically over syllables.

Traditionally, the theory of suprasegmentals has been considered harder than that of segmental phenomena for the following reasons. First, their physical correlates are more elusive: stress is related to amplitude, and tone to frequency, but the relationship is quite indirect (see Lehiste 1970). Second, informant judgments are harder to elicit: native speakers of a language often find it much harder to judge e.g. whether two syllables carry the same degree of stress than to judge whether they contain the same vowel. Finally, until recently, a notation as transparent as the alphabetic notation for phonemes was lacking. In this section, we will deal mainly with tone and tone-like features of speech, leaving the discussion of stress and prosody to Section 4.1.

Starting in the 1970s, phonological theory abandoned the standard string-based theory and notation in favor of a generalization called *autosegmental* theory. Autosegmental theory (so named because it encompasses not only suprasegmental but also subsegmental aspects of sound structure) generalizes the method of using a string over some alphabet P to k-tuples of strings connected by *association relations* that spell out which segments in the two strings are overlapping in time. We will begin with the simplest case, that of *bistrings* composed of two strings and an association relation. First, the strings are placed on *tiers*, which are very much like

Turing-machine tapes, except the number of blank squares between nonblank ones cannot be counted.

Definition 3.3.1 A **tier** is an ordered pair ($\mathbb{Z}$, N) where $\mathbb{Z}$ is the set of integers equipped with the standard identity and ordering relations '$=$' and '$<$' and N is the name of the tier.

The original example motivating the use of separate tiers was *tone*, which is phonologically distinctive in many languages. Perhaps the best-known example of this is Mandarin Chinese, where the same syllable *ma* means 'mother', 'hemp', 'horse', 'admonish', and 'wh (question particle)', depending on whether it is uttered with the first, second, third, fourth, or fifth tone. As the tonology of Chinese is rather complex (see e.g. Yip 2002), we begin with examples from lesser-known but simpler tonal systems, primarily from the Niger-Congo family, where two contrastive level tones called high and low (abbreviated H and L, and displayed over vowels as acute or grave accent; e.g. *má, mà*) are typical, three level tones (high, mid, low, H, M, L) are frequent, four levels (1, 2, 3, 4) are infrequent, and five levels are so rare that their analysis in terms of five distinct levels is generally questionable.

In the study of such systems, several salient (not entirely exceptionless, but nevertheless widespread) generalizations emerge. First, a single syllable may carry not just a single tone but also sequences of multiple tones, with HL realized as falling and LH as rising tones. Such sequences, known as *contour tones*, are easily marked by combining the acute (high) and grave (low) accent marks over the tone-bearing vowel (e.g. *mâ* for falling, *mǎ* for rising tone), but as the sequences get more complex, this notation becomes cumbersome (and the typographical difficulty greatly increases in cases where accent marks such as umlaut are also used to distinguish vowels such as *u* and *ü, o* and *ö*).

Second, sequences of multiple tones show a remarkable degree of stability in that deletion of the tone-bearing vowel need not be accompanied by deletion of the accompanying tone(s) – this 'autonomy' of tones motivates the name *auto*segmental. Third, processes like assimilation do not treat contour tones as units but rather the last level tone of a contour sequence continues, e.g. *mǎ+ta* does not become *mǎtǎ* but rather *mǎtá*. As an example of stability, consider the following example in Lomongo [LOL], where phrase-level rules turn *bàlóngó bǎkáé* 'his book' into *bàlóngǎkáé*: the H on the deleted *o* survives and attaches to the beginning of the LH contour of the first *a* of *(b)ǎkáé*. In the following autosegmental diagram, we segregate the segmental content from the tonal content by placing them on separate *tiers*:

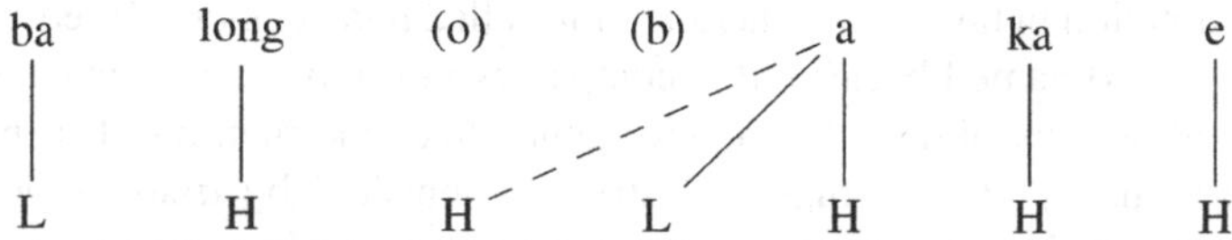

Each tier N has its own **tier alphabet** T_N, and we can assume without loss of generality that the alphabets of different tiers are disjoint except for a distinguished **blank** symbol G (purposely kept distinct from the pause symbol #) that is adjoined to every tier alphabet. Two tiers bearing identical names can only be distinguished by inspecting their contents. We define a tier containing a string $t_0t_1 \dots t_n$ starting at position k by a mapping that maps k on t_0, $k+1$ on $t_1, \dots, k+n$ on t_n, and everything else on G. Abstracting away from the starting position, we have the following definition.

Definition 3.3.2 A tier N **containing** a string $t_0t_1 \dots t_n$ over the alphabet $T_N \cup^* G$ is defined as the class of mappings F_k that take $k+i$ into t_i for $0 \le i \le n$ and to G if i is outside this range. Unless noted otherwise, this class will be represented by the mapping F_0. Strings containing any number of successive G symbols are treated as equivalent to those strings that contain only a single G at the same position. G-free strings on a given tier are called **melodies**.

Between strings on the same tier and within the individual strings, temporal ordering is encoded by their usual left-to-right ordering. The temporal ordering of strings on different tiers is encoded by association relations.

Definition 3.3.3 An **association relation** between two tiers N and M containing the strings n $= n_0n_1 \dots n_k$ and m $= m_0m_1 \dots m_l$ is a subset of $\{0, 1, \dots, k\} \times \{0, 1, \dots, l\}$. An element that is not in the domain or range of the association relation is called **floating**.

Note that the association relation, being an abstract pattern of synchrony between the tiers, is one step removed from the content of the tiers: association is defined on the *domain* of the representative mappings, while content also involves their *range*. By Definition 3.3.3, there are 2^{kl} association relations possible between two strings of length k and l. Of these relations, the *no crossing constraint* (NCC; see Goldsmith 1976) rules out as ill-formed all relations that contain pairs (i, v) and (j, u) such that $0 \le i < j \le k$ and $0 \le u < v \le l$ are both true. We define the **span** of an element x with respect to some association relation A as those elements y for which (x, y) is in A. Rolling the definitions above into one, we have the following definition.

Definition 3.3.4 A **bistring** is an ordered triple (f, g, A), where f and g are strings not containing G, and A is a well-formed association relation over two tiers containing f and g.

In the general case, we have several tiers arranged in a tree structure called the geometry of the representation (see Section 3.2). Association relations are permitted only among those tiers that are connected by an edge of this tree, so if there are k tiers there will be $k-1$ relations. Thus, in the general case, we define a k-**string** as a $(2k-1)$-tuple $(s_1, \dots, s_k, A_1, \dots, A_{k-1})$, where the s_i are strings and the A_i are association relations.

Theorem 3.3.1 The number of well-formed association relations over two tiers, each containing a string of length n, is asymptotically $(6 + 4\sqrt{2})^n$.

Proof Let us denote the number of well-formed association relations with n symbols on the top tier and k symbols on the bottom tier by $f(n, k)$. By symmetry, $f(n, k) = f(k, n)$, and obviously $f(n, 1) = f(1, n) = 2^n$. By enumerating relations according to the pair (i, j) such that no $i' < i$ is in the span of any j' and no $j'' > j$ is in the span of i, we get

$$f(n+1,k+1) = \sum_{i=1}^{k+1} f(n,i)2^{k+1-i} + f(n,k+1) \tag{3.3}$$

From (3.3) we can derive the following recursion:

$$f(n+1,k+1) = 2f(n+1,k) + 2f(n,k+1) - 2f(n,k) \tag{3.4}$$

For the first few values of $a_n = f(n,n)$, we can use (3.4) to calculate forward: $a_1 = 2$, $a_2 = 12$, $a_3 = 104$, $a_4 = 1008$, $a_5 = 10272$, $a_6 = 107712$, and so on. Using (3.4) we can also calculate backward and define $f(0,n) = f(n,0)$ to be 1 so as to preserve the recursion. The generating function

$$F(z,w) = \sum_{i,j=0}^{\infty} f(i,j)z^i w^j \tag{3.5}$$

will therefore satisfy the equation

$$F(z,w) = \frac{1 - \frac{z}{1-z} - \frac{w}{1-w}}{1 - 2z - 2w + 2zw} \tag{3.6}$$

If we substitute $w = t/z$ and consider the integral

$$\frac{1}{2\pi i}\int_C \frac{F(z,t/z)}{z}dz \tag{3.7}$$

this will yield the constant term $\sum_{n=0}^{\infty} f(n,n)t^n$ by Cauchy's formula. Therefore, in order to get the generating function

$$d(t) = \sum_{i=0}^{\infty} a_n t^n \tag{3.8}$$

we have to evaluate

$$\frac{1}{2\pi i}\int_C \frac{1 - \frac{z}{1-z} - \frac{t/z}{1-t/z}}{z(1 - 2z - 2t/z + 2t)}dz \tag{3.9}$$

which yields

$$d(t) = 1 + \frac{2t}{\sqrt{1 - 12t + 4t^2}} \tag{3.10}$$

$d(t)$ will thus have its first singularity when $\sqrt{1 - 12t + 4t^2}$ vanishes at $t_0 = (3 - \sqrt{8})/2$, yielding

$$a_n \approx (6 + 4\sqrt{2})^n \tag{3.11}$$

the desired asymptotics. ■

The base 2 logarithm of this number, $n \cdot 3.543$, measures how many bits we need to encode a bistring of length n. Note that this number grows linearly in the length of the bistring, while the number of (possibly ill-formed) association relations was 2^{n^2}, with the base 2 log growing quadratically. Association relations in general are depicted as bipartite graphs (pairs in the relation are called **association lines**) and encoded as two-dimensional arrays (the incidence matrix of the graph). However, the linear growth of information content suggests that well-formed association relations should be encoded as one-dimensional arrays or strings. Before turning to this matter in Section 3.4, let us first consider two particularly well-behaved classes of bistrings. A bistring is **fully associated** if there are no floating elements and **proper** if the span of any element on one tier will form a single substring on the other tier (Levin 1985). Proper relations are well-formed but not necessarily fully associated.

Let us define $g(i, j)$ as the number of association relations containing no unassociated (floating) elements and define b_n as $g(n, n)$. By counting arguments similar to those used above, we get the recursion

$$g(n+1, k+1) = g(n+1, k) + g(n, k+1) + g(n, k) \tag{3.12}$$

Using this recursion, the first few values of b_n can be computed as 1, 3, 13, 63, 321, 1683, 8989, and so on. Using (3.12) we can calculate backward and define $g(0, 0)$ to be 1 and $g(i, 0) = g(0, i)$ to be 0 (for $i > 0$) so as to preserve the recursion. The generating function

$$G(z, w) = \sum_{i,j=0}^{\infty} g(i, j) z^i w^j \tag{3.13}$$

will therefore satisfy the equation

$$G(z, w) = \frac{1 - z - w}{1 - z - w - zw} = 1 + \frac{zw}{1 - z - w - zw} \tag{3.14}$$

Again we substitute $w = t/z$ and consider the integral

$$\frac{1}{2\pi i} \int_C \frac{G(z, t/z)}{z} dz \tag{3.15}$$

which will yield the constant term $\sum_{n=0}^{\infty} g(n, n) t^n$ by Cauchy's formula. Therefore, in order to get the generating function

$$e(t) = \sum_{i=0}^{\infty} b_n t^n \tag{3.16}$$

we have to evaluate

$$\frac{1}{2\pi i} \int_C \frac{1}{z} + \frac{t}{z(1 - z - t/z - t)} dz = 1 - \frac{t}{2\pi i} \int_C \frac{dz}{(z-p)(z-q)} \tag{3.17}$$

which yields

$$e(t) = 1 + \frac{t}{\sqrt{1 - 6t + t^2}} \tag{3.18}$$

Notice that

$$e(2t) = 1 + \frac{2t}{\sqrt{1 - 6 \cdot 2t + (2t)^2}} = d(t) \tag{3.19}$$

and thus

$$\sum_{i=0}^{\infty} b_n (2t)^n = \sum_{i=0}^{\infty} a_n t^n \tag{3.20}$$

Since the functions $d(t)$ and $e(t)$ are analytic in a disk of radius 1/10, the coefficients of their Taylor series are uniquely determined, and we can conclude that

$$b_n 2^n = a_n \tag{3.21}$$

meaning that fully associated bistrings over n points are only an exponentially vanishing fraction of all well-formed bistrings. In terms of information content, the result means that fully associated bistrings of length n can be encoded using *exactly* one bit less per unit length than arbitrary well-formed bistrings.

Exercise 3.3* Find a 'bijective' proof establishing (3.21) by direct combinatorial methods.

Now, for proper representations, denoting their number by $h(n, k)$, the generating function $H = H(z, w)$ will satisfy a functional equation

$$H - zH - wH - 2zwH + zw^2H + z^2wH - z^2w^2H = r(z, w) \tag{3.22}$$

where $r(z, w)$ is rational. Using the same diagonalizing substitution $w = t/z$, we have to evaluate

$$\frac{1}{2\pi i} \int_C \frac{s(z,t)}{z(1 - z - t/z - 2t + t^2/z + tz - t^2)} dz \tag{3.23}$$

Again, the denominator is quadratic in z, and the radius of convergence is determined by the roots of the discriminant

$$(t^2 + 2t - 1)^2 - 4(t - 1)(t^2 - t) = t^4 + 10t^2 - 8t + 1 \tag{3.24}$$

The reciprocal of the smallest root of this equation, approximately 6.445, gives the base for the asymptotics for c_n, the number of proper bistrings over n points. By taking the base 2 logarithm, we have the following theorem.

Theorem 3.3.2 The information content of a fully associated (proper) well-formed bistring is 2.543 (2.688) bits per unit length.

Exercise 3.4 Count the number of well-formed (fully associated, proper) k-strings of length n assuming each tier alphabet has only one element besides G.

Sets of well-formed (fully associated, proper) bistrings will be called well-formed (fully associated, proper) **bilanguages**. These can undergo the usual set-theoretic operations of **intersection**, **union**, and **complementation** (relative to the 'universal set' of well-formed, fully associated, resp. proper bistrings). **Reversal** (mirror image) is defined by reversing the constituent strings together with the association relation. The concatenation of bistrings is defined by concatenating both the strings and the relations:

Definition 3.3.5 Given two bistrings (f, h, A) and (k, l, B) on tiers N and M, their **concatenation** (fk, hl, AB) is constructed via the tier-alphabet functions $F_0, H_0, K_{|f|}$, and $L_{|g|}$ as follows. $FK_0(i) = F(i)$ for $0 \leq i < |f|$, $K_{|f|}(i)$ for $|f| \leq i < |f| + |k|$, G otherwise. $HL_0(j) = H(j)$ for $0 \leq j < |k|$, $L_{|k|}(j)$ for $|k| \leq j < |f| + |k|$, G otherwise. Finally, $AB = A \cup \{(i + |f|, j + |k|) | (i, j) \in B\}$.

Notice that the concatenation of two connected bistrings will not be connected (as a bipartite graph). This is remedied by the following definition.

Definition 3.3.6 Given two bistrings as in 3.3.5, their **t-catenation** (**b-catenation**) is defined as (fk, hl, AtB) (fk, hl, AbB), where $AtB = AB \cup \{(|f| - 1, |k|)\}$ $(AbB = AB \cup \{(|f|, |k| - 1)\})$.

Using phonological terminology, in t-catenation the last element of the *t*op tier of the first bistring is *spread* on the first element of the bottom tier of the second bistring, and in b-catenation the last element of the *b*ottom tier of the first string is spread on the first element of the top tier of the second bistring.

The only autosegmental operation that is not the straightforward generalization of some well-known string operation is that of **alignment**. Given two bistrings $x = (f, g, A)$ and $y = (g, h, B)$, their alignment $z = x \parallel y$ is defined to be (f, h, C), where C is the relation composition of A and B. In other words, the pair (i, k) will be in C iff there is some j such that (i, j) is in A and (j, k) is in B. Now we are in a position to define projections. These involve some subset S of the tier alphabet T. A **projector** $P_S(h)$ of a string $g = h_0 h_1 \ldots h_m$ with respect to a set S is the bistring (h, h, Id_S), where (i, j) is in Id_S iff $i = j$ and h_i is in S. The **normal bistring** $I(h)$ corresponding to a string h is simply its projector with respect to the full alphabet: $I(h) = P_T(h)$. A **projection** of a string with respect to some subalphabet S can now be defined as the alignment of the corresponding normal bistring with the projector.

The alignment of well-formed bistrings is not necessarily well-formed, as the following example shows. Let f $= ab$, g $= c$, h $= de$, and suppose that the following associations hold: $(0, 0)$ and $(1, 0)$ in x; $(0, 0)$ and $(0, 1)$ in y. By definition, C should contain $(0, 0), (0, 1), (1, 0)$, and $(1, 1)$ and will thus violate the No Crossing Constraint. Note also that a projector, as defined here, will not necessarily be proper. In order to capture the phonologically relevant sense of properness, it is useful to relativize the definition above to 'P-bearing units' (Clements and Ford 1979). We will say that a bistring (f, h, A) is **proper with respect to a subset** S of the tier alphabet T underlying the string h, iff $(f, h, A) \parallel P_S(h)$ is proper.

3.4 Phonological computation

The standard theory of phonology (Chomsky and Halle 1968) enumerates the well-formed strings in a generative fashion (see Section 2.1) by selecting a set E of *underlying forms* and some context-sensitive rules R that manipulate the underlying forms to yield the permissible *surface forms*. Because features are an effective (though imperfect, see Section 3.2) means of expressing natural classes, rules that typically arise in the phonology of natural languages can be stated more economically directly on features, and in fact phonologists rarely have any reason to manipulate strings of phonemes (as opposed to strings of feature bundles). Nevertheless, in what follows we can assume without loss of generality that the rules operate on segments because a rule system employing features can always be replaced by a less economical but equivalent rule system that uses only segments.

Exercise 3.5 Poststress destressing. In our example language there are five unstressed vowels a e i o u and five stressed vowels A E I O U. Whenever two stressed vowels would come into contact, the second one loses its stress: [+stress] → [−stress]/[+stress]_. How many string-rewriting rules are needed to express this regularity without using feature decomposition?

In many problems, such as speech recognition, we are more interested in the converse task of computing the underlying form(s) given some surface form(s). Because of the context-sensitive character of the rules, the standard theory gave rise to very inefficient implementations: although in principle generative grammars are neutral between parsing and generation, the membership problem of CSGs is PSPACE-complete (see Theorem 2.3.2), and in practice no efficient parsing algorithm was found. Context-sensitive phonological rule systems, though widely used for generation tasks (Hunnicutt 1976, Hertz 1982), were too inefficient to be taken seriously as parsers.

The key step in identifying the source of the parsing difficulty was Johnson's (1970) finding that, as long as phonological rules do not reapply within their output, it is possible to replace the context-sensitive rules by finite state transducers (FSTs). That such a condition is necessary can be seen from the following example: $S \rightarrow ab$; $\lambda \rightarrow ab/a\_b$. Starting from S, these rules would generate $\{a^n b^n | n \in \mathbb{N}\}$, a language known not to be regular (see Theorem 3.4.1 below). To show that once the condition is met, context-sensitive rules can be replaced by FSTs, we first need to establish some facts.

We define **regular relations** analogously to the case of regular languages: given two alphabets P, Q, a relation $R \subset P^* \times Q^*$ is regular iff it is finitely generated from finite sets by the operations of union, concatenation, and Kleene closure. (These operations are defined componentwise on relations.) Regular relations are in the same relationship to FSTs as regular languages are to FSAs. In fact, it is convenient to think of languages as unary relations. Similarly, finite state automata will be defined as n-tape automata (transducers) with $n = 1$. In such automata, all tapes are read-only, and the automaton can change internal state when no tape is advanced (λ-move) or when one or more of the tapes is advanced by one square.

Definition 3.4.1 A **finite state transducer (FST)** is a quadruple (S, s, F, T), where S is a finite set of states, $s \in S$ is the starting state, $F \subset S$ is the set of final (accepting) states, and T is a set of **transitions** of the form $(b, a, l_1, \ldots, l_n)$, where b and a are the states *b*efore and *a*fter the move, and the l_j are letters scanned on the jth tape during the move or λ. When for every b and $l_1, \ldots, l_n$ there is at most one $a \in S$ such that $(b, a, l_1, \ldots, l_n) \in T$, the transducer is **deterministic**, and when λ never appears in any transition it is called **length preserving**. Taking $n = 1$, we obtain the important special case of **finite state automata (FSAs)**, which are here viewed as inherently nondeterministic since λ*-moves* (change of state without consuming input) are permitted. For the sake of completeness, we define here **deterministic finite state automata (DFSAs)** as deterministic length-preserving FSTs with $n = 1$, but we leave it to the reader to prove the classic result that FSAs and DFSAs accept the same set of languages and that other minor alterations, such as consuming inputs on states rather than on transitions (Mealy machine rather than Moore machine), have no impact on the class of languages characterized by FSAs.

An n-tuple of words is **accepted** by an FST iff, starting with n tapes containing the n words, the reading heads positioned to the left of the first letter in each word, and the FST in the initial state, the automaton has a sequence of legal moves (transitions in T) that will advance each reading head to the right of the word on that tape, with the automaton ending in a final (accepting) state. When two words (or two n-tuples of words) land an FST in the same state (or in the case of nondeterministic FSTs, the same set of states) starting from the initial state, we call them *right congruent*. Significantly, this notion can be defined without reference to the automaton, based solely on the language it accepts.

Definition 3.4.2 Let L be a language (of n-tuples), x, y are **right congruent** iff for all z either both xz and yz are in L or both of them are outside L. We define **left congruence** analogously by requiring for all z both zx and zy to be (or both not to be) in L. Finally, the **syntactic congruence** (also known as **Myhill-Nerode equivalence**) is defined as the smallest equivalence that is finer than both left and right congruences. The syntactic congruence, as the name suggests, is a primary tool of analysis for syntax (when the alphabet is taken to contain words and the strings are sentences) – we return to its use in Section 5.1.1.

Exercise 3.6 Prove that right congruence, as defined by FST landing sites, is the same relation as defined through the accepted language L. Prove that left and right congruences are equivalence relations. What can we say about the paths through an FST and left congruent elements?

The key property of regular languages (and relations) is that the (right) congruence they define has finitely many equivalence classes (this is also called having **finite index**). If a language (of n-tuples) has finite index, we can use the equivalence classes as states of the accepting FST, with transitions defined in the obvious manner. Conversely, languages accepted by some FST obviously have finite index. Finite languages have finite index, and if two languages have finite index, so will their union, concatenation, and Kleene closure. Thus we have the following theorem.

Theorem 3.4.1 Kleene's theorem. An n-place relation is regular iff it is accepted by an n-tape FST.

Other properties also studied in formal language theory, such as closure under intersection or complementation, will not necessarily generalize from unary to n-ary relations. For example, the binary relations $\{(a^n, b^n c^*)|n \in \mathbb{N}\}$ and $\{(a^n, b^* c^n)|n \in \mathbb{N}\}$ intersect to yield $\{(a^n, b^n c^n)|n \in \mathbb{N}\}$, which is not regular. However, the length-preserving relations are a special case where closure under intersection and set-theoretic difference holds. The method of expressing context-sensitive rules by regular binary relations exploits this fact by adding a new symbol, ϵ, to the alphabet, and using it as padding to maintain length wherever needed.

Although the method of serially composed FSTs and the related method of parallel FSTs (Koskenniemi 1983) play a central role in modern computational phonology, we will not pursue their development here, as it is covered in most newer textbooks on computational linguistics and there are volumes such as Roche and Schabes (1997) and Kornai (1999) devoted entirely to this subject. Rather, we turn our attention to the formalization of the autosegmental theory described in Section 3.3, where regular languages are generalized so as to include the *association relation*. Because of the central role of the finitely generated case, our first task is to identify the family of *regular* or *finite state* bilanguages. We do this based on the finite index property:

Definition 3.4.3 The **syntactic congruence** $\equiv_L$ generated by a bilanguage L contains those pairs of bistrings (α, β) that are freely substitutable for one another; i.e. for which $\gamma\alpha\delta \in L \Leftrightarrow \gamma\beta\delta \in L$. When γ (δ) is fixed as the empty string, we will talk of right (left) congruence. A k-language is **regular** iff it gives rise to a (right) congruence with finitely many classes.

Example 3.4.1 One-member tier alphabets $\{a\} = T_N, \{b\} = T_M$, and empty association relations. Let us denote the bistring $(a^i, b^j, \emptyset)$ by $\langle i, j\rangle$. The bilanguage $B = \{\langle i, i\rangle | i > 0\}$ is not regular because the bistrings $\langle 1, k\rangle$ all belong in different (right)congruence classes for $k = 1, 2, \ldots$, as can be seen using $\delta = \langle l, 1\rangle$: if $k \neq l$, then $\langle 1, k\rangle\langle l, 1\rangle \notin B$ but $\langle 1, l\rangle\langle l, 1\rangle \in B$.

Discussion B can be expressed as the Kleene closure $^+$ of the one-member bilanguage $\{\langle 1, 1\rangle\}$. However, the appropriate closure operation for bilanguages involves not only concatenation but t-catenation and b-catenation as well.

Example 3.4.2 CV-skeleta. Assume that the upper tier alphabet has only two symbols C and V, while the lower has several, and assume further that association relations are limited to binary branching. Symbols on the lower tier linked to a single V (two adjacent Vs) are called *short (long) vowels*, and those linked to a single C (two adjacent Cs) are called *short (long) consonants*. When two adjacent symbols from the lower tier link to a single V, these are called *diphthongs*, when they link to a single C, they are called *affricates*. There are other configurations possible, such as triphthongs, but they are rarely attested.

Discussion CV-skeletal representations formalize a large number of observations that traditional grammars state in somewhat hazy terms, e.g. that diphthongs are the sequence of two vowels 'behaving as a single unit'.

Definition 3.4.4 A **biautomaton** is defined as a 6-tuple (S, U, V, i, F, T), where S is a set of states, U and V are the alphabets of the two tapes, i is the initial state, F is the set of final (accepting) states, and T is the transition function. If we denote the square under scan on the upper tape by x and the square under scan on the lower tape by y, the transition function from a given state will depend on the following factors:

(i) Is there a symbol on square x, and if so, what symbol? (If not, we use the special symbol G introduced in 3.3.)
(ii) Is there a symbol on square y, and if so, what symbol?
(iii) Are the squares associated?
(iv) Are there further association lines from x to some symbol after y?
(v) Are there further association lines from y to some symbol after x?

The transition function T, depending on the present state, the letters under scan, and the presence of association between these letters, will assign a new state and advance the tapes in accordance with the following rule:

> If there are no further association lines from x and y, both tapes can move one step to the right, if there are further association lines from x, only the bottom tape can move, and if there are further association lines from y, only the top tape can move. (3.25)

In other words, the current position of the reading heads can always be added to the association relation without violating the No Crossing Constraint. We will shortly define *coders* as automata in which (3.25) is augmented by the requirement of moving the tapes as fast as possible, but for the moment we will leave the advancement pattern of the tapes nondeterministic. To specify which tape will move, it is best to separate out the transition function into three separate components: one that gives the new state provided a top move t was taken, one that gives the new state provided a bottom move b was taken, and one that gives the new state provided a full move f was taken. Here and in what follows $x[y,t]$ denotes the result state of making a top move from state x upon input y and similarly for $x[y,b]$ (bottom move) and $x[y,f]$ (full move). In general there can be more than one such state, and we do not require that only a top, bottom, or full move be available at any given point – it might still be the case that only one of these moves is available because that is what the association pattern dictates, but there is no general requirement enforcing uniqueness of the next move.

The transition function at state u $\in$ S is **scanning independent** iff for every possible scanning of a string α takes the machine from u to the same state x $=$ u$[\alpha]$. In particular, the machine must be able to perform a full move as a top move followed by a bottom move or as a bottom move followed by a top move. Two full moves should be replaceable by $ttbb$, $bbtt$, $tbtb$, $bttb$, $tbbt$, $btbt$ and similarly for fff

and longer sequences of moves. A biautomaton will be called a **finite autosegmental automaton** iff its transition function is scanning independent at every state. It will be called a **coder automaton** if advancement is by the following deterministic variant of the rule given above:

> If there are no further symbols on either tape, the machine stops. If there are no further symbols on one tape, the other tape is advanced by one. If there are no further association lines from x and y, both tapes move one step to the right; if there are further association lines from x, only the bottom tape moves; and if there are further association lines from y, only the top tape moves, provided the move does not result in scanning G. (The case where there are further lines both from x and y cannot arise since such lines would cross.) (3.26)

Coders can be used to assign a unique linear string, called the **scanning code** in Kornai (1995 Sec. 1.4), to every association relation. Let us denote a top move by t, a bottom move by b, the presence of an association line by 1 and its absence by 0. To assign a unique linear string over the alphabet $\{t, b, 0, 1\}$ to every association relation, it is sufficient to record the top and bottom moves of the coder, leaving full moves unmarked, together with the association lines or lack thereof encountered during the scan. Since scanning a bistring of length n requires at most $2n$ t and b moves, which requires $2n$ bits, and at each step we need to mark 0 or 1, which requires another $2n$ bits, we have a constructive proof that the information content of well-formed bistrings is at most 4 bits per unit length. While this is quite inefficient compared with the optimum 3.543 bits established by Theorem 3.3.1, the constructive nature of the encoding lends further support to the claim in Section 3.3 that multilinear representations are linear (rather than quadratic or other polynomial) data types.

So far we have four families of bilanguages: those accepted by finite autosegmental automata (deterministic or nondeterministic) will be collected in the families **RD** and **RN**, those accepted by biautomata will be collected in the family **BA**, and those accepted by coders will be collected in the family **CA**. Clearly we have **RN** $\subset$ **BA**, **RD** $\subset$ **BA**, **CA** $\subset$ **BA** since both scanning independent and coder automata are special cases of the general class of biautomata. Let us first establish that nondeterminism adds no power – for further "geographic" results, see Theorem 3.4.4.

Theorem 3.4.2 RD = RN.

Proof Same as for the 1-string case. Instead of the state set S of the nondeterministic automaton, consider its power set 2^S and lift the nondeterministic transition function T to a deterministic transition function D as follows: for $i \in$ S, define D($\{i\}$) as $\{$T(i)$\}$, and for $X \subset S$, $D(X) = \bigcup_{i \in X} D(\{i\})$.

Note that the proof will not generalize to coders because different nondeterministic options can lead to different positionings of the heads. However, if the transition function is scanning independent, different positionings of the heads can always be exchanged without altering the eventual state of the machine. Now we can prove

Kleene's theorem that the family of languages **R** characterized by the finite index property is the same as **RN** and **RD**.

Theorem 3.4.3 A bilanguage L is regular iff it is accepted by a regular autosegmental automaton.

Proof ($\Leftarrow$) If L is accepted by a regular autosegmental automaton, it is also accepted by a deterministic regular autosegmental automaton (which can be constructed by the method outlined above), and further it can be accepted by a reduced automaton in which no two states have exactly the same transition function (for such states can always be collapsed into a single state). We claim that there will be as many right congruence classes in $\equiv_L$ as there are states in a minimal (reduced, deterministic, regular) autosegmental automaton $A = (S, U, V, i, F, T)$.

To see this, define $\alpha \equiv_A \beta$ iff for every scanning of α starting in the initial state i and ending in some state j there is a scanning of β starting in i and also ending in j and vice versa. Clearly, $\equiv_A$ is an equivalence relation, and $\alpha \equiv_A \beta \Rightarrow \alpha \equiv_L \beta$. If $\alpha \not\equiv_A \beta$, there must exist a state j such that at least one scanning of one of the bistrings, say α, will lead from i to j, but no scanning of β will ever lead from i to j. Since A is deterministic, scanning β will lead to some state k$\neq$j. We will show that there exists a string δ such that from j we get to an accepting state by scanning δ and from k we get to a nonaccepting state (or conversely), meaning that $\alpha\delta \in L$ but $\beta\delta \notin L$ (or conversely), so in either case $\alpha \not\equiv_L \beta$.

Call two states p and q *distinguishable* iff there exists a string δ such that starting from p, scanning δ leads to an accepting state, but starting from q, scanning δ leads to a rejecting state or vice versa. *In*distinguishability, denoted by I, is an equivalence relation: clearly pIp for every state p, and if pIq, also qIp. For transitivity, suppose indirectly that pIq and qIr, but p and r are distinguishable; i.e. there is a string δ for which p[δ] is accepting but r[δ] is not. Now, q[δ] is either accepting or rejecting: in the former case, qIr was false, and in the latter, pIq was false, a contradiction. Further, in a minimal automaton there can be no two (or more) indistinguishable states, for such states could be collapsed into a single state without altering the accepted bilanguage. Since j and k above are not equal, they are distinguishable by some δ. ■

($\Rightarrow$) To prove the 'only if' part of the theorem, we have to show that if a bilanguage L gives rise to a finite right congruence, it is accepted by some regular autosegmental automaton. We will construct the states of the automaton from the congruence classes of the equivalence relation. Let us denote the congruence class of a bistring α under $\equiv_L$ by (α). The initial state of the machine is the congruence class of the empty bistring, ( ), and the transition function from state (α) is defined for top transitions from (α) as the congruence class $(\alpha t\text{T})$ (t-catenation), and similarly the result state of a bottom transition from (α) will be the congruence class $(\alpha b\text{B})$ (b-catenation). Thus top (bottom) transitions are nondeterministic: there are as many result states as there are congruence classes for each member of the top (bottom) tier alphabet. For ordinary concatenation of a bistring β, the result is defined by the class $(\alpha\beta)$ in order to guarantee scanning independence.

Finally, the accepting states of the automaton are defined as those congruence classes that contain the members of L – this is well-defined because if $\alpha \equiv_L \beta$, both must be members of L or both must be outside L, meaning that L is a union of congruence classes. What remains to be seen is that the bilanguage M accepted by the automaton defined here is the same as the bilanguage L we started with. First let us take a bistring α included in L – since (α) is an accepting state, it follows that α is also in M. Next let us take a bistring β not in L – since (β) is not an accepting state it would follow that β is not in M if we can show that no scanning path would lead to any state other than (β). This can be done by induction on the length (defined as the maximum of the length of the top and bottom strings) of β (see Kornai 1995, ch 1.6).

Discussion The class **R** of regular k-languages is closed under all Boolean operations, while the class of regular k-relations is not. This is due to the fact that finite autosegmental automata can always be determinized (Theorem 3.4.2), while FSTs in general cannot. Determinization creates a machine where complementing the set of accepting states leads to accepting the complement language – if a language is only accepted by a nondeterministic acceptor, this simple method fails, and the complement language may not have an equally simple characterization. Since both families are closed under union, (non)closure under complementation will guarantee, by De Morgan's law, (non)closure under intersection. In this respect, the bilanguage families **BA** and **CA** are closer to regular relations than the family **R**.

Example 3.4.3 Consider the bilanguage $T = \{\langle i, j\rangle | i > j\}\{(a, b, \{(0,0)\})\}$ – it contains those bistrings that have i floating features on the top tier, and j floating features on the bottom tier, followed by an end marker, which is simply a feature a on the top tier associated to a feature b on the bottom tier. Clearly, if $i - j \neq i' - j'$, we have $\langle i, j\rangle \not\equiv_T \langle i', j'\rangle$ so T is not in **R**. However, it is in **CA** since the following automaton will accept it:

in state	from x to y	from x to z>y	from y to w>x	automaton will
0	absent	absent	absent	stay in 0
0	absent	absent	present	go to 1
1	absent	absent	present	stay in 1
1	present	absent	absent	go to 2
2	any	any	any	go to 3

(3.27)

With 2 as the only accepting state, the machine will accept only those strings whose scan puts the machine in 2 but not further. To get into 2, the last thing the machine must encounter is a single association line (the end marker) in state 1. To get into state 1, the machine can make a number of top moves over floating elements (this is the loop over state 1), preceded by a number of full moves over floating elements (this is the loop over state 0). Note that this is not scanning independent: no provision was made for top and bottom moves to replace full moves out of state 0.

What Example 3.4.3 shows is that **CA** is not contained in **R**. It is, of course, contained in **BA**, and the bilanguage B of Example 3.4.1 introduced above shows that the containment is proper. The biautomaton that accepts this bilanguage is trivial: it contains only one state and only full advance is permitted (and that only when no association lines are present). To see that no coder can accept this bilanguage, suppose indirectly that an n-state coder A accepts B. The bistrings $\langle k,k \rangle$ are all accepted ($k = 1,2,3,\ldots,n+1$), so there is at least one accepting state f that accepts both $\langle i,i \rangle$ and $\langle j,j \rangle$, $1 \leq i < j \leq n+1$, by the pigeonhole principle. Let $j-i = p$, and consider the bistring $\langle j,i \rangle$. In the first i steps, we arrive in f, and in the next p steps we make legal top moves (since we are at the end of the bottom string) that are indistinguishable from legal full moves. But p full moves would take us back to f, which is an accepting state, so p top moves also take us back to f, meaning that $\langle j,i \rangle$ is accepted by A, a contradiction. To complete our 'geographic survey', note that **R** is not contained in **CA**. This can be seen e.g. by considering the regular bilanguage $D = \{\langle 1,j \rangle | j > 0\}$. Collecting these results gives us the following theorem.

Theorem 3.4.4 Both **R** and **CA** are properly contained in **BA**, and neither is contained in the other.

To summarize, **R** is closed under union and intersection, as the standard direct product construction shows, and also under complementation, as can be trivially established from the characterization by automata. Boolean operations thus offer no surprises, but string operations need to be revised. If we use concatenation as the only k-string composition operation, there will be an infinite number of further undecomposable structures, such as the bistrings resulting from the spreading of a single element on the bottom (top) tier. These structures, and many others, have no structural break in them if indeed concatenation was the only possibility: that is why we introduced t-catenation and b-catenation above. Regular expressions also work:

Theorem 3.4.5 Every bilanguage accepted by an autosegmental automaton can be built up from the elementary bistrings $(x,y,\emptyset)$ and $(x,y,\{(0,0)\})$ by union, t-catenation, b-catenation, concatenation, and Kleene closure.

The proof is left as an exercise for the reader. Once these operations are available for creating larger bistrings from two successive bistrings, **Kleene closure** will include them as well. This way the oddness of the bilanguage B introduced in Example 3.4.1 above disappears: $B = \{\langle i,i \rangle | i \rangle 0\}$ is *not* the Kleene $^+$ of $\langle 1,1 \rangle$ because the closure means arbitrary many catenation operations *including* t-catenations and b-catenations. The Kleene $^+$ of $\langle 1,1 \rangle$ is really the set of all well-formed bistrings (over one-letter tier alphabets), which is of course regular. From the characterization by automata, it easily follows that the concatenation, t-catenation, b-catenation, and Kleene closure of regular bilanguages is again regular. Standard proofs will also generalize for closure under (inverse) homomorphisms and (inverse) transductions.

As we discussed earlier, transductions play a particularly important role in replacing context-sensitive phonological rules by finite state devices. For this reason, we combine the notions of k-strings and n-place regular relations and state this last result (for the proof, modeled after Salomaa 1973 Ch. IV, see Kornai 1995 Sec. 2.5.2), as a separate theorem.

Theorem 3.4.6 If L is a regular bilanguage and $B = (S,I,O,i,F,T)$ a generalized bisequential mapping, the image $(L)B$ of L under B is also a regular bilanguage.

Autosegmental rules can be grouped into two categories: rules affecting the strings stored on the various tiers and rules affecting the association relations. The two categories are not independent: rules governing the *insertion* and *deletion* of symbols on the various tiers are often conditioned on the association patterns of the affected elements or their neighbors, while rules governing the *association* and *delinking* of elements of different tiers will often depend on the contents of the tiers.

Repeated insertion is sufficient to build any string symbol by symbol – deletion rules are used only because they make the statement of the grammar more compact. Similarly, repeated association would be sufficient for building any association relation line by line, and delinking rules are only used to the extent that they simplify the statement of phonological regularities. A typical example would be a *degemination* rule that will delink one of the two timing slots (a notion that we will develop further in Section 4.1) associated to a segment k, if it was preceded by another segment s, and also deletes the freed slot. In pre-autosegmental theory the rule would be something like $k[+long] \rightarrow k[-long]/s\_$ (the features defining k and s are replaced by k and s to simplify matters), and in autosegmental notation we have

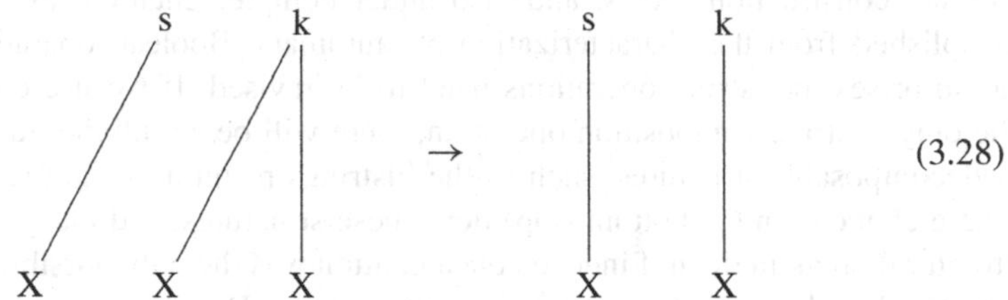

(3.28)

The regular bitransducer corresponding to this rule will have an *input bistring*, an *output bistring*, and a *finite state control*. The heads scanning the bistrings and the association relations are read-only. If we denote the bistring on the left-hand side of the arrow by β and that on the right-hand side by β', the bitransducer accepts those pairs of bistrings that have the form $(\alpha\beta\gamma, \alpha\beta'\gamma)$ and those pairs (δ, δ) that meet the rule vacuously (β is not part of δ).

Exercise 3.7 Verify that the regular bitransducer defined with the finite state control in (3.26) accepts those and only those pairs of bistrings (ρ, σ) in which σ is formed from ρ by applying (3.29).

in state	input bistring	output bistring	heads move	result state
0	s1X	s1X	f/f	1
0	q!=s1X	q	m/m	0
0	q!=s1X	r!=q	m/m	-
1	k1X	k1X	b/0	2
1	q!=k1X	q	m/m	0
1	q!=k1X	r!=q	m/m	-
2	k1X	k1X	m/m	0
2	q!=k1X	r!=q	m/m	-

(3.29)

What makes the theory of regular autosegmental languages, relations, automata, and transducers presented here central to the study of phonology is the view shared by many, though by no means all, phonologists that in spite of a superficial context-sensitivity, there is actually no need to go beyond the regular domain. For segmental rules, this was already established by Johnson (1970), but suprasegmentals, in particular tone and stress, remained a potential threat to this view. To the extent that autosegmental theory provides our current best understanding of tone, the foregoing imply that counterexamples will not be found among tonal phenomena since the basic regularities of tone, in particular the handling of floating elements and spreading, are clearly expressible by regular means. This still leaves stress phenomena as a potential source of counterexamples to the regularity thesis, particularly as hierarchical bracketing of the sort familiar from context-free grammars is at the heart of the so-called cyclic mode of rule application, a matter we shall turn to in Section 4.2.

3.5 Further reading

The issue of whether signs are truly arbitrary still raises its head time and again in various debates concerning *sound symbolism, natural word order*, and the *Sapir-Whorf hypothesis* (SWH). That there is some degree of sound symbolism present in language is hard to deny (Allott 1995), but from a frequency standpoint it is evident that the 'conventional' and 'arbitrary' in language vastly outweigh the 'motivated' or 'natural'. We will return to natural word order and the SWH in Chapters 5 and 6. We note here that the SWH itself is clearly conventionally named since neither Sapir's nor Whorf's published work shows much evidence that either of them ever posed it in the strong form 'language determines thought' that came to be associated with the name SWH.

We owe the program of eliminating references to meaning from operational definitions of linguistic phenomena to the structuralist school, conventionally dated to begin with de Saussure (1879). The first detailed axiom system is that of Bloomfield (1926). The version we present here relies heavily on later developments, in particular Harris (1951).

Distinctive features provide a vector decomposition of phonemes: whether the field underlying the vector space is taken to be continuous or discrete is a matter strongly linked to whether we take an empiricist or a mentalist view of phonemes. The first position is explicitly taken by Cherry (1956) and Stevens and Blumstein (1981), who assume the coordinates in feature space to be directly measurable properties of the sounds. The second is implicit in the more abstract outlook of Trubetzkoi (1939) and Jakobson et al. (1952) and explicit in the single most influential work on the subject, Chomsky and Halle (1968).

The issue of characterizing the set of well-formed phoneme strings in P^* is impacted more severely by the appropriate notion of 'well-formed' than by the choice of generative tools (context-sensitive, context-free, or finite state): as we shall see in Chapter 4, the key question is how we treat forms that have not been attested but sound good to native speakers. To get a taste of the debate at the time, the reader may wish to consult Chomsky and Halle (1965a), Householder (1965, 1966), and Kortlandt (1973) – in hindsight it is clear that the *mentalist* view propounded by Chomsky and Halle took the field (see Anderson (2000) for a nuanced overview of the debate). Anderson (1985) describes the history of modern phonological theory from a linguistic perspective.

Readers who wish to get some hands-on experience with the kinds of problems phonologists routinely consider should consult Nida (1949) or Halle and Clements (1983) – the latter includes several problems like that of Kikuyu [KIK] tone shift (Clements and Ford 1979) that served to motivate autosegments. Autosegmental theory initially grew out of research on African tone languages (Welmers 1959, Leben 1973, Goldsmith 1976, Williams 1976) and prosody (Firth 1948) – the standard monographic treatment is Goldsmith (1990). The basic insight that tones and segments need not be fully aligned with one another but rather must be placed on separate tiers was soon generalized from tone to vowel harmony (Clements 1977), aspiration (Thráinsson 1978) nasality (Hyman 1982), and eventually, by placing all distinctive features on separate tiers, to the theory of feature geometry (for a review, see McCarthy 1988; for a current proposal, see Padgett 2002).

For further discussion of the Pāṇinian system of describing natural classes, see Petersen (2004). The modern treatment of natural classes begins with Trubetzkoi (1939) and Jakobson et al. (1952), who assumed that the defining properties of natural classes are all orthogonal (see also Cherry 1956). Distinguishing the marked member of an opposition has its roots in morphology, where it is often the case that only one member of an opposition (such as singular vs. plural) is marked explicitly, the other member being known from the absence of marking (e.g. we know *boy* is singular because it does not have the plural suffix *-s*). Extending this notion to phonological oppositions begins with Trubetzkoi, but the full theory of markedness developed only gradually with the work of Jakobson, Halle, and Chomsky – for an overview, see Hyman (1975).

The key ideas of decomposing context-sensitive rules into simple regular (and, when needed, length-preserving) steps such as the introduction and eventual deletion of temporary brackets that keep track of the locus of rule application were discovered independently by Johnson (1970) and Kaplan and Kay (an influential unpublished manuscript eventually published in 1994). Kiraz (2001) presents a full system based on n-way relations (n-tape automata) in the context of Semitic morphology. A different *intersective* approach is introduced in Bird and Ellison (1994).

The cyclic mode of rule application was first introduced in Chomsky and Halle (1968 Sec. 2.3–2.5). The formalism proposed there relied heavily on the use of explicit boundary markers, a device that is no longer viewed as appropriate by most phonologists, but the central idea is well-preserved in lexical phonology and morphology (LPM, see Kiparsky 1982), which no longer uses boundary markers.

4
Morphology

Morphology, the study of the shape and structure of words, is a field that brings into sharp relief what are perhaps the most vexing aspects of linguistics from a mathematical perspective: radical typological differences, flexible boundaries, and near-truths. Mild typological differences are common to most fields of study. For example, the internal organs of different primates are easily distinguished by experts yet differ only mildly, so that a person who knows something about gorillas and knows human anatomy well can make a reasonable guess about the position, shape, size, and functioning of gorilla livers without ever having seen one. *Radical typological differences* are much less common. Continuing with the analogy, one knowledgeable about the internal sex organs of males but not of females would have a hard time guessing their position, shape, size, or functioning. In morphology, radical typological differences abound: no amount of expert knowledge about Modern English is sufficient to make a reasonable guess e.g. about the case system of Modern Russian, in spite of the fact that the two languages descended from the same Indoeuropean origins. Mathematics, on the whole, is much better suited for studying mild (parametric) typological differences than radical ones. We exemplify the problem and discuss a possible solution in Section 4.1, which deals with prosody in general and the typology of stress systems in particular.

In mathematics, *flexible boundaries* are practically unheard of: if in one case some matter depends on arithmetic notions, we are unlikely to find other cases where the exact same matter depends on topological notions. It is easier to find examples in computer science, where the same functionality (e.g. version control of files) may be provided as part of the operating system in one case or as part of the text editor in another. In morphology, flexible boundaries are remarkably common: the exact same function, forming the past tense, may be provided by regular suffixation (*walk* → *walked*), by ablaut (*sing* → *sang*), or by suppletion (*go* → *went*). We exemplify the problem in Section 4.2, where we introduce the notions of derivation and inflection.

Finally, we call *near-truths* those regularities that come tantalizingly close to being actually true yet are detectably false with the available measurement techniques. A well-known example is Prout's law that atomic weights are integer multiples of that of hydrogen. For example, the helium/hydrogen atomic weight ratio is 3.971 and that of nitrogen/hydrogen 13.897. Near-truths are so powerful that one is inclined to disregard the discrepancies: from a chemistry-internal perspective, everything would be so much simpler if atomic weights were truly subject to the law. In fact, we have to transcend the traditional boundaries of chemistry and gain a good understanding of isotopes and nuclear physics to see why the law is only nearly true. Unfortunately, there is no good candidate for a deeper theory that can clean up linguistic near-truths. As we shall see repeatedly, all forms of 'cognitive' and 'functional' explanations systematically fall short. Therefore, in Section 4.3 we look at a solution, *optimality theory*, which builds near-truths into the very architecture of linguistics.

A central task that is shared between phonology and morphology is characterizing the set of words. The operational definition, *maximal pause-free stretch between potential pauses*, has a number of drawbacks. First, it is restricted to the subset of words that are attested, while it is clear that new words are added to the language all the time. A list of attested forms fails to characterize either the individual ability to serve as an oracle capable of rendering well-formedness judgments on word candidates or the collective ability of the speakers to introduce new words in their language. Second, it fails to assign structure to the words, substituting a simple accept/reject decision for detailed analysis. This failure is especially frustrating in light of the fact that every attempt at capturing the theoretically more interesting notion of *potential* words inevitably proceeds from structural considerations. Third, it offers no help in assigning meaning to words, except perhaps for the rare subclass of onomatopoeic words whose meaning can be inferred from their sound.

Finally, the operational definition leaves open the issue of word frequency. Part of the task of characterizing the set of words is to describe their frequency either in corpora or, more interestingly, in the populations of which the corpora are samples. Typical populations of interest include the set of utterances a person in a given language community is likely to encounter or the set of texts some natural language software (spellchecker, machine translator, information retrieval system, etc.) should be prepared for. Especially for software, it is desirable to make the system less likely to fail on the frequently encountered cases than on some rare or marginal cases. We defer introducing the mathematical machinery of weighted languages and automata required for dealing with frequencies to Chapter 5, but introduce the basic empirical regularity, *Zipf's law*, in Section 4.4.

As far as the acoustic content of potential word forms is concerned, the task of characterizing this (weighted) set can be subdivided into two parts: first, characterizing the segmental content as a set of well-formed phonemic strings, and second, characterizing the suprasegmental content (tone and stress) associated to the segments. The former task is called *phonotactics*, and it has been traditionally recognized that the central object of inquiry is not the full word but considerably smaller *syllabic* units that show very strong combinatorical restrictions internally, and only much weaker restrictions externally. As we shall see in Section 4.1, such units play a central role in characterizing the suprasegmental patterns of words as well. As far as the meaning of words is concerned, the pivotal units, called *morphemes*, are again smaller than the word and show much less predictability than the words composed of them. There are some languages, most notably Chinese, where the correlation between syllables and morphemes is quite strong, but most languages show evident mismatches in the form of both polysyllabic morphemes such as *country* and polymorphemic syllables such as *knives*. As we shall see in Section 4.2, the description of morpheme combinations, called *morphotactics*, has practically no commonalities with phonotactics, a fact often referred to as the *double articulation* or *duality of patterning* of language (Martinet 1957, Hockett 1960).

4.1 The prosodic hierarchy

The marking of certain substructures as belonging to a certain prosodic domain such as the *mora*, *syllable*, *foot*, or *prosodic word* is an essential part of phonological representations for three interrelated reasons. First, a great number of phonological processes or constraints make reference to such domains; for the syllable (in English), see Kahn (1976), and for the foot (in Japanese), see Poser (1990). Second, the domains themselves can carry feature information that cannot properly be attributed to any smaller constituent inside the domain; *stress* and *boundary tones* provide widely attested examples, though some readers may also be familiar with *emphasis* (pharyngealization) in Arabic, Aramaic [CLD], and Berber [TZM] (Jakobson 1957, Hoberman 1987, Dell and Elmedlaoui 1985, 1988). Finally, the shape of words is largely determined by the prosodic inventory of the language; for example, if a name ends in a vowel, we can be virtually certain it does not belong in the Anglo-Saxon layer of English.

It should be said at the outset that our understanding of the prosodic hierarchy is not yet sufficient. For example, it is not clear that moras are constituents of the syllables in the same way syllables are constituents of feet, whether notions such as extrametricality or ambisyllabicity are primitive or derived, whether abstract units of stress can be additively combined (as in the *grid-based* theories starting with Liberman 1975), and so on. Even so, there is no doubt that the prosodic hierarchy plays an absolutely pivotal role in phonology and morphology as the guardian of well-formedness in a manner broadly analogous to the use of *checksums* in digital signal transmission. Arguably, the main function of phonology is to repair the damage that morphological rules, in particular concatenation, would cause.

4.1.1 Syllables

Syllables are at the middle of the prosodic hierarchy: there are higher units (feet, and possibly superfeet, the latter also called *cola*), and there are lower units (onset, nucleus, rhyme, mora), but we begin the discussion with the syllable since the other prosodic units are less likely to be familiar to the reader. There are three different approaches one may wish to consider for a more precise definition of the syllable. First, we can introduce a *syllabic alphabet* or *syllabary* that is analogous to the phonemic alphabet introduced in Section 3.1, but without the requirement that distinct syllables show no similarity (segmental overlap). This approach is historically the oldest, going back at least a thousand years to the syllabic writing systems such as found in Japanese Hiragana and arguably much earlier with Brahmi (5th century BCE) and the Cypriot syllabary (15th century BCE), though in fact many of the early scripts are closer to being mora-based than syllable-based.

Second, we can introduce explicit boundary markers such as the - or · used in dictionaries to indicate syllable boundaries, a notation that goes back at least to the 19th century. An interesting twist on the use of boundary markers is to permit *improper* parentheses, e.g. to denote by $[a(b]c)$ the case where b is said to be *ambisyllabic* (belonging to both syllables ab and bc). The notion of ambisyllabicity receives a slightly different formulation in autosegmental theory (see Section 3.3). Note that improper parentheses could describe cases like $[a(bc]d)$, where more than one element is ambiguously affiliated, while the autosegmental well-formedness conditions would rule this out.

Finally, we can use tree structure notation, which is more recent than the other two and has the advantage of being immediately familiar to contemporary mathematicians and computer scientists, but is incapable of expressing some of the subtleties, such as ambisyllabicity, that the earlier notations are better equipped to handle. One such subtlety, easily expressible with autosegmental notation or with improper parentheses, is the notion of *extrametricality*, meaning that the parse tree simply fails to extend to some leaves.

The unsettled notation is much more a reflection of the conceptual difficulties keenly felt by the linguist (for an overview, see Blevins 1995) than of practical difficulties on the part of the native speaker. In fact, most speakers of most languages have clear intuitions about the syllables in their language, know exactly where one ends and the next one begins, and can, with little or no formal training, draw up an inventory of syllables. It is precisely the confluence of these practical properties that makes the syllable such a natural building block for a script, and when new scripts are invented, such as Chief Sequoyah of the Cherokee [CER] did in 1819, these are often syllabaries. The ease with which native speakers manipulate syllables is all the more remarkable given the near impossibility of detecting syllable boundaries algorithmically in the acoustic data.

For our purposes it will be sufficient to group phonemes into two broad classes, called *vowels* (V) and *consonants* (C), based on whether they can appear in isolation (i.e. flanked by pauses on both sides) or not. We mention here that in phonology consonants are generally subdivided into other *major classes* like stops such as *p t k b d g*, nasals such as *n m*, liquids such as *l r*, and glides such as *y* (see Chomsky and Halle 1968 Sec. 7.3). We will assume a separate CV tier where autosegments correspond to elementary timing units: a consonant or short vowel will take a single timing unit, and a long vowel will take two. To build a *syllable*, we start with a *nucleus*, which can be a short vowel V or a long vowel VV, and optionally add consonants at the beginning (these will be called the *onset*) and/or at the end (these will be called the *coda*). This way, no syllable composed of Cs alone will ever get built.

Different languages put different constraints on the number and type of consonants that can appear in the onset and the coda. On the whole, codas tend to be more restricted: it is easy to find languages with CCC onsets but only CC codas, or CC onsets but only C (or no) codas, but it is hard to find languages with more complex codas than onsets. Combinatorical restrictions within the onset are common. For example, in English, if the onset is CCC, the first C must be *s* and the second must be a voiceless stop. Similarly, the *sr* onset is common and **rs* is impossible, while in the coda it is the other way around. Generally, the nucleus serves as a barrier through which combinatorical restrictions do not propagate.

The inventory of V sounds that can serve as syllabic nuclei is not constant across languages: vowels always can, but many languages, such as Czech, treat liquids as syllable-forming; and some, like Sanskrit, also permit nasals. At the extreme, we find Berber, where arguably every sound, including stops, can form a syllable, so no sound is truly C-like (incapable of serving as a syllabic nucleus).

It is not evident whether the onset and the coda are truly symmetrical or whether it makes sense to group the nucleus and the coda together in a constituent called the *rhyme.* Many observations concerning the combinatorical restrictions can be summarized in terms of *sonority*, a linear scale based somewhat loosely on the vocal energy (see Section 8.1) contained in the sound – those sounds that can be heard farther away are considered more sonorous. Formally, the **sonority hierarchy** groups the sounds in discrete, ranked sonority classes, with low vowels and voiceless stops being the most and least sonorous, respectively. The overall generalization is that syllables are constructed so that sonority rises from the margins toward the nucleus. In English and many other languages, sibilants such as *s* are a known exception since they are more sonorous than stops yet can appear farther from the nucleus. There are many mechanisms available to the phonologist to save the overall law from the exception that 'proves' it (Lat. *provare*, to test). One is to declare sibilants extrametrical, and another one is to treat sibilant+stop clusters as a single consonant. We return to this question in Section 4.3.

Exercise 4.1† English graphemotactics. Take a large list of written (lowercase) words, and define vowels **V** as *aeiouy* and consonants as the rest. Define as **on**sets (**co**das) those consonantal strings that begin (end) words, including the empty string in both sets. How much of the word list is matched by the regular expression **syll**$*$, where **syll** is defined by **on V co** $\cup$ **on VV co**? Does the grammar overgenerate? How? Why?

4.1.2 Moras

Syllables often come in two, and sometimes in three, sizes (weights): *light* syllables are said to contain one mora, *heavy* syllables contain two moras, and *superheavy* syllables contain three moras. It is surprisingly hard to define **moras**, which are some kind of abstract syllable weight unit, any better than by this simple listing since the containment of moras in syllables is like a dime containing two nickels: true for the purposes of exchanging equal value but not in the sense that the nickels could be found inside the dime. To see this, compare the typical light syllable, CV, to the typical heavy syllable, CVV or CVC. If the second mora were contributed by the vowel length or by the coda consonant, we would expect CVVC syllables to be superheavy, but in fact only CVVCC syllables generally end up trimoraic.

The exchange of equal value is best seen in the operation of various rules of stress and tone assignment. A familiar example involving stress is classical Latin, where the last syllable is extrametrical (ignored by the rule of stress placement) and stress always falls on the penultimate mora before it. For tone, consider the Kikuria [KUJ] example discussed in Odden (1995). In Bantu languages, tense/aspect is often marked by assigning high tone to a given position in the verb stem: in Kikuria, the high tone's falling on the first, second, third, or fourth mora signifies remote past, recent past, subjunctive, and perfective, respectively. To quote Odden,

> Regardless of how one counts, what is counted are vowel moras, not segments or syllables. Stated in terms of mora count, high tone is simply assigned to the fourth mora in the perfective, but there is no consistent locus of tone assignment if one counts either syllables or segments.

An entirely remarkable aspect of the situation is that in some other languages, such as Lardil [LBZ] (see Wilkinson 1988), the whole phenomenon of rules being sensitive to the number of moras is absent: it is the number of syllables, as opposed to the number of moras, that matters. The distinction is known in linguistics as *quantity-sensitive* vs. *quantity-insensitive* languages, and for the most part it neatly divides languages into two typological bins. But there are nagging problems, chief among them the existence of typologically *split* languages such as Spanish, where the verbal system is quantity insensitive but the nominal system is quantity sensitive.

4.1.3 Feet and cola

One step up from syllables we find *metrical feet*, groupings that contain one strong and one weak syllable. Such feet account nicely for the long observed phenomenon that syllable stress generally appears as a pulse train, with stressed and unstressed syllables alternating quite predictably. When two stressed syllables meet, one of them generally gets destressed: compare Italian *cittA* 'city' (stress is indicated by capitalized vowels here), and *vEcchia* 'old' to the combination *citta vEcchia* 'old city' (stress retraction, see Nespor and Vogel 1989 – for a symmetrical rule see Exercise 3.5). Another way of resolving such *clashes* is by the insertion of unstressed material such as a pause (Selkirk 1984).

Other feet constructions include *unbounded feet*, a flat structure incorporating an arbitrary number of syllables, *degenerate feet*, containing just one syllable, and even *ternary feet*. There is no doubt that in the vast majority of cases, feet are binary (they contain exactly two syllables), Kiribati [GLB] being the best, perhaps the only, counterexample that resists reanalysis in terms of binary feet (Blevins and Harrison 1999). In some cases, especially for the study of secondary, tertiary, and weaker levels of stress, it may make sense to join feet in a higher structure called a *colon* (see Hammond 1987).

4.1.4 Words and stress typology

Segmentally, utterances can be parsed into words, and the words can be parsed into syllables (barring extrametrical material). The suprasegmental shape of the resulting syllable stream gives strong cues to where the word boundaries are located: in any given word, there is exactly one primary stress. Some complications arise because certain particles, called *clitics*, are adjoined to an adjacent word prosodically even though the relation between the elements is morphologically undefined. The existence of such particles requires some dissociation between the phonological and the syntactic definitions of 'word' – the former are called *prosodic words*, the latter just *words*. For example, in Arabic, where conjunctions and prepositions are proclitic (attach to the following word), the written word unit (delimited by whitespace) is the prosodic, rather than the syntactic, word. In Latin-based orthography, the apostrophe is often used to separate clitics from their hosts (but this is not a reliable criterion, especially as the apostrophe is used for many other purposes as well). Ignoring this complication for the moment (we take up this matter in Section 4.2), there are as many words in an utterance as there are primary stresses, and to find out where the word boundaries fall, all that is required is an understanding of where the stress falls within the words.

At the top of the prosodic hierarchy we find the *prosodic word*, composed of cola or feet and carrying exactly one primary stress. Locating the syllable with the main stress as well as those syllables that carry lesser (secondary, tertiary, etc.) stress is a primary concern of the *metrical theory* of phonology. (Although in many ways related to the generative theory of *metrics*, metrical theory is an endeavor with a completely different focus: metrics is a branch of *poetics*, concerned with poetic meter (see e.g. Halle and Keyser 1971), while metrical theory is a branch of linguistics proper.) When the problem is solvable in the sense that the location of the primary stress is rule-governed, linguists speak of *fixed* stress. When the location of stress cannot be predicted either on the basis of the phonological composition of the word (e.g. number and weight of syllables) or on the basis of morphological composition, linguists speak of *free* stress, and make recourse to the purely descriptive method of marking in the lexicon where the stress should fall.

While this last recourse may strike the mathematician as pathetically inept, bordering on the ridiculous, it is nothing to be sneered at. First, languages provide many examples of genuinely unpredictable features: as any foreign learner of German will know from bitter experience, the gender of German nouns *must be* memorized, as any heuristic appeal to 'natural gender' will leave a large number of exceptions in its wake. Second, the procedure is completely legitimate even in cases where rules are available: tabulating a finite function can define it more compactly than a very complex formula would. The goal is to minimize the information that needs to be tabulated: we will begin to develop the tools to address this issue in Chapter 7. The reader who feels invincible should try to tackle the following exercise.

Exercise 4.2^M Develop rules describing the placement of accents in Sanskrit. For general information, see Whitney (1887) §80–97, 128 130, 135a; for nouns §314–320; for numerals §482g, 483a–c, 488a; for verbs §591–598; for adverbs §1111g, 1112e, 1114d; for personal suffixes §552–554; with other parts of the system discussed in §556, 945, 1073e, 1082–1085, 1144, 1205, 1251, 1295, and elsewhere. An overview of Pāṇini's system, which also treats accents by rules scattered throughout the grammar, is presented in Cardona (1988 Sec. 2.8).

For the purposes of the typology, the interesting cases are the ones where stress is fixed. For example, in Hungarian, primary stress is always on the first syllable; in French, it is on the last syllable; in Araucanian [ARU], it is on the second syllable; in Warao [WBA], it is on the next to last (penultimate) syllable; and in Macedonian [MKJ], it is on the antepenultimate syllable. Interestingly, no example is known where stress would always fall on the third syllable from the left or the fourth from the right.

Quantity-sensitive languages offer a much larger variety. For example, in Eastern Cheremis [MAL], stress falls on the rightmost heavy syllable, but if all syllables are light, stress is on the leftmost syllable (Sebeok and Ingemann 1961). In Cairene Arabic [ARZ], stress is on the final syllable if it is superheavy or on the penultimate syllable if heavy; otherwise it is on the rightmost nonfinal odd-numbered light syllable counting from the nearest preceding heavy syllable or from the beginning of the word (McCarthy 1979). The reader interested in the full variety of possible stress rules should consult the StressTyp database (Goedemans et al. 1996), which employs the following overall categorization scheme:

```
1 Unbounded systems
   1.1 Quantity-sensitive
   1.2 Quantity-insensitive
   1.3 Count systems
2 Binary Bounded systems
   2.1 Quantity-insensitive
   2.2 Quantity-sensitive
3 Special systems
   3.1 Broken-window systems
   3.2 n-ary weight distinctions
     3.2.1 Superheavy syllables
     3.2.2 Prominence systems
4 Bounded ternary systems
```

Although from the outside it is somewhat haphazard, the system above is typical in many respects; indeed, it is among the very best that linguistic typology is currently capable of producing. First, the categories are rather sharply separated: a system that fits in one will not, as a rule, fit into any other. There are very few split cases, and these tend to involve major subsystems (such as the verbal and nominal systems) rather than obscure corner cases. Given that stress rules are generally riddled with exceptions, this is a remarkable achievement. Second, the system is based on several hundred languages analyzed in great depth, giving perhaps a 10% sample of known languages and dialects. That certain categories are still instantiated by only a handful of examples is a cause for concern, but again, the depth of the research is such that these examples are fairly solid, rarely open to major reanalysis. Third, the empirical correlates of the categories are rather clear, and it requires only a few examples and counterexamples to walk down the decision tree, making the problem of classification much less formidable than in those cases of typology where the operative categories are far more elusive.

The modern theory of stress typology begins with Hayes (1980), who tried to account for the observed variety of stress patterns in terms of a few simple operations, such as building binary trees over the string of syllables left to right or right to left – for a current proposal along the same lines, see Hayes (1995). Here we will discuss another theory (Goldsmith and Larson 1990), which, in spite of its narrower scope (it applies to quantity-insensitive systems only), has taken the important step of divesting stress typology from much of its post hoc character.

Suppose we have n syllables arranged as a sequence of nodes, each characterized at time T by two real parameters, its current activation level $a_k(T)$ and the bias b_k that is applied to it independent of time T. The two parameters of the network are the leftward and rightward feeding factors α and β. The model is updated in discrete time: for $1 < k < n$, we have $a_k(T+1) = \alpha a_{k+1}(T) + \beta a_{k-1}(T) + b_k$. At the edges, we have $a_1(T+1) = \alpha a_2(T) + b_1$ and $a_n(T+1) = \beta a_{n-1}(T) + b_n$. Denoting the matrix that has ones directly above (below) the diagonal and zeros elsewhere by $\mathbf{U}$ ($\mathbf{L}$) and collecting the activation levels in a vector $\mathbf{a}$, and the biases in $\mathbf{b}$, we thus have

$$\mathbf{a}(T+1) = (\alpha\mathbf{U} + \beta\mathbf{L})\mathbf{a}(T) + \mathbf{b} \tag{4.1}$$

The iteration will converge iff all eigenvalues of $\mathbf{W} = \alpha\mathbf{U} + \beta\mathbf{L}$ lie within the unit disk. Gershgorin's Circle Theorem provides a simple sufficient condition for this: if $|\alpha| + |\beta| < 1$, $\mathbf{W}^n$ will tend to zero and (4.1) yields a stable vector $\mathbf{a} = (\mathbf{I} - \mathbf{W})^{-1}\mathbf{b}$ no matter where we start it. Here we will not analyze the solutions in detail (see Prince 1993), but just provide a sample of the qualitatively different cases considered by Goldsmith and Larson.

If we set $\alpha = -0.8, \beta = 0$, and a bias of 1 on the last node, we obtain an alternating pulse train proceeding from the right to left, corresponding to the stress pattern observed in Weri [WER], with primary stress falling on the last syllable and secondary stresses on the third, fifth, etc., syllables counting backward from the right edge. With the bias set at -1, we obtain the stress pattern of Warao [WBA], with primary stress on the penultimate syllable and secondary on the fourth, sixth, etc., counting backward. The mirror image of the Weri pattern, obtained by setting the bias to 1 on the first syllable, is observed in Maranungku [ZMR].

What is remarkable about this situation is that the bewildering typological variation appears as a consequence of the model, different regions of the parameter space show typologically different patterns, without any particular effort to reverse-engineer rules that would provide the attested patterns. Chomsky (1981) called this the *explanatory depth* of the theory, and on the whole there is little to recommend models lacking in it.

Exercise 4.3 Provide a better bound on the convergence of $\mathbf{W}$ than what follows from the Gershgorin Circle Theorem. Investigate which qualitative properties of the solutions are independent of the number of syllables n.

4.2 Word formation

The basic morphological unit is the **morpheme**, defined as a minimal (atomic) sign. Unlike phonemes, which were defined as minimum concatenative units of sound, for morphemes there is no provision of temporal continuity because in many languages morphemes can be discontinuous and words and larger units are built from them by processes other than concatenation. Perhaps the best known examples are the triconsonantal roots found in Arabic, e.g. *kataba* 'he wrote', *kutiba* 'it was written', where only the three consonants *k-t-b* stay constant in the various forms. The consonants and the vowels are put together in a single abstract *template*, such as the perfect passive CaCCiC (consider *kattib* 'written', *darris*, 'studied' etc.), even if this requires a certain amount of stretching (here the second of the three consonants gets lengthened so that a total of four C slots can be filled by three consonants) or removal of material that does not fit the template. The phenomenon of discontinuous morphemes extends far beyond the Semitic languages, e.g. to Penutian languages such as Yokuts [YOK], Afroasiatic languages such as Saho [SSY], and even some Papuan languages.

In phonotactics, matters were greatly simplified by distinguishing those elements that do not appear in isolation (consonants) from those that do (vowels). In morphotactics, the situation is more complex, requiring us to distinguish at least six categories of morphemes: roots, stems, inflectional affixes, derivational affixes, simple clitics, and special clitics; as these differ from one another significantly in their combinatorical possibilities. When analyzing the prosodic word from the outside in, the outermost layer is provided by the *clitics*, elements that generally lack independent stress and thus must be prosodically subordinated to adjacent words: those attaching at the left are called *proclitic*, and those at the right are *enclitic*. A particularly striking example is provided by Kwakiutl [KWK] determiners, which are prosodically enclitic (attach to the preceding word) but syntactically proclitic (modify the following noun phrase).

Once the clitics are stripped away, prosodic and morphological words will largely coincide, and the standard Bloomfieldian definition becomes applicable as our first criterion for wordhood (1): *words are minimal free forms*, i.e. forms that can stand in isolation (as a full utterance delimited by pauses) while none of their constituent parts can. We need to revise (1) to take into account not just clitics but also *compounding*, whereby words are formed from constituent parts that themselves are words: Bloomfield's example was *blackbird*, composed of *black* and *bird*. To save the minimality criterion (1), Bloomfield noted that there is more to *blackbird* than *black+bird* inasmuch as the process of compounding also requires a rule of *compound destressing*, which reduces (or entirely removes) the second of the two word-level stresses that were present in the input. Since *bird* in isolation has full word-level stress (i.e. the stress-reduced version cannot appear in isolation), criterion (1) remains intact.

Besides compounding, the two most important word-formation processes are *affixation*, the addition of a bound form to a free form, and *incorporation*, which will generally involve more than two operands, of which at least two are free and one is bound. We will see many examples of affixation. An English example of incorporation would be *synthetic compounds* such as *moviegoer* (note the lack of the intermediate forms **moviego*, **goer* and **movier*). If affixation is concatenative (which is the typical case, often seen even in languages that have significant nonconcatenative morphology), affixes that precede (follow) the stem are called **prefixes (suffixes)**. If the bound morpheme is added in a nonconcatenative fashion (as e.g. in English *sing, sang, sung*), traditional grammar spoke of *infixation*, *umlaut*, or *ablaut*, but as these processes are better described using the autosegmental theory presented in Section 3.3, the traditional terminology has been largely abandoned except as descriptive shorthand for the actual multitiered processes. Compounded, incorporated, and affixed stems are still single prosodic words, and linguists use (2): *words have a single main stress* as one of many confluent criteria for deciding wordhood.

An equally important criterion is (3): *compounds are not built compositionally* (semantically additively) from their constituent parts. It is evident that *blackbird* is not merely some black bird but a definite species, *Turdus merula.* Salient characteristics of blackbirds, such as the fact that the females are actually brown, cannot be inferred from *black* or *bird*. To the extent the mental lexicon uses words and phrases as access keys to such *encyclopedic knowledge,* it is desirable to have a separate entry or *lexeme* for each compound. We emphasize here that such considerations are viewed in morphology as heuristic shortcuts: the full theory of lexemes in the mental lexicon can (and, many would argue, must) be built without reference to encyclopedic knowledge. What is required instead is some highly abstract grammatical knowledge about word forms belonging in the same lexeme, where lexemes are defined as a **maximal set of paradigmatically related word forms**.

But what is a *paradigm?* Before elaborating this notion, let us briefly survey the remaining criteria of wordhood: (4) *words are impervious to reordering* by the kinds of processes that often act freely on parts of phrases (cf. *the bird is black* vs. **this is a birdblack*). (5) *words are opaque to anaphora* (cf. *Mary used to dare the devil but now she is afraid of him* vs. **Mary used to be a daredevil but now she is afraid of him*). Finally, (6) *orthographical separation* also provides evidence of wordhood, though this is a weak criterion – orthography has its own conventions (which include hyphens that can be used to punt on the hard cases), and there are many languages such as Chinese and Hindi, where the traditional orthography does not include whitespace.

Given a finite inventory M of morphemes, criteria (1–6) delineate a formal language of **words** W as the subset of $w \in M^*$ for which either (i) w is free and no $w = w_1 w_2$ concatenation exists with both w_1 and w_2 free or (ii) such a concatenation exists but the meaning of w is not predictable from the meanings of w_1 and w_2. By ignoring the nonconcatenative cases as well as the often considerable phonological effects of concatenation (such as stress shift, assimilation of phonemes at the concatenation boundary, etc.), this simple model lets us concentrate on the internal syntax of words, known as *morphotactics*. Words in $W \cap M$ ($M^2, \ldots$) are called *monomorphemic (bimorphemic, ...)*. Morphemes in $M \setminus W$ are called **bound**, and the rest are **free**.

The single most important morphotactic distinction is between *content* morphemes such as *eat* or *food* and *function* morphemes such as *the*, *of*, *-ing*, *-ity*, the former being viewed as truly essential for expressing any kind of meaning, while the latter only serve to put expressions in a grammatically nicer format. In numerical expressions, the digits would be content, and the commas used to group digits in threes would be function morphemes. Between free or bound, content or function, all four combinations are widely attested, though not all four are simultaneously present in every language. Bound content morphemes are known as **roots**, and free content morphemes are called **stems**. Bound function morphemes are known as **affixes**, and free function morphemes are called *function words* or **particles**. On occasion, one and the same morpheme may appear in multiple classes. Take for example Hungarian case endings such as *nAk* 'dative' or *vAl* 'instrumental'. (Here *A* is used to denote an *archiphoneme* or partially specified phoneme whose full specification for the backness feature (i.e. whether it becomes *a* or *e*) depends on the stem. This phenomenon, called *vowel harmony*, is a typical example of autosegmental spreading, discussed in Section 3.3.) These morphemes generally function as affixes, but with personal pronouns they function as roots, as in *nekem* '1SG.DAT' and *velem* '1SG.INS'.

Affixes (function morphemes) are further classified as *inflectional* and *derivational* – for example, compare the English plural *-s* and the noun-forming adjectival *-ity*. As we analyze words from the outside in, in the outermost layer we find the inflectional prefixes and suffixes. After stripping these away, we find the derivational prefixes and suffixes surrounding the stem. To continue with the example, we often find forms such as *polarities* in which first a derivational and subsequently an inflectional suffix is added, but examples of the reverse order are rarely attested, if at all.

Finally, and at the very core of the system, we may find *roots*, which require some derivational process (typically templatic rather than concatenative) before they can serve as stems. By the time the analysis reaches the root level, the semantic relationship between stems derived from the same root is often hazy: one can easily see how in Arabic *writing, to write, written, book*, and *dictate* are derived from the same root *k-t-b*, but it is less obvious that *meat, butcher* is similarly related to *battle, massacre* and to *welding, soldering, sticking* (as in the root *l-h-m*). In Sanskrit, it is easy to understand that *desire* and *seek* could come from the same root *iSh*, but it is something of a stretch to see the relationship between *narrow* and *distressing (aNh)* or between *shine* and *praise (arch)*.

The categorization of the morphemes as roots, stems, affixes, and particles gives rise to a similar categorization of the processes that operate on them, and the relative prevalence of these processes is used to classify languages typologically. Languages such as Inuktitut [ESB], which form extremely complex words with multiple content morphemes, are called *incorporating*. At the opposite end, languages such as Vietnamese, where words typically have just one morpheme, are called *isolating*. When the majority of words are built recursively by affixation, as in Turkish or Swahili, we speak of *agglutinating* languages. Because of the central position that Latin and ancient Greek occupied in European scholarship until the 19th century, a separate typological category is reserved for *inflecting languages* where the agglutination is accompanied by a high degree of morphological *suppletion* (replacement of affix combination by a single affix). Again we emphasize that the typological differences expressed in these labels are radical: the complexities of a strongly incorporating language cannot even be imagined from the perspective of agglutinating or isolating languages.

The central organizational method of morphology, familiar to any user of a (monolingual or bilingual) dictionary, is to collect all compositionally related words in a single class called the *lexeme*. On many occasions, we can find a single distinguished form, such as the nominative singular of a noun or the third-person singular form of a verb, that can serve as a representative of the whole collection in the sense that all other members of the lexeme are predictable from this one *citation form*. On other occasions, this is not quite feasible, and we make recourse to a theoretical construct called the *underlying form* from which all forms in the lexeme are generated. The citation form, when it exists, is just a special case where a particular surface form (usually, but not always, the one obtained by affixing zero morphemes) retains all the information present in the underlying form.

Within a single lexeme, the forms are related *paradigmatically*. Across lexemes in the same class, the abstract structure of the lexeme, the *paradigm*, stays constant, meaning that we can abstract away from the stem and consider the shared paradigm of the whole class (e.g. the nominal paradigm, the verbal paradigm, etc.) as structures worth investigating in their own right. A paradigmatic contrast along a given morphological dimension such as *number*, *gender*, *person*, *tense*, *aspect*, *mood*, *voice*, *case*, *topic*, *degree*, etc., can take a finite (usually rather small) number of values, one of which is generally **unmarked** (i.e. expressed by a zero morpheme). A typical example would be the number of nouns in English, contrasting the singular (unmarked) to plural *-s*, or in classical Arabic, contrasting singular, dual, and plural. We code the contrasts themselves by abstract markers (diacritics, see Section 3.2) called *morphosyntactic features*, and the specific morphemes that realize the distinctions are called their *exponents*. The abstract markers are necessary both because the same distinction (e.g. plurality) can be expressed by different overt forms (as in English *boy/boys*, *child/children*, *man/men*) and because the exponent need not be a concatenative affix – it can be a templatic one such as reduplication.

The full **paradigm** is best thought of as a direct sum of direct products of morphological contrasts obtaining in that category. For example, if in a language verbs can be inflected for person (three options) and number (two options), there are a total of six forms to consider. This differs from the feature decomposition used in phonology (see Section 3.2) in three main respects. First, there is no requirement for a single paradigmatic dimension to have exactly two values, and indeed, larger sets of values are quite common – even *person* can distinguish as many as four (singular, dual, trial, and plural) in languages such as Gunwinggu [GUP]. Second, it is common for paradigms to be composed of direct sums; e.g. a whole subparadigm of infinitivals plus a subparadigm of finite forms as in Hungarian. Third, instead of the subdirect construction, best characterized by a subset (embedding) in a direct product, we generally have a fully filled direct product, possibly at the price of repeating entries. For example, German adjectives have a maximum of six different forms (e.g. *gross*, *grosses*, *grosse*, *grossem*, *grossen*, and *grosser*), but these are encoded by four different features, case (four values), gender (three values), number (two values), and determiner type (three values), which could in principle give rise to 72 different paradigmatic slots. Repetition of the same form in different paradigmatic slots is quite common, while missing entries (a combination of paradigmatic dimensions with no actual exponent), known as *paradigm gaps*, are quite rare, but not unheard of.

For each stem in a class, the words that result from expressing the morphosyntactic contrasts in the paradigm are called the *paradigmatic forms* of the stem: a lexeme is thus a stem and all its paradigmatic forms. This all may look somewhat contrived, but the notion that stems should be viewed not in isolation but in conjunction with their paradigmatic forms is so central to morphology that the whole field of study (morphology, 'the study of forms') takes its name from this idea. The central notational device in the linguistic presentation of foreign language data, *morphological glosses*, also has the idea of paradigmatic forms built in: for example, *librum* is glossed as *book. ACC*, thereby explicitly translating not only the stem *liber*, which has a trivial equivalent in English, but also the fact that the form is accusative, an idea that has no English equivalent.

The processes that generate the paradigmatic forms are called **inflectional**, while processes whose output is outside the lexeme of the input are called **derivational** – a key issue in understanding the morphotactics of a language is to distinguish the two. The most salient difference is in their degree of automaticity: inflectional processes are highly automatic, virtually exceptionless both in the way they are carried through and in the set of stems to which they apply, while derivational processes are often subject to alternations and/or are limited in their applicability to a subclass of stems. For example, if *X* is an adjective, *X-ity* will be a noun meaning something like 'the property of being X', but there are different affixes that perform this function and there is no clear way to delineate their scope without listing (compare *absurdity*, *redness*, *teenhood*, **redity*, **teenity*, **redhood*, **absurdhood*, **absurdness*, **teenness*). When we see an affix with limited combining ability, such as *-itis* 'swelling/inflammation of' (as in *tendinitis*, *laryngitis*, *sinusitis*), we can be certain that it is not inflectional.

Being automatic comes hand in hand with being compositional. It is a hallmark of derivational processes that the output will not necessarily be predictable from the input (the derived stems for *laHm* 'meat', *malHama* 'battle', and *laHam* 'welding' from the same root *l-h-m* are a good example). The distinction is clearly observable in the frequency of forms that contain a given affix. For example, the proportion of plural forms ending in *-s* is largely constant across various sets of nouns, while the proportion of nouns ending in *-ity*, *-ness*, and *hood* depends greatly on the set of inputs chosen; e.g. whether they are Latinate or Germanic in origin. The same distinction in *productivity* is also observable in whether people are ready to apply the generative rule to stems they have not encountered before.

By definition, inflection never changes category, but derivation can, so whenever we have independent (e.g. syntactic or semantic) evidence of category change, as in English *quick* (adjective) but *quickly* (adverb), we can be certain that the suffix causing the change is derivational. Inflectional features tend to play a role in rules of syntactic agreement (concord) and government, while derivational features do not (Anderson 1982). Finally, inflection generally takes place in the outermost layers (last stages, close to the edges of the word, away from the stem or root), while derivation is in the inner layers (early stages, close to the stem or root).

Although in most natural languages M is rather large, 10^4–10^5 entries (as opposed to the list of phonemes, which has only 10^1–10^2 entries), the set of attested morphemes is clearly finite, and thus does not require any special technical device to enumerate. This is not to say that the individual list entries are without technical complexity. Since these refer to (sound, meaning) pairs, at the very least we need some representational scheme for sounds that extends to the discontinuous case (see Section 3.3), and some representational scheme for meanings that will lend itself to addition-like operations as more complex words are built from the morphemes. We will also see the need for marking morphemes *diacritically* i.e. in a way that is neither phonological nor semantic. Since such diacritic marking has no directly perceptible (sound or meaning) correlate, it is obviously very hard for the language learner to acquire and thus has the greatest chance for faulty transmission across generations.

The language W can be infinite, so we need to bring some variant of the generative apparatus described in Chapter 2 to bear on the task of enumerating all words starting from a finite base M. The standard method of generative morphology is to specify a base inventory of morphemes as (sound, diacritic, meaning) triples and a base inventory of rules that form new triples from the already defined ones. It should be said at the outset that this picture is something of an idealization of the actual linguistic practice, inasmuch as no truly satisfactory formal representation of (word-level) meanings exists, and that linguistic argumentation generally falls back on natural language paraphrase (see Section 5.3) rather than employing a formal theory of semantics the same way it uses formal theories of syntax and phonology. While this lack of *lexical semantics* is keenly felt in many areas of knowledge representation, in morphology we can make remarkable progress by relying only on some essential notions about whether two words mean, or do not mean, the same thing. Here we illustrate this on the notion of **blocking**, which says that if s_1s_2 is a derived (generated) form with meaning m and s is a primitive (atomic) form with the same meaning m, the atomic form takes precedence over the derived one:

$$(s,m) \in W \Rightarrow (s_1s_2,m) \notin W \tag{4.2}$$

To see blocking in action, consider the general rule of past tense formation in English, which (keeping both the sound and the meaning sides of the morphemes loosely formulated, and ignoring the issue of diacritics entirely) says

(Verb, m) + (*-ed*, PAST) = (Verb.ed, m.PAST)

i.e. the past tense of a verb *Verb* is formed by concatenation with the morpheme *ed*. Clearly, this is a general rule of English, even though it fails in certain cases: when the verb is *go*, the past tense form is *went* rather than **goed*. It would be awkward to reformulate the rule of English past tense formation to say "add the morpheme *-ed* except when the verb is *go*, *eat*, ...". By assuming a higher principle of blocking, we are freed of the need to list the exceptions to each rule; the mere existence of (went, go.PAST) in the lexicon of English will guarantee that no other form with semantics go.PAST need be considered. For a more subtle example, not immediately evident from (4.2), consider the rule of agentive noun formation: for (Verb, m), Verb.er is the agent of m (e.g., *eat/eater, kill/killer*). Much to the chagrin of children learning English, there is no **bicycler* – such a person is called a *bicyclist*. Here the mere existence of the *-ist* form is sufficient to block the *-er* form, though there is no general rule of *-ist* taking precedence over *-er* (cf. *violinist/*violiner, *fiddlist/fiddler*).

Kiparsky (1982b) treats blocking as a special case of the more general Elsewhere Principle going back to Pāṇini: rules with narrower scope (affecting fewer items) take precedence over rules with broader scope. By treating lexical entries like *bicyclist* as identity rules, blocking appears at the very top of a precedence hierarchy running from the most specific (singleton) rules to the most generic rules. Since much of morphology is about listing (diacritically marking) various elements, it may appear attractive to do away with general rules entirely, trying to accomplish as much by tabulation as possible. But it is clear that children, from a relatively early age, are capable of performing morphological operations on forms they have never encountered before. The classic "wug" test of Berko (1958) presents children with the picture of a creature and the explanation "this is a wug". When you add a second picture and say "now there are two of them. These are two ...", by first grade children will supply the correct plural (with voiced /wugz/ rather than unvoiced /wugs/). Since the evidence in favor of such *productive* use of morphological rules is overwhelming, it is generally assumed that general (default) rules or constraints are part of the adult morphological system we wish to characterize.

In this sense, the set of content morphemes is not just large but infinite (open to the addition of new stems like *wug*). In contrast, the set of function morphemes is small (10^2–10^3 entries) and closed. For each function morpheme m_i, we can ask what is the set $S_i \subset M$ for which affixation with m_i is possible; e.g. $\{s \in S | s.m_i \in W\}$. Perhaps surprisingly, we often find $S_i = S_j$. In other words, the distributions of different function morphemes are often identical or near-identical. For example, in English, every stem that can take *-ed* can also take *-ing* and conversely – exceptions such as strong verbs that do not take their past tense in *-ed* can be readily handled by the blocking mechanism discussed above. This clustering gives rise to the fundamental set of diacritics known as *lexical categories* or *parts of speech* (POS). Given that (two and a half millennia after Pāṇini) the actual format of morphological rules/constraints and their interaction is still heavily debated, it may come as something of a surprise that there is broad consensus on the invisible (diacritical) part of the apparatus, and in fact it would be next to impossible to find a grammar that does not employ lexical categories.

In most, perhaps all, languages, the largest category is that of *nouns* (conventionally denoted N), followed by the class A of *adjectives*, the class V of *verbs*, Adv *(adverbials)*, and Num *(numerals)*. Again, multiple membership is possible, indeed typical. For each lexical category there is an associated paradigm, sometimes trivial (as e.g. for adverbs, which generally have no paradigmatically related forms) but on occasion rather complex (as in verbal paradigms, which often express distinctions as to voice, aspect, tense, mood, person, and number of subject, direct and even indirect object, and gender). Paradigms are such a powerful descriptive tool that in an ideal agglutinating/inflecting language, the task of characterizing W would be reduced to providing a few tabular listings of paradigms, plus some diacritic marking (lexical categorization) accompanying the list of content morphemes. As an added bonus, we will see in Section 5.2 that *syntax*, the study of the distribution of words in sentences, is also greatly simplified by reference to lexical categories and paradigmatic forms. To be sure, no language is 100% agglutinating (though many come remarkably close), but enough have large and complex paradigms to justify architecting the rest of the morphology around them.

A central issue of morphology is whether the use of diacritics can be eliminated from the system. For example, it is often assumed that lexical categories are semantically motivated, e.g. nouns are names of things, verbs denote actions, etc. Yet in truth we do not have a theory of lexical semantics that would sustain the category distinctions that we see, and many things that could equally well be one or the other (the standard example is *fire* noun vs. *burn* verb) actually end up in different categories in different languages. There has been better progress in eliminating diacritics by means of phonological marking. For example Lieber (1980) replaces the paradigm classes of traditional Latin descriptive grammar with phonologically motivated triggers. Yet some hard cases, most notably Sanskrit and Germanic (Blevins 2003), still resist a purely phonological treatment, and the matter is by no means settled.

4.3 Optimality

Formal descriptions of morphology standardly proceed from a list of basic entries (roots and stems) by means of inflectional rules that manipulate strings or multitiered representations to produce the paradigmatic forms and derivational rules that produce stems from roots or other stems. While this method is highly capable of expressing the kind of regularities that are actually observed in natural language, very often we find rule systems that *conspire* to maintain some observable regularity without actually relying on it. As an example, consider a language (such as Tiberian Hebrew) that does not tolerate complex codas. Whenever some process such as compounding or affixation creates a complex coda, we find some secondary cleanup process, such as the insertion of a vowel (a process called *epenthesis*) or the deletion of a consonant, that will break up the complex coda.

Optimality theory (OT) is built on the realization that in these situations the observable (surface) regularity is the cause, rather than the effect, so that to state the grammar one needs to state the regularities rather than the processes that maintain them. There are two main technical obstacles to developing such a theory: first, that the regularities may contradict each other (we will see many examples of this), and second, that regularities may have exceptions. In the previous section, we saw how individual (lexical) exceptions can be handled by means of the Elsewhere Condition, and this mechanism extends smoothly to all cases where one subregularity is entirely within the domain of a larger (super)regularity. But there are many cases where the domains overlap without either one being included in the other, and for these we need some prioritization called *constraint ranking* in OT.

An OT grammar is composed of two parts: a relation GEN between abstract (underlying) representations and candidate surface realizations, and a function EVAL that ranks the candidates so that the highest-ranked or *optimal* one can be chosen. The ranking of the candidates is inferred from the ranking of the constraints, which is assumed to be absolute and immutable for a given language at a given synchronic stage but may vary across languages and across different synchronic stages of the same language. Optimality theory is closely linked to the kind of linguistic typology that we exemplified in Section 4.1 with stress typology in that the constraints are assumed to come from a universal pool: variation across languages is explained by assuming different rankings of the same constraints rather than by assuming different constraints.

The (universal) family of constraints on GEN requires *faithfulness* between underlying and surface representations. These can be violated by epenthesis processes that create elements present in the surface that were missing from the underlying form, by deletion processes that remove from the surface elements that were present in the underlying form, and by reordering processes that change the linear order of elements on the surface from the underlying order. The constraints on EVAL are known as *markedness* constraints. These prohibit configurations that are exceptional from the perspective of universal grammar. For example, complex syllabic onsets or codas are more marked (typologically more rare) than their simple counterparts, so a universal prohibition against these makes perfect sense, as long as we understand that these prohibitions can be overridden by other (e.g. faithfulness) considerations in some languages.

Once an ordering of the constraints is fixed, it is a mechanical task to assess which constraints are violated by which input-output pair. This task is organized by the *tableau* data structure introduced in the founding paper of OT (Prince and Smolensky 1993), which places constraints in columns according to their rank order, and candidate forms in rows. Each form (or, in the case of faithfulness constraints, each input-output pair) will either *satisfy* a constraint or *violate* it. (In early OT work, the severity of a violation could be a factor, for a modern view, see McCarthy 2003.) In some cases, it is possible for a form (pair) to violate a constraint more than once; e.g. if the constraint requires full syllabification and there is more than one unsyllabified element. Given the infinite set of candidates produced by GEN, the *selection* problem is finding an optimal candidate ω such that for any other candidate χ the first constraint (in rank order) that distinguishes between ω and χ favors ω.

Standard phonology (Chomsky and Halle 1968) used context-sensitive rules not just to describe how the various morphosyntactic features are expressed (often constrained by reference to other invisible diacritic elements) but also to clean up the output of earlier rules so that the results become phonologically well-formed. In some grammars, most notably in Pāṇini, the cleanup is left to a single postprocessing stage (the last three sections of the grammar following 8.2, known as the *Tripādī*), but in many cases (often collected under the heading of *boundary strength* or *level ordering*) it seems advisable to mix the phonological rules with the purely morphological ones. To the extent that the morphological constituent structure dictates the order of rule application, a theory that strongly couples morphological and phonological well-formedness has little freedom in ordering the phonological rules. But to the extent that weaker (or as in the Ashṭādhyāyī, no) coupling is used, there remains a certain degree of freedom in ordering the rules, and this freedom can be exploited to describe near-truths.

Context-sensitive rules of the form $B \rightarrow C/A\_D$ are called *opaque* if we encounter strings ABD, suggesting the rule should have applied but somehow did not, or if C appears in an environment other than A_D, suggesting that C came about by some other means. To continue with the Tiberian Hebrew example, the language has a rule of deleting *?* (glottal stop) outside syllable onsets, which should be ordered *after* the rule of epenthesis that breaks up complex codas. This means that a form such as *deš?* first undergoes epenthesis to yield *deše?* but the final *?* is deleted, so from the surface it is no longer evident what triggered the epenthesis in the first place.

In a constraint-based system the notion of opacity has to be reinterpreted slightly – we say that a constraint is not *surface true* if there are exceptions beyond the simple lexical counterexamples and not *surface apparent* if the constraint appears true but the conditions triggering it are not visible. Generally, such situations put such a high burden on the language learner that over the course of generations the rule/constraint system is replaced by one less opaque, even at the expense of not transmitting the system faithfully.

4.4 Zipf's law

We define a **corpus** simply as any collection of texts. Linguists will often impose additional requirements (e.g. that all texts should originate with the same author, should be about the same topic, or should have consistent spelling), but for full generality we will use only the more liberal definition given above. When a collection is exhaustive (e.g. the complete works of Shakespeare), we speak of *closed* corpora, and when it can be trivially extended (e.g. the issues of a newspaper that is still in publication), we speak of *open* corpora.

Perhaps the simplest operation we can perform on a corpus is to count the words in it. For the sake of concreteness we will assume that the texts are all ascii, that characters are lowercased, and all special characters, except for hyphen and apostrophe, are mapped onto whitespace. The terminal symbols or *letters* of our alphabet are therefore $\Sigma = \{a, b, \ldots z, 0, 1, \ldots 9, ', -\}$, and all word types are strings in Σ^*, though word tokens are strings over a larger alphabet including capital letters, punctuation, and special characters. Using these or similar definitions, counting the number of tokens belonging in the same type becomes a mechanical task. The results of such *word counts* can be used for a variety of purposes, such as the design of more efficient codes (see Chapter 7), typology, investigations of style, authorship, language development, and statistical language modeling in general.

Given a corpus S of **size** $L(S) = N$ (the number of word tokens in S), we find V different types, $V \leq N$. Let us denote the **absolute frequency** (number of tokens) for a type w by $F_S(w)$ and the **relative frequency** $F_S(w)/N$ by $f_S(w)$. Arranging the w in order of decreasing frequency, the rth type, w_r, is said to have **rank r**, and its relative frequency $f_S(w_r)$ will also be written f_r. As Estoup (1916) and Zipf (1935)

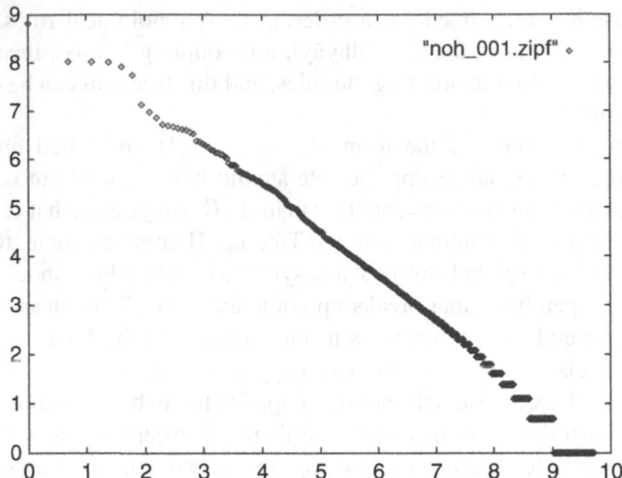

Fig. 4.1. Plot of log frequency as a function of log rank for a newspaper issue (150k words)

noted, the plot of log frequencies against log ranks shows a reasonably linear relation. Figure 4.1 shows this for a single issue of an American newspaper, the *San Jose Mercury News*, or *Merc* for short. Historically, much of our knowledge about word frequencies has come from corpora of this size: the standard Brown Corpus (Kucera and Francis 1967) is about a million words. Today, corpora with size 10^7–10^8 are considered medium-sized, and only corpora above a billion words are considered large (the largest published word count is based on over 10^{12} words). The study of such corpora makes the utility of the log scale evident: perfectly ordinary words like *dandruffy* or *uniform* can have absolute frequency 0.000000001 or less. Denoting the slope of the linear portion by $-B$, B is close to unity, slightly higher on some plots and slightly lower on others. (Some authors, such as Samuelsson (1996), reserve the term "Zipf's law" to the case $B = 1$.) As a first approximation, Zipf's law can be given as

$$\log(F_r) = H_N - B\log(r) \tag{4.3}$$

where H_N is some constant (possibly dependent on S and thus on N, but independent of r). This formula is closely related to, but not equivalent with, another regularity, often called Zipf's second law. Let $V(i, N)$ be the number of types that occur i times. Zipf's second law is usually stated as

$$\log(i) = K_N - D_N \log(V(i, N)) \tag{4.4}$$

The status of Zipf's law is highly contentious, and the debate surrounding it is often conducted in a spectacularly acrimonious fashion. As an example, we quote here Herdan (1966:88):

> The Zipf law is the supposedly straight line relation between occurrence frequency of words in a language and their rank, if both are plotted

> logarithmically. Mathematicians believe in it because they think that linguists have established it to be a linguistic law, and linguists believe in it because they, on their part, think that mathematicians have established it to be a mathematical law. ... Rightly seen, the Zipf law is nothing but the arbitrary arrangement of words in a text sample according to their frequency of occurrence. How could such an arbitrary and rather trivial ordering of words be believed to reveal the most recondite secrets, and the basic laws, of language?

Given the sheer bulk of the literature supporting some Zipf-like regularity in domains ranging from linguistic type/token counts to the distribution of wealth, it is natural that statisticians sought, and successfully identified, different genesis mechanisms that can give rise to (4.3, 4.4), and related laws.

The first results in this direction were obtained by Yule (1924), working on a version of (4.3) proposed in Willis (1922) to describe the number of species that belong to the same genus. Assuming a single ancestral species, a fixed annual probability s of a mutation that produces a new species, and a smaller probability g of a mutation that produces an entirely new genus, Yule shows that over time the distribution for the number of genera with exactly i species will tend to

$$1/i^{1+g/s} \tag{4.5}$$

This is not to say that words arise from a single undifferentiated ancestor by a process of mutation – the essential point of Yule's work is that a simple uniform process can give rise, over time, to the characteristically nonuniform 'Zipfian' distribution. Zipf himself attempted to search for a genesis in terms of a "principle of least effort", but his work (Zipf 1935, 1949) was never mathematically rigorous and was cut short by his death. A mathematically more satisfying model specifically aimed at word frequencies was proposed by Simon (1955), who derived (4.3) from a model of text generation based on two hypotheses: (i) new words are introduced by a small constant probability, and (ii) old words are reused with the same probability that they had in earlier text.

A very different genesis result was obtained by Mandelbrot (1952) in terms of the classic "monkeys and typewriters" scenario. Let us designate an arbitrary symbol on the typewriter as a word boundary and define "words" as maximum strings that do not contain it. If we assume that new symbols are generated randomly, Zipf's law can be derived for $B > 1$. Remarkably, the result holds true if we move from a simple Bernoulli experiment (zero-order Markov process; see Chapter 7) to higher-order Markov processes.

In terms of content, though perhaps not in terms of form, the high point of the Zipfian genesis literature is the Simon-Mandelbrot debate (Mandelbrot 1959, 1961a, 1961b, Simon 1960, 1961a, 1961b). Simon's genesis works equally well irrespective of whether we assume a closed ($B < 1$) or open ($B > 1$) vocabulary. For Mandelbrot, the apparent flexibility in choosing any number close to 1 is a fatal weakness in Simon's model. While we side with Mandelbrot for the most part, we believe his critique of Simon to be too strict in the sense that explaining too much is not as fatal a flaw as explaining nothing. Ultimately, the general acceptance of Mandelbrot's genesis as the linguistically more revealing rests not on his attempted destruction of Simon's model but rather on the fact that we see his model as more assumption-free.

Although Zipf himself held that collecting more data about word frequency can sometimes distort the picture, and there is an "optimum corpus size" (for a modern discussion and critique of this notion, see Powers 1998), here we will follow a straight *frequentist* approach that treats corpora as samples from an underlying distribution. To do this, we need to normalize (4.3) so that its fundamental content, linearity on a log-log scale, is preserved independent of sample size. Although in corpus linguistics it is more common to study sequences of dependent corpora $S_N \subset S_{N+1}$, here we assume a sequence of *independent* corpora satisfying $L(S_N) = N$. On the y axis, we divide all values by N so that we can work with relative frequencies f, rather than absolute frequencies F, and on the x axis we divide all values by $V(N)$ so that we can work with relative ranks $0 \leq x = r/V(N) \leq 1$, rather than absolute ranks r. Accordingly, (4.3) becomes

$$\log(f(xV(N))) = H_N - \log(N) - B_N \log(x) - B_N \log(V(N)) \tag{4.6}$$

The Zipf line intersects the x axis at $x = 1$, where the relative frequency of the least frequent item is just $1/N$. This is because at the low end of the distribution, we find a large number of **hapax legomena**, words that appear only once in the corpus, and **dis legomena**, words that appear only twice. (For large corpora, typically about 40% to 60% of all word types appear only once and another 10% to 15% only twice.) Since $\log(1)$ is zero, we have $H_N = B_N \log(V(N))$ i.e. that the Zipf line is always shifted up from the origin by $B_N \log(V(N))$. We can reasonably call a population Zipfian only if the B_N will converge to a Zipf constant B, an empirical requirement that seems to be met by sequences of medium to large corpora – our current computational ability to analyze corpora without special hardware limits us to about 10^{11} words.

Using $r = 1$ in (4.3) we obtain $\log(F_1) = H_N = B \log(V(N))$ and by subtracting $\log(N)$ from both sides $\log(f(1)) = B \log(V(N)) - \log(N)$. Here the left hand side is a constant, namely the log frequency of the most frequent word (in English, *the*), so the right-hand side must also tend to a constant with increased N. Thus we have obtained $B \log(V(N)) \sim \log(N)$, known as the *power law of vocabulary growth.* This law, empirically stated by many researchers including Guiraud (1954) (with $B = 2$), Herdan (1960), and Heaps (1978), becomes the following theorem.

Theorem 4.4.1 In corpora taken from populations satisfying Zipf's law with constant B, the size of the vocabulary grows with the $1/B$-th power of corpus length N,

$$V(N) = cN^{1/B} \tag{4.7}$$

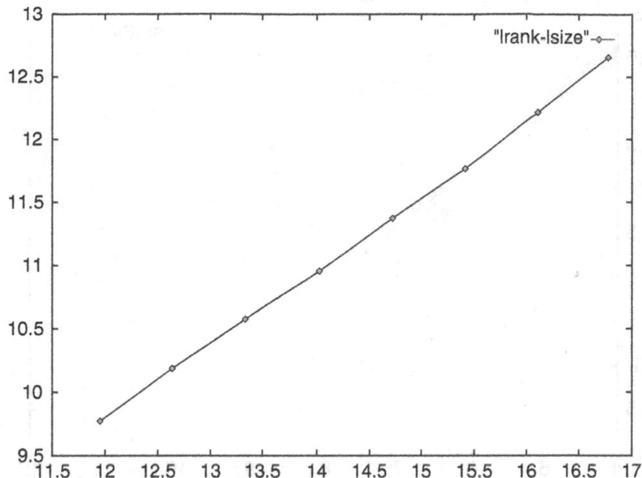

Fig. 4.2. Growth of vocabulary size *V(N)* against corpus size *N* in the *Merc* on log-log scale

Here c is some fixed multiplicative constant (not to be confused with the constant C of Herdan (1964:157) and later work, which corresponds to the exponent $1/B$ in our notation). We illustrate this law on a corpus of some 300 issues of the *Merc* totaling some 43 millon words (see Figure 4.2). Increasingly larger *independent* samples were taken so as to guard against the effects of diachronic drift (new words get added to the vocabulary). Although we derived (4.7) as a theorem, it should be emphasized that it still has the status of an approximate empirical law (since it was derived from one), and in practice some slower patterns of infinite growth such as $V(N) = N^{D/\log(\log(N))}$ would still look reasonably linear for $N < 10^{11}$ at log-log scale, and would be just as compatible with the observable data. The same holds for Zipf's second law (see Figure 4.3).

Theorem 4.4.2 In corpora taken from populations satisfying Zipf's law with parameter B, (4.4) is satisfied with parameter $D = B/(1 + B)$.

Proof Zipf's first law gives $f(r) = x^{-B}/N$, so the probability of a word is between i/N and $(i + 1)/N$ iff $i \leq x^{-B} \leq i + 1$. Therefore, we expect $V(i, N) = V(N)(i^{-1/B} - (i + 1)^{-1/B})$. By Rolle's theorem, the second term is $1/B\theta^{-1/B-1}$ for some $i \leq \theta \leq i + 1$. Therefore, $\log(V(i, N)) = \log(V(N)) - \log(\theta)(B + 1)/B - \log(B)$. Since $\log(B)$ is a small constant, and $\log(\theta)$ can differ from $\log(i)$ by no more than $\log(2)$, rearranging the terms we get $\log(i) = \log(V(N))B/(B + 1) - \log(V(i, N))B/(B + 1)$. Since $K_N = \log(V(N))B/(B + 1)$ tends to infinity, we can use it to absorb the constant terms.

Discussion The normalization term K_N is necessitated by the fact that second law plots would otherwise show the same drift as first law plots. Using this term, we can state the second law in a much more useful format. Since $\log(i) = \log(V(N))B/(B + 1) - \log(V(i, N))B/(B + 1)$ plus some additive constant,

$$V(i, N) = mV(N)/i^{1+1/B} \tag{4.8}$$

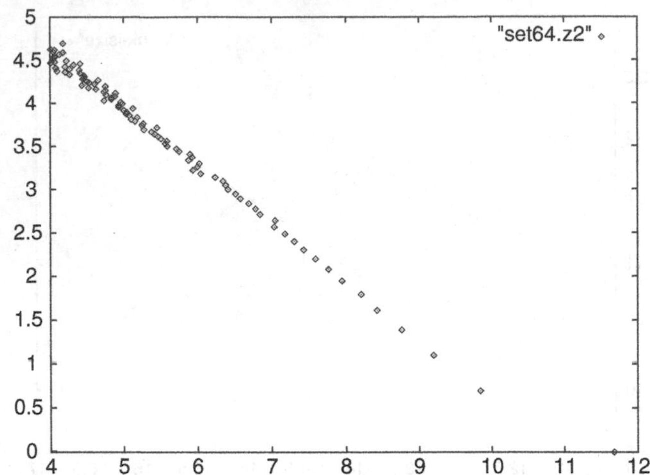

Fig. 4.3. Number-frequency law on the Merc (10m words)

where m is some multiplicative constant. If we wish $\sum_{i=1}^{\infty} V(i,N) = V(N)$ to hold we must choose m to be $1/\zeta(1+1/B)$, which is the reason why Zipfian distributions are sometimes referred to as ζ distributions. Since this argument assumes Zipf's second law extends well to high-frequency items where the empirical fit is not particularly good (see Section 7.1 for further discussion), we find Mandelbrot's (1961c) criticism of $B = 1$ to be somewhat less compelling than the case he made against $B < 1$. Recall from the preceding that B is the reciprocal of the exponent $1/B$ in the vocabulary growth formula (4.6). If we choose a very 'rich' corpus (e.g. a table of logarithms), virtually every word will be unique, and $V(N)$ will grow faster than $N^{1-\varepsilon}$ for any $\varepsilon > 0$, so B must be 1. The following example sheds some light on the matter.

Example 4.4.1 Let $L = \{0, 1, \ldots, 9\}$ and our word tokens be the integers (in standard decimal notation). Further, let two tokens share the same type if their smallest prime factors are the same. Our size N corpus is constructed by N drawings from the exponential distribution that assigns frequency 2^{-i} to the number i. It is easy to see that the token frequency will be $1/(2^p - 1)$ for p prime and 0 otherwise. Therefore, our corpora will not satisfy Zipf's law, since the rank of the ith prime is i but from the prime number theorem $p_i \sim i\log(i)$ and thus its log frequency $\sim -\,i\log(i)\log(2)$. However, the corpora will satisfy Zipf's second law since, again from the prime number theorem, $V(i,N) = N/i^2(\log(N) - \log(i))$ and thus $\log(V(N))/2 - \log(V(i,N))/2 = \log(N)/2 - \log(\log(N))/2 - \log(N)/2 + \log(i) + \log(\log(N) - \log(i))/2$, which is indeed $\log(i)$ within $1/\log(N)$.

Discussion Example 4.4.1 shows that Theorem 4.4.2 cannot be reversed without additional conditions (such as $B > 1$). A purist might object that the definition of token/type relation used in this example is weird. However, it is just an artifact of the Arabic system of numerals that the smallest prime in a number is not evident. If we used the canonical form of numbers, everything after the first prime could simply be discarded as mere punctuation.

Zipf's laws, including the power law of vocabulary growth, are near-truths in three different senses. First, the domain of the regularity needs to be circumscribed: it is clear that the empirical fit is not nearly as good at the high end as for lower frequencies. Second, these laws are only true in the asymptotic sense: the larger the sample the better the fit. Finally, these laws are true only as first-order approximations. In general it makes perfect sense to use not only linear, but also quadratic and higher-order terms to approximate the function of interest. On a log-log scale, however, it is not evident that the next term should come from this kind of approach; there are many other series that would work equally well.

Corollary 4.4.1 Vocabulary is infinite. Since there is no theoretical limit on the size of the corpus we can collect, and $N^{1/B}$ tends to infinity with N, it is a trivial corollary of the Zipf/Herdan laws that there is no theoretical limit to vocabulary size.

4.5 Further reading

Morphology is historically the oldest layer of linguistics: most of the early work on Sanskrit (Pāṇini, circa 520–460 BCE), Greek (Dionysius Thrax, circa 166–90 BCE), and Latin (Stilo, circa 152–74 BCE, Varro, 116–27 BCE) concerns morphological questions. For a clear exposition of the structuralist methods, see Nida (1949). The idea that paradigms can freely repeat the same form has been raised to a methodological axiom, known as the *Principle of the maximally differentiated paradigm*, by one of the founding fathers of structuralism, Hjelmslev (1961) [1943]. It is not always trivial to distinguish purely phonological from purely morphological concerns, and readers interested in morphology alone should still consult Anderson (1985). The most encyclopedic contemporary source is Spencer and Zwicky (1998). The use of (sound, diacritic, meaning) triples is discussed e.g. in Mel'čuk (1993–2000) – Kracht (2003) speaks of *exponent* rather than sound and *category* rather than diacritic when elucidating the same idea of signs as ordered triples.

The prosodic hierarchy is discussed in detail in Hammond (1995); the approach we follow here is that of Clements and Keyser (1983). How the abstract picture of segmental duration assumed in this theory is reconciled with the actual length variation observed in speech is discussed in Kornai (1995 Ch. 3). The sonority hierarchy originates with Jespersen (1897); for a detailed application to Tashlhiyt Berber [SHI], see Dell and Elmedlaoui (1985, 1988). Moras constitute a still unresolved area; see e.g. Broselow (1995) and Hayes (1995). For the relationship of typology and diachrony see Aristar (1999). The modern theory of Arabic roots starts with McCarthy (1979); see also Heath (1987). For clitics, see Zwicky (1985), and for bound words, see Nevis (1988).

For the purely phonological part of Pāṇini's system, see Buiskool (1939). Level ordering in standard phonology is motivated in Kiparsky (1982); for OT-internal arguments, see Rubach (1977, 2000); for a synthesis of OT and LPM, see Kiparsky

(2006). Opacity was introduced in Kiparsky (1968), for a syntesis of level ordering and OT, see Kiparsky (2006). For an influential modern treatment of blocking, see Aronoff (1976), and for a more detailed discussion of blocking in Pāṇini's system, see Kiparsky (2002). Note that blocking is not necessarily absolute since doublets like *bicycler/bicyclist* can coexist, but (according to Kroch 1994) these are "always reflections of unstable competition between mutually exclusive grammatical options. Even a cursory review of the literature reveals that morphological doublets occur quite frequently, but also that they are diachronically unstable".

Methods based on character and word frequency counts go back to the Middle Ages: in the 1640s, a Swedish sect was deemed heretical (relative to Lutheran orthodoxy) on the basis of a larger than expected frequency of forms such as *Christ bleeding, Christ suffering, Christ crucified* found in its Sion Psalmbook. The same power law distribution as Zipf's law has been observed in patterns of income by Pareto (1897), and there is again a large body of empirical literature supporting Zipf's law, known in economics as Pareto's law. Champernowne (originally in 1936, but not fully published until 1973) offered a model where the uneven distribution emerges from a stochastic process (Champernowne 1952, 1953, 1973; see also Cox and Miller 1965) with a barrier corresponding to minimum wealth. The modern genesis is due to Mandelbrot (1952) and Miller (1957), though Li (1992) is often cited in this regard. Mitzenmacher (2004) is a good survey. The fundamental observation behind Herdan's law is due to Herdan (1960), though Heaps (1978) is often credited. Theorem 4.4.1 is from Kornai (1999a); for a more recent discussion, see van Leijenhorst and van der Weide (2005).

Corpus size has grown dramatically in the past half-century. In the 1960s and 1970s the major corpora such as the Brown Corpus, the London-Lund Corpus, or the Lancaster-Oslo-Bergen (LOB) Corpus had $N = 10^6$ or less. By the 1980s corpora with $N = 10^7$–10^8 were widely disseminated by the Linguistic Data Consortium, and by the 1990s billion-word (10^9) probabilistic language models were commonly used; e.g. in speech recognition. Today, monolingual segments of large search engine caches provide access to corpora such as the Google 5-gram corpus with $N > 10^{12}$ – inspecting the tail end of such caches makes it clear that infinite vocabulary growth is fueled by compounding and incorporation, as well as certain kinds of affixation (see Kornai 2002 for further discussion).

5

Syntax

The theory of syntax addresses three strongly interconnected ranges of facts. The first of these is the combinatorical possibilities of words. It is very clear that *The boys ate* is a sentence of ordinary English, while four other permutations of these three elements, **The ate boys, *Ate boys the, *Boys the ate*, and **Boys ate the*, are outside the bounds of ordinary English. The remaining one, *?Ate the boys*, is harder to pass judgment on, but it seems clear that its stylistic value is very different from that of the first sentence. Similarly, speakers of English will strongly agree that *The boys eat* and *The boy eats* are ordinary sentences of the language, while **The boy eat* and **The boy ates* are not, a highly generalizable observation that justifies the statement, familiar to all from school grammar, that predicates agree with their subjects in person and number.

In most, though not necessarily all, cases it is relatively easy to construct pairs of sentences, one grammatical and the other not, that bring into sharp relief how a particular rule or constraint operates or fails to operate. There are areas of grammar such as 'weak crossover' or 'heavy NP shift' which are *weak* in the sense that the contrast is less visible and obvious than in the examples above, but even if we raise the bar very high, there are plenty of significant contrasts left for a theory of syntax to account for. On the whole, the development of syntax is not crucially impacted by the weaker examples, especially as there are generally other languages where phenomena marginal in one language, such as resumptive pronouns in English, can be observed in unequivocal examples, and often in far richer detail.

The second range of facts concerns the internal structure of sentences. Generally there is a verb, which takes a subject, and often an object, sometimes a second (indirect) object, and there can be other dependents, modifiers, and so on. Here the problem is twofold: first, it is not at all evident what primitive notions one should consider (though the grammatical tradition of millennia offers some rather clear guidelines), and second, even if a set of primitive grammatical notions is agreed upon, it is far from clear which word or combination of words fits which notion in any given sentence.

The third range of facts concerns the fit, or lack thereof, between what is being said and what is seen in the world. In Chapter 6, we will view sentences as broadly analogous to formulas and inquire about their interpretation in various model structures. Here we begin with a word of caution: clearly, people are often capable of assigning meaning to grammatically ill-formed utterances, and conversely, they can have trouble interpreting perfectly well-formed ones. But even with this caveat, there is a great deal of correlation between the native speaker's judgments of grammaticalness and their ability to interpret what is being said, so a program of research that focuses on meaning and treats grammaticality as a by-product may still make sense.

In fact, much of modern syntax is an attempt, one way or the other, to do away with separate mechanisms in accounting for these three ranges of facts. In Section 5.1, we will discuss *combinatorical* approaches that put the emphasis on the way words combine. Although in formal theories of grammar the focus of research activity historically fell in this category, the presentation here can be kept brief because most of this material is now a standard part of the computer science curriculum, and the reader will find both classical introductions such as Salomaa (1973) and modern monographic treatments such as Kracht (2003). In Section 5.2, we turn to *grammatical* approaches that put the emphasis on the grammatical primitives; prominent examples are *dependency grammar* (Tesnière 1959), *tagmemics* (Brend and Pike 1976), *case grammar* (Fillmore 1968), and *relational grammar* (Perlmutter 1983), as well as classical Pāṇinian *morphosyntax* and its modern variants. These theories, it is fair to say, have received much less attention in the mathematical literature than their actual importance in the development of linguistic thought would warrant. In Section 5.3, we discuss *semantics-driven* theories of syntax, in particular the issues of frame semantics and knowledge representation. In Section 5.4, we take up weighted models of syntax, which extend the reach of the theory to another range of facts, the *weight* a given string of words has. We will present this theory in its full generality, permitting as special cases both standard formal language theory, where the weights 1 and 0 are used to distinguish grammatical from ungrammatical, the extension (Chomsky 1967) where weights between 1 and 0 are used to represent intermediate degrees of grammaticality, and the *probabilistic* theory, which plays a central role in the applications. In Section 5.5, we discuss weighted regular languages, giving an asymptotic characterization of these over a one-letter alphabet. Evidence of syntactic complexity that comes from external sources such as difficulty of parsing or acquisition is discussed in Section 5.6.

5.1 Combinatorical theories

Given some fixed set of vocabulary items Σ, a central question is to characterize the subset $L \subset \Sigma^*$ that speakers take to be 'ordinary' or 'grammatical' or 'well-formed' (we use these notions interchangeably) by means of a formal system. One significant practical difficulty is that Σ, as traditionally understood, collects together all words from the language, and this is a very large set. For English, we already have $|\Sigma| \gg 10^6$ if we restrict ourselves to words attested in print, and potentially infinite, if we permit all well-formed, but not necessarily attested, words. In Section 5.1.1, we discuss how to define classes of words by distributional equivalence and thereby replace the set Σ of words by a considerably smaller set C of (strict) *categories*. In Section 5.1.2, we introduce *categorial grammar* as a means of characterizing the set of well-formed expressions, and in Section 5.1.3, we turn to constituent structure.

5.1.1 Reducing vocabulary complexity

Since the vocabulary of natural languages is large, potentially infinite (see Corollary 4.4.1), grammarians have throughout the ages relied on a system of *lexical categories* or *parts of speech* to simplify matters. Originally, these were defined by similarity of meaning (e.g. *adze* and *ax*) or similarity of grammatical function (e.g. both can serve as the subject or object of a verb). Both these methods were deemed inadequate by the structuralists:

> The school grammar tells us, for instance, that a noun is 'the name of a person, place, or thing'. This definition presupposes more philosophical and scientific knowledge than the human race can command, and implies, further, that the form-classes of a language agree with the classifications that would be made by a philosopher or scientist. Is fire, for instance, a thing? For over a century, physicists have believed it to be an action or process rather than a thing: under this view, the verb burn is more appropriate than the noun fire. Our language supplies the adjective hot, the noun heat, and the verb to heat, for what physicists believe to be a movement of particles in a body. ... Class meanings, like all other meanings, elude the linguist's power of definition, and in general do not coincide with the meanings of strictly defined technical terms. To accept definitions of meaning, which at best are makeshifts, in place of an identification in formal terms, is to abandon scientific discourse. (Bloomfield 1933 Sec. 16.2)

The method preferred by the structuralists, formalized in contemporary terms by Myhill (1957) and Nerode (1958), relies on similarity of combinatorical potential. Given a word w, we call its **distribution** the set of all pairs of strings α, β such that $\alpha w \beta \in L$. Two words u, v have the same distribution iff

$$\alpha u \beta \in L \Leftrightarrow \alpha v \beta \in L \tag{5.1}$$

Strict lexical categories are defined as the equivalence classes C of (5.1). In Section 3.4 we already introduced a variant of this definition with u, v arbitrary strings (multiword sequences), and we note here that for any strings x, x' from class c_1 and y, y' from class c_2, their concatenation xy will belong in the same distributional equivalence class as $x'y'$. Therefore, concatenation is a well-defined binary operation among the equivalence classes, which is obviously associative since concatenation was associative. This operation turns C into a semigroup and, with the addition of the empty word that serves as the identity, a monoid over C called the **syntactic semigroup** or **syntactic monoid** associated with L.

The algebraic investigation of syntax is greatly facilitated by the introduction of these structures, which remain invariant under trivial changes such as renaming the lexical categories. Needless to say, algebraic structures bring with themselves the whole apparatus of universal algebra, in particular **homomorphisms**, which are defined as mappings that preserve the operations (not just the semigroup product, but in monoids also the identity, conceptualized as a nullary operation); **substructures**, defined as subsets closed under the operations; and **direct products**, defined componentwise. We shall also avail ourselves of some notions and theorems more specific to the theory of semigroups and (semi)automata, in particular **divisors**, which are defined as homomorphic images of subsemigroups/sub(semi)automata (denoted $A \prec B$ – when A is a divisor of B, we also say B is a **cover** of A), *cascade products*, and the powerful Krohn-Rhodes theory. But before we turn to the general case in Section 5.6, here we concentrate on the special case where u, v are words, here construed as atomic symbols of the alphabet rather than as having (phonological or morphological) internal structure.

By **lexical categorization (category assignment)** we mean the coarsest one to one (one to many) mapping $f : \Sigma \rightarrow C$ that respects distributional equivalence. The system of strict categories C employed here is slightly different from the traditional linguistic notion of lexical categories, in that we do not permit words to have multiple categories. Where linguists would assign a word such as *run* two categories, Noun (as in *This was a successful run*) and Verb (as in *Run along!*), f maps *run* to the unique category NounVerb. Also, the finer distinctions relegated in linguistics to the level of subcategories, in particular differences among the range of optional and obligatory complements, are here seen as affecting the category. Verbs with a single obligatory argument slot are called **intransitive**, those with two are called **transitive**, and those with three are called **ditransitive**. As far as (5.1) is concerned, these are clearly distinct: compare *She ate something* (transitive verb) to **She hiccuped something* (intransitive verb used in a transitive context). Finally, changes in inflection, viewed by linguists as leaving the category unchanged, have the effect of changing the strict categories defined by (1): for example, *f(eat)* $\neq$ *f(eats)* because e.g. *John eat* $\notin L$ while *John eats* $\in L$.

Altogether, the system C of strict categories used here is obtained from the traditionally defined set of lexical categories or *parts of speech* (POS) by (i) making inflection part of the category, (ii) elevating subcategorial distinctions to full categorial status, and (iii) taking the Boolean atoms of the resulting system. Linguistics generally uses a set of only a dozen or so basic lexical categories, presumed to be common to all languages: the major classes are Noun, Verb, Adjective, Adverb, and perhaps Preposition (or Postposition, depending on language) – given the prominence assigned to these in school grammars, these will all be familiar to the reader. There is less agreement concerning minor categories such as Determiner, Pronoun, Quantifier, etc., and many grammarians permit *syncategorematic* elements that are exempted from the domain of f. By extending the discussion to inflectional distinctions, we blow up this basic set to a few hundred or a few thousand, depending on the complexity of the inflectional morphology of the language in question (Spencer and Zwicky 1998). While in principle the move to Boolean atoms could blow this up exponentially, in practice only very few combinations are actually attested, so this step is far less important than the (additive) effect of using strict subcategories.

For English, well-developed sets of categories (usually called *tags* or *POS tags* in corpus linguistics) include the Brown Corpus tagset comprising 226 categories (Greene and Rubin 1971), the LOB Corpus tagset comprising 153 categories (Johansson et al. 1986), and the CLAWS2 and C6 tagsets with over 160 categories (Leech et al. 1994). These numbers are slightly extended by adding subcategory distinctions within nouns (such as count noun vs. mass noun, simple vs. relational, etc.), adjectives (attributive vs. predicative), and adverbs (manner, time, place, cause, etc.). A far more significant extension comes from adding subcategory distinctions for verbs. For example, Levin (1993) distinguished 945 subcategories for English verbs, with no two of these having exactly the same distribution. The differences are often expressed in terms of *selectional restrictions*. For example *murder* and *kill* differ in the requirement imposed on the subject: the former requires a human (compare *The assassin murdered him* to **The accident murdered him*), while the latter imposes no such restriction (*The assassin/accident killed him*).

But when all is said and done, the move from the specific lexical items in Σ to the strict categories in C brings a great deal of simplification, especially as it collapses some very large (potentially infinite) sets such as that of numbers or personal names. Altogether, the method brings down the size of the alphabet to that of an abridged dictionary, perhaps 10^3–10^4 strict categories, sometimes called *preterminals* in recognition of the fact that they do not correspond well to either the terminals or the nonterminals in string-rewriting systems. In this regard, we will follow the practice of contemporary linguistics rather than that of early formal language theory and think of terminals as either words or (strict) categories, whichever is more convenient.

5.1.2 Categorial grammar

Once we collapsed words with the exact same distribution, the next task is to endow the set of (pre)terminals with some kind of grammatical structure that will regulate their combinatorical potential. One particularly attractive method would be to assign each (pre)terminal $p \in C$ a word $s(p)$ from a free group G with generators $g_0, g_1, \ldots, g_k$ and define a string of preterminals $p_{i_1} p_{i_2} \ldots p_{i_r}$ to belong in L iff $s(p_{i_1}) s(p_{i_2}) \ldots s(p_{i_r})$, as a group-theoretical product, is equal to a distinguished generator g_0 (cf. Example 2.2.2). For reasons that will become clear from the following discussion, this method is more attractive in the Abelian than in the general (noncommutative) case, and the latter will require conventions that go beyond what is familiar from group theory.

Example 5.1.1 Arithmetical expressions. Let us collect arithmetic expressions that contain both variables and numbers in the formal language A. For example $(1.3 - x) * (1.33 + x) \in A$, but $9)(x \notin A$. The major categories, with infinitely many elements, will be Var and Num, and there will be smaller categories Bin for binary operators $+, -, :, *$ and Uni for the unary operator $-$. In addition, we need one-member categories OpenPar and ClosePar. (To eliminate the multiple category assignment of minus, instead of Bin and Uni we could use StrictBin for $+, :, *$ and BinUni for $-$, but this would only complicate matters here.)

Since Num and Var give well-formed expressions as they are, we take s(Num) = s(Var) = g_0. Members of StrictBin take well-formed expressions both on the left and the right, so we take s(StrictBin) = g_0^{-1} – this will work out nicely since $g_0 \cdot g_0^{-1} \cdot g_0 = g_0$. (Notice that the system slightly overgenerates, permitting the operator to come in Polish or reverse Polish order as well – we will return to this matter later.) The case of the minus sign is more complicated: when it is a binary operator, it is behaving the same way as the other operators and thus should get the signature s(−) = g_0^{-1}, but when it is a unary operator it can be prefixed to well-formed expressions and yields a well-formed expression, making it s(−) = I_L (left unit). Finally, OpenPar creates an expression that turns well-formed expressions into ill-formed expressions requiring a ClosePar. To handle this requires a second generator g_1, and we take s(OpenPar) = $g_1 g_0^{-1}$ and s(ClosePar) = $g_1^{-1} g_0$.

Discussion This toy example already displays many of the peculiarities of categorial grammar writing. First, whenever a preterminal w belongs to multiple categories, we need to assign it multiple values s(w). In the language of arithmetic, this is a rare exception, but in natural language the situation is encountered quite frequently. The categorial signature assignment s thus has to be taken as a one-to-many relation rather than a function, and the key definition of grammaticality has to be modified so that as long as there is at least one permitted assignment that reduces to g_0, the string is taken to be grammatical. In a string of r words that are all d-way ambiguous as far as their categorial assignment is concerned, we therefore need to examine d^r products to see if any of them reduce to g_0.

The main difficulty concerns the left unit. In a group, left units are indistinguishable from right units, and the unit will commute with every element even if the group itself is not commutative. However, in syntax, elements that preserve grammaticality can still have order restrictions: consider, for example, the discourse particle *well*. Ordinarily, if α is a sentence of English, so is *Well,* α, but the string α, *well* will not be grammatical. To address this issue, categorial grammars formally distinguish two kinds of **cancellation**, **left** and **right**. While in arithmetic $3 * (5/3)$ is indistinguishable from $(5/3) * 3$, in categorial grammar we will make the distinction between A/B, a syntactic unit that requires a *following* B to make it A, and $B\backslash A$, a syntactic unit that requires a *preceding* B to make it A. If the category of a full sentence is T, the category of *well* is T/T, meaning that it requires something of category T following it to yield T. This is in sharp contrast with group theory, where $g_i g_i^{-1}$ and $g_i^{-1} g_i$ would cancel equally (see Section 6.3.2 for further details). By a **cancellation structure** we shall mean the minimal set of formal expressions generated from a finite set of basic categories P and closed under the operations $(p\backslash q)$ and (p/q).

Once cancellation order is controlled, we are in a good position to capture syntactic conventions about arithmetic operators. In standard notation, strictly binary operators, such as +, take one argument on the left and one on the right, so their type is $(g\backslash g)/g$. In reverse Polish notation, their type is $g\backslash(g\backslash g)$, and in Polish notation $(g/g)/g$. Left units, such as the unary minus operator, can no longer attach anywhere, so expressions like (23.54−) are now ruled out as long as the type of the unary minus is g/g, demanding a well-formed expression to its right. But the syntax of natural language still presents some difficulties, even with unary operators like *well*. Sentences like *Well, well, ..., well, John went home* look increasingly strange as the number of *wells* increases. This effect has no counterpart in (directional) multiplication: if I_L is a left unit, multiplying by it many times will not change the outcome in any way. We return to this matter in Section 5.4, where we discuss *valuations* e.g. in terms of degree of grammaticalness, but here we ignore this complication and summarize the preceding discussion in two definitions.

A **unidirectional categorial grammar** is given by a set of preterminals P and a mapping s: $P \to 2^G$, where G is a free Abelian group with a distinguished generator g. We say that the string $p_1 p_2 \dots p_n$ is accepted by the grammar iff there are elements of G $a_1, a_2, \dots a_n$ such that $a_i \in s(p_i)$ and $a_1 a_2 \dots a_n = g$.

A **bidirectional categorial grammar** or just **categorial grammar** is given by a set of preterminals P and a mapping s: $P \to 2^G$, where G is a cancellation structure with a distinguished generator g. We say that the string $p_1 p_2 \dots p_n$ is accepted by the grammar iff there are elements of G $a_1, a_2, \dots a_n$ such that $a_i \in s(p_i)$ and there is a bracketing of $a_1 a_2 \dots a_n$ that cancels to g by a series of left- and right-cancellations sanctioned by the rules $q(q\backslash p) = (p/q)q = p$.

Categorial grammar (CG) offers an extremely flexible and intuitive method for sorting out complex syntactic situations: schoolchildren spend years mastering the complexities that we can now succinctly capture in a handful of categorial signature assignments. What is particularly attractive about the system is that the syntax ties to the semantics in exactly the desired manner: the order of cancellations corresponds exactly to the order of evaluation. Take for example $3/2/2$. The categorial signatures are $g, g\backslash g/g, g, g\backslash g/g, g$ so the product is g, making the expression well-formed. If we begin by evaluating from the left, we first get $3/2 = 1.5$, and subsequently $1.5/2 = 0.75$. If we begin by evaluating from the right, we get $2/2 = 1$, and subsequently $3/1 = 3$. Remarkably, the same kind of ambiguity can be observed in natural language. Consider, following Montague (1970a),

$$\text{Every man loves a woman} \tag{5.2}$$

To keep the notation close to Montague's original, the distinguished generator g_0 will be written T, and the second generator g_1 is written E. (Anticipating developments in Section 6.3, T is mnemonic for *t*ruth value and E for *e*ntity.) The categorial signatures s are given as follows. Common nouns such as *man* and *woman* are assigned the signature $T\backslash E$. Quantifiers and determiners have signature $E/(T\backslash E)$, so that noun phrases come out as $E/(T\backslash E) \cdot (T\backslash E) = E$. Finally, transitive verbs such as *loves* have signature $E\backslash T/E$ (i.e. they require an E on both sides to produce the distinguished generator T). As in the arithmetic example, the product s(*every*)s(*man*)s(*loves*)s(*a*)s(*woman*) $= E/(T\backslash E) \cdot T\backslash E \cdot E\backslash T/E \cdot E/(T\backslash E) \cdot T\backslash E$ can be computed in two essentially different orders, and these correspond to the two *readings* of (5.2), namely

$$\forall x \text{man}(x) \exists y \text{woman}(y) \text{loves}(x, y) \tag{5.3}$$

$$\exists y \text{woman}(y) \forall x \text{man}(x) \text{loves}(x, y) \tag{5.4}$$

It would be trivial to extend this grammar fragment[1] to cover proper nouns such as *John* by taking s(*John*) = E, and intransitive verbs such as *sleeps* by taking s(*sleeps*) = $E\backslash T$. This would correctly derive sentences such as *Every man sleeps* or *A woman loves John.* To add prepositions and prepositional phrases requires another generator P. By assigning prepositions such as *to* the signature P/E and to ditransitives such as *gives* the signature $E\backslash T/E/P$, we obtain derivations for sentences like *John gave a book to every woman.* Prefixing adjectives like *lazy* to common nouns leaves the category of the construction unchanged: *lazy man* has essentially the same distribution as *man*. For this reason, we assign adjectives the signature $(T\backslash E)/(T\backslash E)$. The same method is available for *adadjectives* like *very*. Since these leave the category of the following adjective unchanged, we assign them to the category $((T\backslash E)/(T\backslash E))/((T\backslash E)/(T\backslash E))$.

Exercise 5.1 Consider a programming or scripting language with which you are familiar, such as Fortran, C, Perl, or Python. Write a categorial grammar that captures its syntax.

5.1.3 Phrase structure

It is a remarkable fact about natural language that some of the more complex (in linguistic terminology, *derived*) constructions from Σ^* clearly fit in some of the equivalence classes of lexical items defined by (5.1). For example, proper nouns such as *John*, and noun phrases such as *every man*, can be substituted for one another in virtually every context such as *Bill visited* _. Comparing *John didn't go home yesterday evening* to *?Every man didn't go home yesterday evening* (to which speakers strongly prefer *No man went home yesterday evening*) makes clear that there are exceptions, but the distributional similarity is so strong that our preference is to assume identity and fix the exceptions by a subsequent rule or a more highly ranked constraint.

Not only do proper nouns show near-perfect distributional equivalence with noun phrases, but the same phenomenon, the existence of distributionally equivalent lexical entries, can be observed with most, if not all, derived constructions. The phenomenon is so pervasive that it has its own name, *lexicality*. Informally, the principle says that for any grammatical position or role that can be filled by a phrase, there is a single-word equivalent that can serve in the same position or role. In the special case where the equivalence is between a construction and one of its components, this component is called the **lexical head** of the construction. A good example is adjectival modification: *lazy old man* is equivalent to its lexical head *man*. Constructions that contain their own lexical heads are called **endocentric** and the rest are called **exocentric**. Noun phrases formed by determiners or quantifiers are a good example of the latter. For example, neither *every* nor *woman* shows distributional equivalence, or even broad distributional similarity, to *every woman* since the full construction can appear freely in subject or object position, while **Every sleeps,*

[1] The presentation here is not faithful to the classic Montague (1973) 'PTQ fragment' – see Section 6.2 for details.

Woman sleeps, *John saw every* or **John saw woman* are ungrammatical, and conversely, nouns can undergo adjectival modification while noun phrases cannot (lazy every woman*). When quantifiers undergo similar modification, as in *nearly every*, the semantics makes it clear that these do not attach to the whole construction, we have *(nearly every) woman* rather than *nearly (every woman)*.

In endocentric constructions, it is nearly always the case that the nonhead constituents are entirely optional. This phenomenon, though not quite as general as lexicality, has its own name: *optionality*. For constructions composed of two parts, optionality follows from the definition: if uv is equivalent to v, this means u was optional. For larger constructions, the only exceptions seem to be due to agreement phenomena: for example, *Tom, Dick, and Harry* is equivalent to *Dick and Harry* but not to any smaller substring since the construction (e.g. in subject position) demands plural agreement. Even exocentric constructions, where there is no clear candidate for head or for deletable elements, show evidence for not being composed of parts of equal importance, and we commonly find the term *head* extended to the most important component.

To the extent that there seems to be something of an Artinian condition (no infinite descending chain) on natural language constructions, we may want to inquire whether there is a Noetherian condition (no infinite ascending chain) as well. A construction whose head is some lexical category c is said to be a **projection** of c: the idea is that we obtain more and more complex constructions by successively adjoining more and more material to the head lexical entry. Can this process terminate in some sense? Linguistics has traditionally recognized **phrases** as maximal projections (i.e. as constructions that can no longer be extended in nontrivial ways). The most important example is the noun phrase, which is effectively closed off from further development by a determiner or quantifier. Once this is in place, there is no further adjective, numeral, or other modifier that can be added from the left (compare **three the books, *three every books* to *the three books, every three books*) and only relative clauses are possible from the right (*the three books that I saw yesterday*). Once such a *that*-clause is in place, again there is no room for different kinds of modifications. Further relative clauses are still possible (*the three books that I saw yesterday that you bought today*), but no other kind of element is. Other notable examples include the verb phrase (VP), the prepositional phrase (PP), the adjectival phrase (AP), and the adverbial phrase (AdvP) – since this covers all major categories, it is commonly assumed that every construction is part of a maximal (phrasal) construction that can be further extended only by the trivial means of coordination.

Another observation connects nonheads (also called *dependents*) to phrases: in most constructions, dependents are freely substitutable for their maximal projections. Thus, where a construction has a dependent adjective, such as *traditional* in *the traditional dish*, it can also have a full adjectival phrase, as in *the exceedingly traditional dish*. The phenomenon that dependents are maximal projections is common enough to have a name of its own: *maximality*.

There are two formally rather different systems of syntactic description built on these insights. The first, *dependency grammar* (DG), uses directed graphs where arcs always run from dependents to heads (in some works, the opposite convention is used), and the second, *X-bar theory*, uses string-rewriting techniques. Historically, dependency grammar has not been very concerned with word order, its main focus being the relationships that can obtain between a head and its dependents. String rewriting, on the other hand, does not really have the apparatus to deal with grammatical function, valence, case, and similar matters occupying the dependency grammarian, its primary focus being the hierarchical buildup of sentence structure. In the remainder of this section, we concentrate on X-bar theory, leaving case, agreement, valence, and related issues to Section 5.2.

The key to hierarchical structure is provided by optionality, which enables us to gradually reduce complex constructions to simpler ones. This is the method of *immediate constituent analysis* (ICA), which relies on extending the relation (5.1) from Σ to Σ^*. Recall Definition 3.4.2: two strings γ, δ are distributionally equivalent whenever

$$\alpha\gamma\beta \in L \Leftrightarrow \alpha\delta\beta \in L \tag{5.5}$$

The key insight, due to Chomsky (1956), was to notice that the typical reduction step in ICA does not require distributional equivalence, just substitutability in positive contexts:

$$\alpha\gamma\beta \in L \Rightarrow \alpha\delta\beta \in L \tag{5.6}$$

To make this concrete, consider the original example from Wells (1947): *The king of England opened Parliament*. We wish to find how this sentence is built from parts and recursively how the parts are built from smaller parts, etc. ICA begins with locating the structural break between *England* and *opened* (rather than, say, between *of* and *England*) based on the observation that *opened Parliament* can be substituted by *slept* without loss of grammaticality and *The king of England* can be similarly substituted by *the king*. It is clear that the substitution works well only in one direction, that of simplification. For example, the construction *a bed rarely slept in* will not tolerate substitution in the other direction. The famous rule

$$\text{S} \rightarrow \text{NP VP} \tag{5.7}$$

captures exactly this step of the ICA. The sentence is analyzed in two constituents, a (subject) NP *The king of England* and a verb phrase *opened Parliament*.

In the original notation of Wells (1947), the constituents were described by boundary symbols |, ||, |||, ... of ever-increasing strength, so that the sentence would come out *The* | *king* || *of* | *England* ||| *opened* | *Parliament*, but this has been superseded by *bracketing*, which would make the example come out as *(((The king) (of England)) (opened Parliament))*. In general, to formalize ICA by string rewriting, we need to adjoin new *nonterminal* symbols to the (pre)terminals found in C or Σ so as to cover constructions (multiword constituents). Lexicality means that the new symbols can be patterned after the ones used for categories. Traditionally, the same symbols, with superscript bars (hence the name X-bar), are reused so that e.g. nouns are N, bare noun phrases are $\overline{\text{N}}$, and full noun phrases are $\overline{\overline{\text{N}}}$. We leave open the issue of whether there is exactly one bar level between the lexicon and maximal phrases and use the notation X^M to denote a maximal projection of X. (For many minor lexical categories $X^M = X$; i.e. no complex phrases of that type can be built.) If we attach the nonterminal to the opening parenthesis of each constituent, we speak of **labeled bracketings**. Just as scanning codes (see Section 3.4) offer a simple linearized version of an association relation, (labeled) bracketings offer a simple linearization of parse trees, and many questions about tree structures can be recast in terms of the code strings (Chomsky and Schützenberger 1963).

Since (5.5) defines an equivalence relation, L itself is the union of some equivalence classes. To the extent that sentences are freely substitutable for one another, L is covered by a single symbol S, the start symbol of string-rewriting systems. Lexicality in this case is traditionally taken to mean that S is a projection of the main verb – in more modern accounts it is often some abstract property of the verb, such as its tense marking, that is taken to be the head of the entire sentence. Either way, rule (5.7) amounts to the statement that the NP on the left (the subject) completes the VP and thus S has one more bar than VP.

In a string-rewriting formalism, maximality means that only rules of the type $X^n \rightarrow Y_1^M \dots Y_l^M X^{n-1} Z_1^M \dots Z_r^M$ are possible. Counterexamples are not easy to find but they exist; consider infinitival clause complements to verbs. Some verbs, such as *want*, can take both subjectful and subjectless clauses as their complement: compare *John wanted Bill to win* and *John wanted to win*. Other verbs, such as*try*, only take subjectless clauses: *John tried to win, *John tried Bill to win*. By analogy to tensed sentences, the tenseless *Bill to win* has one more bar level than *to win* since the subject *Bill* adds one bar. If this is so, the rule expanding *try*-type VPs has one fewer bar on the clausal complement than the rule expanding *want*-type verbs, meaning that maximality is violated in the former. Altogether, maximality remains one of the fascinating near-truths about syntax, and trying to come to grips with this and similar counterexamples remains an active area of research.

Given the special role that coordination plays in the system, one technique of great practical significance is **extending** context-free grammars by permitting rules of the form $A \rightarrow r$, where r is any regular expression over the alphabet composed of both terminals and nonterminals: for any string w that matches r, we say that A can be rewritten as w. It is easy to see that this extension does not change the generative capacity of the system (if a language can be described by means of an extended CFG, it can also be defined by an ordinary CFG), yet the perspicuity of the system is significantly increased.

Exercise 5.2 Prove that extended CFGs only generate CFLs.

All in all, (extended) CFGs offer an extremely transparent and flexible method of syntactic analysis. Several rather different-looking grammatical traditions are equivalent, at least in some formulation, to CFGs: Postal (1964) argued this for tagmemics (Pike 1967), and Bar-Hillel et al. (1960) made the argument for categorial grammars. Yet CFGs can be faulted both for over- and undergeneration. On the one hand, CFGs can easily handle the problem of balanced parentheses of arbitrary depth. This is obviously helpful inasmuch as CFGs could not have gained their prominence as the primary means of syntactic analysis for programming languages without the ability to correctly handle parenthesized expressions of arbitrary depth, but in the analysis of natural languages, arbitrary depth constructions are rarely encountered, if at all. On the other hand, there is a class of *copying* phenomena, ranging from cross-serial dependencies (see Section 5.2.2) to constructions such as *Freezing cold or no freezing cold (we will go barefoot)*, that is outside the CFL domain, see Manaster-Ramer (1986).

Exploring combinatorical mechanisms that transcend the limitations of CFGs has been, and continues to be, a major avenue of research in mathematical linguistics. The general problem of *discontinuous constituents*, i.e. constituents that are interrupted by material from other constituents, was noted early by the developers of ICA – for a current summary, see Ojeda (2006b). The fact that cross-serial dependencies are problematic for CFGs was noted by Postal (1964), who cited Mohawk [MOH] in this regard – later work concentrated on Germanic languages, in particular Dutch and Swiss German. The first formalism to address the issue was *tree adjoining grammar* (TAG – for a modern summary, see Joshi 2003) and, perhaps even closer to the original spirit of ICA, *head grammars* (Pollard 1984). Among categorial grammars, *combinatory categorial grammar* (CCG) provided the same kind of extension to discontinuities; see Steedman (2001). (Note that where it does not clash with established mathematical usage, we distinguish *combinatorial*, 'pertaining to combinators', from *combinatorical* 'pertaining to combinatorics' – CCG uses combinators.) Readers interested in further details are directed to Chapter 5 of Kracht (2003), which discusses all key developments in this area.

5.2 Grammatical theories

The approaches that put the emphasis on string, tree, or category manipulation are rather new (less than a hundred years old) compared with traditional notions such as subject, object, or case, which grammarians have invoked for millennia in accounting for syntactic regularities. The primary impulse of the combinatorical theories is to do away with these and similar notions entirely; for example, Chomsky (1981:59) uses (5.7) to define subject as "NP of S" (i.e. as the NP that appears in the rule rewriting S), predicate as "VP of S", and object as "NP of VP". Such definitions imply that there is no explanatory role for these notions in grammar, that they are purely epiphenomenal, serving at best as convenient abbreviations for constituents in some frequently seen combinatorical configurations. Yet grammarians not only persist in using these notions but in fact the class of theories that rely crucially on them, what we call 'grammatical theories' for lack of a better umbrella term, has undergone intense development in the past forty years.

The first generative model to give primacy to a grammatical notion of case over the combinatorical notions of constituency and category was *case grammar* (Fillmore 1968), and we will cover *role and reference grammar* (Valin 2005), and *relational grammar* (Perlmutter 1983), as well as classical Pāṇinian *morphosyntax* and its modern variants (Ostler 1979, Kiparsky 1987, Smith 1996). Over the years, even mainstream combinatorical theory (Chomsky 1981, 1995) came to incorporate modules of *case theory* and *theta theory* that employ such grammatical devices.

With the notable exception of *lexical functional grammar* (LFG; see Bresnan 1982) and a variety of unification grammars, which came mathematically fully articulated, the formal theory has not always kept up with the insights gained from the grammatical work. In Section 5.2.1, we take the first steps toward formalizing some of the key ideas, starting with dependency, agreement, surface case, and government. Deep cases and direct linking are discussed in Section 5.2.2, and grammatical functions and indirect linking are discussed in Section 5.3.

5.2.1 Dependency

Classical grammar has imbued our culture to such an extent that many of its technical notions, such as *subject, object*, or *predicate*, are applied almost unthinkingly in scientific discourse. So far, we have used these terms without much reflection, essentially in the same somewhat vague but nevertheless widely approved sense as they are used by all people past grammar school. To see what motivates the use of such devices, consider first sentence pairs such as

The farmer killed the duckling (5.8)

The duckling was killed by the farmer (5.9)

Clearly there is a very close paraphrase relationship between (5.8), known as an *active* sentence, and (5.9), known as a *passive*: it seems impossible to imagine a state of affairs in which one is true and the other is false. There is more to active/passive pairs than semantic relatedness; the constructions themselves show deeper parallelism. Whatever selectional relationship obtains between the active verb and its object is carried over to the passive verb and its subject, and whatever relationship obtains between the active verb and the subject is replicated between the passive verb and the agentive *by*-phrase.

For Chomsky (1957) and much of the subsequent tradition of transformational generative syntax, these observations were taken as indicative of a need to treat (5.8) as basic and (5.9) as derived, obtained from (5.8) by means of a *passive transformation* that rearranges the constituents of the active construction and supplies the requisite grammatical formatives to yield the passive. Unfortunately, the derivational analysis creates as many problems as it solves. In particular, we find actives with no passive counterparts, such as *John resembles Bill* and **Bill is resembled by John.* These require some special mechanism to block the passive transformation. Even more problematic are passives with no active counterparts, such as **Everyone said John to be honest/John was said to be honest by everyone*, since these call for an obligatory application of the transformation and thus rely on an abstract *deep structure*. This is extremely challenging from the perspective of language acquisition since the learner has to reverse-engineer the path from deep structure to surface based on an opaque surface structure.

That something of a more abstract sort than surface word order or constituency is required can hardly be doubted, especially if we consider a wider range of alternations. Compare, for example, *The butcher cuts the meat/The meat cuts easily* to *Kelly adores French fabrics/*French fabrics adore easily* or *Jack sprayed the wall with paint/Jack sprayed paint on the wall* to *June covered the baby with a blanket/*June covered the blanket over the baby*. (The examples are from (Levin 1993), where a broad range of similar alternations are discussed.) Traditional grammatical theories, having been developed on languages that show overt case marking, generally use an abstract version of *case* as the key mechanism to deal with the subtleties of syntax.

What is case? Structuralist theories of syntax such as Jakobson (1936) took case to be at the confluence of three major domains: morphology, semantics, and syntax. From the morphological standpoint, *surface* cases are noun affixes common to all but the most isolating of languages. The details of case systems vary greatly, from languages such as Hindi that have only two, nominative and oblique (plus a vestigial vocative), to languages such as Hungarian with seventeen and Tsez [DDO] with over sixty morphologically distinct cases. The typical *nominative/accusative* system will include locative cases, accusative, dative, and perhaps some more specialized cases, such as instrumental and genitive (by convention, all nonnominative cases except the vocative are called oblique). Latin, which lacks an instrumental case but has a vocative and two locatives, or Russian, are good examples of the typical pattern.

From the semantic standpoint, cases correlate with the role the nouns or NPs marked by them play in the sentence. In particular, vocative marks the one the sentence is addressed to, locatives describe where the action or event described by the sentence takes place, the dative is associated to the recipient or beneficiary, the instrumental to instruments, and genitive to possession. The correlation between morphological marking and semantic content is far from perfect. For example, in many languages, the same dative morphology will express both recipients and experiencers. Much of traditional syntax is concerned with enumerating the discrepancies, e.g. the 'dative of separation' that occurs with some verbs instead of the expected ablative, the 'genitive of material' which denotes a relationship of 'being composed of' as in *a bar of gold* rather than a relationship of possession, and so forth – we will return to this matter in Section 5.3.

Finally, from the syntactic side, cases correlate with grammatical function: subjects are typically marked with the nominative case, objects with the accusative case, indirect objects with the dative case, and so forth. Again it has been recognized from the earliest times that the correlation between grammatical function and case marking is imperfect. On the one hand, we find what appears to be the same functional relation expressed by different cases, and on the other, we find that one and the same case (e.g. the accusative) can serve in different functions. Even so, the syntactic impact of case is undeniable. In particular, it is quite clear that in languages with overt case marking, the noun phrases that carry these marks can generally be permuted much more freely than in languages such as English that lack overt case marking. For example, in Latin, we can have *anaticulam agricola occisit* or any of the other five permutations of *duckling.ACC farmer.NOM kill.PAST* and the meaning of the sentence remains the same as that of (5.8).

Here we formalize some essential aspects of grammatical theories, while leaving many issues open. From a combinatorical perspective, the easiest of these is *agreement (concord)*, which obtains between two elements of a construction. The minimum requirement for agreement is a set of two words u and v and a paradigmatic dimension D that can take at least two values D_1, D_2. For example, the words can be the subject and the predicate, and the dimension can be number, taking the values singular and plural. With the standard *glossing* notation (see Section 2.3), where forms are given as $x.D$ with x being a stem and D some paradigmatic value, we have the following definition.

Definition 5.2.1 If for all values D_i the strings $\alpha u.D_i \beta v.D_i \gamma \in L$ while for all $i \neq j$ we have $\alpha u.D_i \beta v.D_j \gamma \notin L$, we say that u and v **agree** in D.

In a string-rewriting formalism, agreement is handled by means of complex symbols: instead of (7) we write $S \rightarrow NP.D_i\, VP.D_i$ (the . is used here as part of the morphemic gloss rather than as a pure concatenation marker). Since words can agree in more than one dimension (e.g. the subject and the predicate will generally agree both in number and in person), a rule schema like this can be thought of as subsuming as many rules as there are D_i. This brings to light the phenomenon of *internal agreement*, when a construction as a whole agrees with some part of it; e.g. *boys* and *lazy boys* are both plural, and *boy* and *lazy boy* are both singular. Since constructions typically agree with their heads, the head feature convention (HFC; see Gazdar et al. 1985) stipulates that internal agreement between constructions and their heads is automatic and it is suspension, rather than enforcement, of this rule that requires a special rule.

Note that more than two words can participate in agreement. For example, in Georgian, the predicate agrees not just with the subject but with the object (and in certain circumstances the indirect object) as well – for a fuller discussion, see Anderson (1992 Sec. 6.1). The definition naturally extends to situations where one of the words carries the feature inherently rather than through explicit morphological marking. For example, in Russian, adjectives agree in gender with the nouns they modify. The nouns do not exhibit gender variation: they are feminine, masculine, or neuter, and have no means of assuming a different value for gender as they could for number. In the typical case of concord it is not clear which of the two D_i is the cause and which is the effect. Traditionally we say that the choice of e.g. first-person subject forces first-person verbal morphology on the predicate, but we could equally well say that the choice of first-person verbal morphology forces the use of a first-person subject. In the Russian case described above, the direction of the value assignment is clear: only the adjective has a range of values, so agreement comes from the noun imposing its value on the adjective and not the other way around. Such cases are therefore also called *government*.

Definition 5.2.2 If in a construction $\alpha u.D_i \beta v.D_i \gamma \in L$ the value D_i is an inherent lexical property of u (v), we say that u governs v (v governs u) if $\alpha u.D_i \beta v.D_j \gamma \notin L$ ($\alpha u.D_j \beta v.D_i \gamma \notin L$) for any $j \neq i$.

In dependency grammar, the key diagnostic for assuming a dependency link between two items is agreement. In the special case of government, the direction of the link is set by convention so that it runs from the dependent to the head (governor). Since verbs can impose case marking on their complements, verbs are the heads, and not the other way around, a conclusion that in phrase structure grammar would be reached from a different fact, the optionality of the dependents. The characterization of the various dependency relations that can obtain between verbs and their dependents is a primary concern of both *case grammar*, which uses an inventory of abstract *deep cases* to describe them, and *relational grammar*, which uses the grammatical functions subject (called '1' in RG), object ('2'), indirect object ('3'), as well as more deep case-like functions such as Benefactive and Locative. Other relations of note are *modification*, as obtains between an adjective and a noun or an adverbial and a verb, *possession* between nouns and nouns, and *coreference* between pronouns and nouns.

Once an inventory of relations is fixed, it is natural to formulate **dependency grammar** in terms of directed graphs: words or higher constituents are vertices and the relations are the edges. A closely related formalism is that of *tagmemics*, which uses abstract construction types (called tagmemes) that have empty *slots* that require *fillers*. To this simple dependency apparatus, **relational grammar** adds an ordered set of *strata*: conceptually each stratum corresponds to a stage in the derivation. Here we go further, permitting edges to run directly between nodes and other edges as in traditional *sentence diagrams* (Reed and Kellog 1878, Kolln 1994, Klammer and Schultz 1996). We take the graph visualization as secondary, and the primary formalism is given as an algebraic system. For reasons of wider applicability (see in particular Section 5.3) rather than defining a single system, we define a metasystem MIL that can be instantiated differently depending on the choice of atoms and operations.

Definition 5.2.3 (i) The atomic elements of MIL form a finite set A. (ii) The primitive operations of MIL are &, =, and perhaps also finitely many binary operations $P_1, P_2, \ldots, P_n$. & will also be denoted by P_0, and we use H as a (metalanguage) variable ranging over the $P_i\,(i = 0, 1, \ldots, n)$.

(iii) If p and q are arbitrary elements, Hpq will be an element, and $p = q$ is an (elementary) statement. The only predicate of MIL is '='. $x, y, z, \ldots$ will be (metalanguage) variables ranging over elements.

(iv) The system is defined inductively: the only elements, operations, and statements of MIL are those resulting from the iterated application of (iii).

(v) The axioms of MIL (using prefix notation for operations and infix for equality) are

$$\begin{aligned} x &= x \\ Hx\&yz &= \&HxyHxz \\ \&xx &= x \\ H\&xyz &= \&HxzHyz \\ \&xy &= \&yx \end{aligned}$$

(vi) The rules of deduction are

$$\frac{x = y}{y = x} \qquad \frac{x = y, y = z}{x = z} \qquad \frac{x = y}{Hxz = Hyz} \qquad \frac{x = y}{Hzx = Hzy}$$

The elements of the algebraic structure can be thought of as **dependency diagrams**. These are directed labelnode hypergraphs of a special kind, differing from ordinary graphs only in that edges can also run to/from other edges, not just nodes. What the operations specify are not the diagrams, just formulaic descriptions of them. Different formulas can describe the same object, depending on the order in which the graph was built up. The equivalence '=' makes it possible to define a conjunctive normal form with respect to &. To put it in other words, MIL is a free algebra over a finite set A generated by the binary operations $P_0, P_1, \ldots, P_n$ satisfying the equations (v). Since the rules of deduction in (vi) make = compatible with the operations, = is a congruence, and its classes can be represented by terms in conjunctive normal form. & corresponds to union of subgraphs, with concomitant unification of identical atomic nodes.

To apply the formalism to dependency grammar, we must take words (or morphemes) as the primitive elements and the set of deep cases and other relations as the primitive operations. To apply it to relational grammar, we need to use a different primitive operation P_{i,c_j} for each relation i and each stratum c_j. To connect MIL formulas in normal form to the more standard graph notation, depict P_iab as a directed graph with vertices a and b and a directed edge labeled P_i running from a to b. Note that a nested formula such as $P_ia(P_jbc)$ does not have a trivial graph-theoretic equivalent since the edge P_i now runs from the vertex a to the *edge* P_jab as in (5.10.i). This frees the formalism from the spurious nodes introduced in Gaifman (1965), making it perhaps more faithful to the original grammatical ideas.

(i) $b \xrightarrow{P_j} c$ (vertical edge from b down to c), with $a \xrightarrow{P_i}$ running into that edge

(ii) $a \xrightarrow{P_i} b$, $c \xrightarrow{P_j} d$

(iii) $a \xleftarrow{P_i} b \xrightarrow{P_j} c$

(iv) $a \longrightarrow b$, $a \longrightarrow d$ (5.10)

To build more complex dependency graphs, the operation & is interpreted as (graph-theoretic) union: expressions like $\&P_iabP_jcd$ are as depicted in (5.10.ii). Expressions like $\&P_iabP_jac$ correspond to structures like (5.10.iii) – since atoms have to be unique, structures like (5.10.iv) cannot be formed. The axioms in (v) and the rules of deduction in (vi) serve to make indistinguishable those structures that differ only in the order in which they were built up. The temporal aspect of the derivation, to the extent it is taken to be relevant (in particular in relational grammar), is modeled through explicit temporal indexing of the edges.

An intuitively very appealing way of formulating dependency theories is to invoke a notion of *valences* in analogy with chemical valences. Under this conception, grammatical constructions are viewed as molecules composed of atoms (words or morphemes) that each have a definite number of slots that can be filled by other atoms or molecules. One key issue is the strength of such valences. Consider for example

$$[\text{John}]_1 \text{ rented } [\text{the room}]_2 \text{ [from a slumlord]}_3 \text{ [for a barbershop]}_4 \\ \text{[for the first three months]}_5 \text{ [for fifteen hundred dollars]}_6 \quad (5.11)$$

There are a total of six complements here, and most of them are optional: leaving their slots unfilled leaves a less informative but still evidently grammatical sentence such as *John rented.* In English at least, the subject (complement 1) is obligatory *(*rented a room)*, and we find many verbs such as *admit* or *transcend* that positively require an object (complement 2) as well: consider **John admits* or **John transcended.* Ditransitive verbs require an indirect object (complement 3) as well; compare **John gave, *John gave Bill,* and **John gave a last chance* to *John gave Bill a last chance.* These constructions are perhaps analogous to radicals such as CH_3, which are not found in nature (because they are so highly reactive that the valence gets filled almost immediately). One point where the chemical analogy may break down is the existence of situations where one slot is filled by more than one filler. Superficially, this is what appears in (5.11) with complements 4–6, which are all prepositional *for*-phrases – the notion of deep cases is introduced precisely with the goal of separating such complements from one another, see Section 5.2.3 for further details. Pāṇini distinguishes six deep cases: Agent, Goal, Recipient, Instrument, Locative, and Source (see Staal 1967). The exact inventory is still heavily debated.

5.2.2 Linking

In the Western tradition of grammar, action sentences like (5.8 and (5.9) are viewed as prototypical, and for the moment we will stay with them. Such sentences always have a main verb, and here we assume that verbs come with a **case frame** that specifies (i) the grammatical relation that the verb has to its dependents, (ii) the case or other marking that expresses this relation, (iii) the syntactic category of the complements, (iv) the obligatory or optional nature of the complements, and (v) the subcategorization restrictions the head places on the complements. The primary goal of syntax, at least in regard to the prototypical class of sentences, is to specify how the various entities (NPs) named in the sentence are linked in the argument slots.

Well-developed grammatical theories that rely on case frames include classical DG, head-driven phrase-structure grammar (HPSG; see Pollard and Sag 1987), lexical-functional grammar (LFG; see Bresnan et al. 1982), case grammar (Fillmore 1968), role and reference grammar (Foley and van Valin 1984, van Valin 2005), relational grammar (Perlmutter 1983), as well as classical Pāṇinian morphosyntax and its modern variants (Ostler 1979, Kiparsky 1987, Smith 1996). This is not to say that all these theories are identical or even highly similar. All that is claimed here is that a linking mechanism, expressed one way or another, is an integral part of their functioning. (Note also that many would take exception to calling this structure a *case* frame, preferring, often for reasons that make a lot of sense internal to one theoretical position or another, names such as *subcategorization frame, lexical form*, or *thematic grid*.)

Consider again the sentences (5.8) and (5.9). In the active case, we simply say that the transitive frame shared by *kill* and many transitive verbs requires a subject and an object NP (or perhaps an agent and a patient NP, depending on whether our terminology is more grammatically or semantically inspired). As long as we accept the notion that in Latin the relationship of subjecthood (or agency) that holds between *the farmer* and the act of killing is expressed by overt case marking, while in English the same relationship is expressed by word order, the essence of the analysis remains constant across languages. This is very satisfactory, both in terms of separating meaning from form (e.g. for the purposes of machine translation) and in terms of capturing variation across languages in typological terms.

In the passive, the central part of the construction is the verbal complex *was killed*, which also has two slots in its frame, but this time the second one is optional. The first slot is that of the subject (a passive experiencer rather than an active agent) and must be filled by an NP, while the second, that of the agent, can be left open or be filled by a *by*-phrase. There is no sense that (5.9) is derived from (5.8); these are independent constructions related only to the extent that their main verbs are related. Unlike syntax, which is expected to be fairly regular in its operation, the lexicon, being the repository of all that needs to be memorized, is expected to be storing a great deal of idiosyncratic material, so the appearance of idiom chunks such as *was said to be* that have their own case frames (in this instance, that of a subject NP and a predicative AP, both obligatory) comes as no surprise.

We still want to say that *resemble* is an exception to the lexical (redundancy) rule that connects transitive verbs *V-ed* to their passive form *was V-en*, and the case frame mechanism offers an excellent opportunity to do so. Instead of taking the verb as a regular transitive (which demands an accusative object), we treat the second complement as nonaccusative (e.g. goal) and thereby exempt the verb from the passive rule that operates on true transitives. Such an analysis is strongly supported cross-linguistically in that in many other languages with overt case endings, the second complement is indeed not accusative.

If we connect any specific theory of syntax (among the many variants sketched so far) to a theory of morphological realization rules (see Section 4.2), we obtain something approaching an end-to-end theory of action sentences, starting with their logical form and going ultimately to their phonetic realization. In a somewhat confusing manner, both the overall directionality from meaning to form and a specific theory of syntax within transformational grammar that uses this direction are called *generative semantics*. (The reverse direction, progressing from syntactic form to some kind of meaning representation, is called *interpretative semantics* and will be discussed in Chapter 6.) Many theories, such as HPSG (Pollard and Sag 1987) or role and reference grammar (RRG; see van Valin 2005), are purposely kept neutral between the two directions.

To get a richer description of syntax, we need to supplement case frames by a number of other mechanisms, in particular by rules for dealing with free adverbials and other material outside the case frame. In (5.11), many grammarians would argue that the goal adverbial complement 4, *for a barbershop*, is outside the core argument structure of *rent*: everything can be said to happen with some goal in view so this is a free adverbial attached to the periphery rather than the core of the structure. Other supplementary mechanisms include language-particular or universal word order rules, rules that govern the placement of dependents of nouns (adjectives, possessors, determiners), the placement of their dependents, and so forth.

Even with these additions, the case frame system has a clearly finitistic flavor. For every slot, we need to specify (i) the grammatical relation, taken from a small (typically ≤ 6) inventory, (ii) the case or other marking that expresses the relation, again taken from a small (typically ≤ 20) inventory, (iii) the syntactic category, typically a maximal major category, again taken from a small (typically ≤ 5) inventory, (iv) whether the complement is obligatory or optional, a binary choice, and (v) subcategorization restrictions, generally again only a few dozen, say ≤ 50, choices. Since a verb can never have more than five slots, the total number of case frames is limited to $(6 \cdot 20 \cdot 5 \cdot 2 \cdot 50)^5$, a number that at $7.7 \cdot 10^{23}$ is slightly larger than Avogadro's number but still finite. The size of this upper bound (which could be improved considerably e.g. by noting that only a handful of verbs subcategorize for four or five complements) gives an indication of the phenomenon we noted in Section 5.1.1 that nominal, adjectival, and other categories have only a few subcategories, while verbs have orders of magnitude more.

One case of particular interest to modern syntax is whether the frame can become recursive: what happens when verbs take other verbs as their complement? The most frequent cases, auxiliary verbs and modals such as *have* or *can*, are relatively easy to handle because they require main verbs (rather than auxiliaries) to complement them, which blocks off the recursion in short order. There is a less frequently seen class of verbs (sometimes called *control* or *raising* verbs) that take infinitival complements such as *try* and *continue* that recurse freely: both *John tries to continue to run* and *John continues to try to run* are perfectly grammatical (and mean different things). Some of these verbs, such as *persuade* and *want*, take sentential infinitival clauses that have subjects (*John persuaded/wanted Bill to run*), and the others are restricted to infinitival VPs (**John tried/continued Bill to run*). To resolve the apparent violation of maximality (see Section 5.1.3), it is tempting to pattern the latter after the former, assuming simply that the subjectless cases already have their subject slot filled by the subject of the matrix verb by some process of sharing arguments (dependents). The issue is complicated by the fact that the behavior of such verbs is not homogeneous in regard to the use of dummy subjects: compare *There continues to be an issue* to **There tries to be an issue* – we return to the matter in Section 5.2.3.

Clause (v) of the case frame definition expresses the fact that verbs can 'reach down' to the subcategory of their complements (we have seen an example in Section 5.1, *kill* vs. *murder*) and if some verbal complements are themselves endowed with a case frame, as must be the case for infinitival complements, it is a question for recursive constructions of this sort as to how far the matrix verb can control the complement of a complement of a complement. The answer is that such control can go to arbitrary depth, as in *Which books did your friends say your parents thought your neighbor complained were/*was too expensive?* This phenomenon, as Gazdar (1981) has shown, is still easily within the reach of CFGs. There are other significant cases where this is no longer true, the most important being *cross-serial dependencies*, which we will illustrate here on Dutch using a set of verbs such as *see*, *make*, *help*, ... that all can, in addition to the subject NP, take an infinitival sentence complement (Huybregts 1976). To make the word order phenomena clear, we restrict ourselves to subordinate clauses beginning with *dat* 'that' and leave it to the reader to supply a main clause such as *It is impossible* _. The verbs in question can thus participate in constructions such as

...dat Jan de kinderen zag zwemmen
that Jan the child.PL see.PAST swim.INF
that Jan saw the children swim

...dat Piet de kinderen hielp zwemmen
that Piet the child.PL help.PAST swim.INF (5.12)
that Piet helped the children swim

...dat Marie de kinderen liet zwemmen
that Marie the child.PL make.PAST swim.INF
that Marie made the children swim

Once we begin to recursively substitute these constructions in one another, two things become evident. First, that there is no apparent limit to this process. Second, the insertion proceeds at two different sites: both subjects and verbs get stacked left to right.

...dat Jan Piet de kinderen zag helpen zwemmen
that Jan Piet the child.PL see.PAST help.INF swim.INF
that Jan saw Piet help the children swim

(5.13)

...dat Jan Piet Marie de kinderen zag helpen laten zwemmen
that Jan Piet Marie the child.PL see PAST help.INF make.INF swim.INF
that Jan saw Piet help Marie make the children swim

The case frames of verbs like *see* contain a subject slot, to be filled by a nominative NP, and an infinitival complement slot to be filled by an S.INF, which again is subject to the rule (5.7) that rewrites it as a subject NP and an (infinitival) VP.INF, and the latter can be filled by an intransitive V.INF. (Mentioning the INF in these rules is redundant since the Head feature Convention will enforce the inheritance of

INF from S to VP to V.) In the DG formalism, the innermost arrow runs from the dependent (subject) *child.PL* to the head *swim.INF*. This arrow, corresponding to the innermost VP, is a dependent of the matrix verb *make*, which has another dependent, the subject *Marie*, yielding the complete infinitival sentence *Marie make the children swim* by (7). Progressing inside out, the next matrix verb is *help*, with subject *Piet* and complement *Marie make the children swim*, yielding *Piet help Marie make the children swim*, and the process can be iterated further. In a standard CFG, recursive application of the rule S.INF $\rightarrow$ NP V S.INF, terminated by an application of S.INF $\rightarrow$ V.INF, would yield the bracketing [NP V [NP V [NP V]]] as in *Piet help Marie make children swim*, which would be entirely satisfactory for English. But Germanic word order does not cooperate: we need to supplement the phrase structure rules either by some verb second (V2) transformation (den Besten 1985), by a *wrapping* mechanism (Bach 1980, Pollard 1984), or by a separate linear precedence mechanism to which we turn now.

Here the fact that DG is generally silent on word order is helpful: all that needs to be said is stating the typologically relevant generalization (namely that Dutch is V2 in subordinate clauses) to get the required word order. One way to formulate generalizations about word order is to state them as constraints on linear precedence. In the Dutch case at hand, we can simply say $NP \prec V$ (noun phrases precede verbs) within structures dominated by an infinitival S. This technique, called *immediate dominance/linear precedence*, or ID/LP for short, is the one used in generalized phrase structure grammar (GPSG) (Gazdar et al. 1985). Another possible technique is based on *nearness*, a weakening of the relation of immediate adjacency. If no morphological marking is available to signal the fact that two words are strongly related, the best stand-in is to insist that they appear next to each other, or at least as close as possible. English uses this device for two rather different functions. In *parataxis*, adjacency signals coordination, as in *Tom Dick and Harry*, where logical calculi would require *Tom and Dick and Harry*. The other case is signaling subjecthood: the only structural position where a subject can appear is the one immediately before the verb.

Modern theories of morphosyntax (Ostler 1979, Kiparsky 1987, Smith 1996), which differ from Pāṇini chiefly in their ambition to also handle analytic languages such as English, have enlarged the inventory of *linkers* from morphological devices such as case marking, prepositions, and agreement to include positionally defined relationships as well. *Direct* theories of linking offer a straightforward mechanism to capture the basic intuition with which we started, that cases mediate between semantic, morphological, and syntactic generalizations. Their central device is the deep case *(kāraka)*, which serves to link complements to the verb both in sentences and in nominal constructions derived from verbs. The derivation starts by selecting a verb with the appropriate tense marking, some nominals with the appropriate number marking, and the deep cases – the latter will be realized (*abhihita*, 'spoken') by morphological and structural devices biuniquely but heterogeneously across verbs. We have biuniqueness in any fixed construction since every deep case that appears there will be realized by a single linker and every linker realizes some deep case, and we have heterogeneity in that there is no requirement for the realization to be constant across verbs. For example, the accusative (surface) case can serve to realize Goal or Patient (but not both in any given sentence or nominalization), and Goal can also be realized by the instrumental case.

Such theories are 'direct' because they proceed from the arguments of the verb, defined semantically, to the linkers, which are visible on the surface, using only one intermediary, the deep cases. In contrast, *indirect* theories invoke two sets of intermediaries, deep cases and grammatical functions (subject, object, etc.), as well. Relational grammar does not use a separate apparatus for these two (grammatical functions and deep cases are intermingled) but does permit derivational strata, which make the overall theory indirect. The same is true of case grammar, which in its original form (Fillmore 1968) was clearly intended as a direct theory but used transformations, which brought strata with them. The case frame definition left room for using *both* grammatical functions and deep cases, and so far there is little to recommend the traditional notion of grammatical function. Yet there are some remarkable phenomena that point at differences not easily explained without reference to subjects and objects. Consider first the behavior of reflexives: *John shaved himself* is obviously grammatical and **Himself shaved John* is obviously not, yet in both cases the Agent and the Patient of the shaving are both *John*, so the difference in acceptability can hardly be attributed to a difference in the semantics.

To further complicate matters, there is a third set of primitives, called *thematic roles* or *theta roles*, that are regularly invoked in classifying verbal arguments. These are intended as fully semantical, expressing generalizations that follow from the meaning of verbs. For example, if V is an action and NP refers to the Agent of this action, then NP intends V to happen. Comparing *John accidentally killed the pedestrian* to **John accidentally murdered the pedestrian* shows that under this definition of Agent (the names used for thematic roles largely overlap the names used for deep cases), *kill* does not require an Agent but *murder* does. The theories of linking discussed so far distinguish between deep and surface cases and keep both of these distinct from both grammatical functions (subject, object, indirect object) and thematic roles to the extent they admit them. Direct theories that refer to thematic roles use them purely as abbreviatory devices to distinguish different classes of verbs with different lexical entailments, while indirect theories, to which we now turn, permit combinatorical statements that refer to more than one set of primitives.

5.2.3 Valency

In Section 5.1 we already considered the informal notion of *valency* as a means of stating grammatical regularities. Here we consider a more formal, and in key respects more general, mechanism for syntactic computations, the use of algebraic structures as a means of regulating the computation. We illustrate the way these formal systems are intended to be used by well-known linguistic phenomena, such as the possessor-possessed relationship, but we do not intend these illustrations to be exemplary in the linguistic sense. In many cases, different analyses of the same phenomenon are also available, often within the confines of the same formal systems.

In many languages, such as Latin or German, the *genitive* case is affixed to a noun to indicate that it is the possessor of another noun (or NP) within the same sentence. The relationship of possession is to be construed as a rather loose one, ranging from 'being a physical part of', as in *John's hand*; to 'ownership', as in *John's book*; to 'being closely associated to', as in *John's boss*; and to 'being loosely associated to', as in *John's cop*, who can be the cop that always tickets John at a particular intersection, the cop that John always talks about, the cop John always calls when a fight breaks out in the bar, and so on.

Many linguists, starting with Pāṇini, would argue that the genitive is not even a case, given that it expresses a relation between two nouns rather than a noun and verb. It is also true that the same possessor-possessed relation is expressed in many languages by suffixation on the head of the construction, the possessed element, rather than on the modifier (the possessor). For our purposes, the main fact of note is that the possessor can show agreement with the possessed in head-marking languages. For example, Hungarian *az én könyvem, a te könyved, az ő könyve* 'my book, your book, his book'. By Definition 5.2.1, this means that in the pure case we have some forms $u.D_i$ and $v.D_i$ that cooccur in some context α_β_γ, while for $i \neq j, \alpha u.D_i \beta v.D_j \gamma \notin L$. (Not every case is pure since often the paradigmatic distinction is not entirely visible on the form, such as English *you*, which can be second-person singular or plural, or Hungarian *az ő*, which in the possessive construction can be third-person singular or plural possessor alike. As there are many languages where all paradigmatic forms are distinct, we can safely ignore this complication here.)

Since there are only a finite number of cases to consider, almost any method, including tabulation (listing), would suffice to handle agreement phenomena like this. For example, if the paradigm has four slots and we denote $u.D_i$ ($v.D_i$) by *a,b,c,d* (resp. a', b', c', d'), the four forms $\alpha a \beta a' \gamma, \alpha b \beta b' \gamma, \alpha c \beta c' \gamma, \alpha d \beta d' \gamma$ are admissible, while $\alpha a \beta b' \gamma$ and the other eleven nonagreeing forms are not. Yet linguistics rejects this method since the intrinsic complexity of such a list would be the exact same as that of listing the cyclic permutation $\alpha a \beta b' \gamma, \alpha b \beta c' \gamma, \alpha c \beta d' \gamma, \alpha d \beta a' \gamma$, and the latter is clearly unattested among natural languages. The preferred method is to formulate a rough rule that generates $\alpha u \beta v \gamma$ and supplement this by a further condition stipulating agreement between u and v.

One important method of stipulating such conditions is to map preterminals onto some algebraic structure that offers some direct means of checking them. For example, given a free group F over four generators *a, b, c, d* and their inverses $a' = a^{-1}, b' = b^{-1}, c' = c^{-1}, d' = d^{-1}$, if α, β, and γ are mapped on the unit e of F and $a, \ldots, d, a', \ldots, d'$ onto themselves, the acceptable forms, the ones with proper agreement, will be exactly the ones whose image is mapped onto e. What makes cancellation in a group such an attractive model of valency is that it is easily typed. When talking about valence informally, we always mean finding an *appropriate* filler for some slot. By assigning each slot an independent group element and each appropriate filler its inverse, we can guarantee both that slots will require filling and that different slots receive different fillers.

Given that agreement in natural languages is independent of word order, the fact that in a group we always have $aa^{-1} = a^{-1}a = e$ is useful, and so is the fact that we can take the group used for checking agreement to be the direct product of simpler cyclic groups used for checking agreement in any particular paradigmatic dimension. But in situations where the filling of valences is order-dependent, using structures other than groups and using decomposition methods other than direct products may make more sense. Before turning to these, let us first consider other potential pitfalls of modeling valences with cancellation.

A strong argument against a strict cancellation model of valency may be a case where a single slot (negative valence) is filled by more than one filler (positive valence), as in coordinated constructions, such as *John and Mary won*, or conversely, cases where a single filler fills more than one slot. Consider, for example, the accusativus cum infinitivo (ACI) construction in Latin, as in *Ad portum se aiebat ire* 'He said he (himself) was going to the harbor'. Evidently the reflexive pronoun *se* 'himself.ACC' acts both as the subject of *say* and the subject of *go*. On the whole, the scope of ACI in Latin is remarkably large: almost every verb that can take an object can also take an infinitival construction as an object. Unlike in English or Dutch, where *He wanted/helped/watched the children to swim* are possible but **He said/judged/delighted the children to swim* are not, in Latin sentences such as *Thales dixit aquam esse initium rerum, Karthaginem delendam esse censeo*, and *Gaudeo te salvum advenisse* are easily formed. Another potential case of a single filler filling multiple slots comes from the class of control verbs discussed in Section 5.2.2: in *John tried to run*, arguably the same filler, *John*, acts the subject of *try* and *run* at the same time. Even more interesting are cases of *object control* such as *John asked Bill to run*, where *Bill* is both the object of *ask* and the subject of *run* – compare *John promised Bill to run.*

To extend valency to coordination, we need a device that will act as a multiplexer of sorts, taking two syntactic objects with one positive valence each and returning a single compound object with one positive valence. Problems arise only in cases where the compound filler can appear in positions that are closed off for elementary fillers: clearly *John and Mary hated each other* is grammatical, while neither **John hated each other* nor **Mary hated each other* are, and it is hard to conceive of the former as being in any sense the coordination of the latter two. Once a multiplexer is available, it or its dual could be used for unifying slots as well, and expressing general statements like 'in object control verbs, the object of the matrix verb is shared with the subject of the embedded verb' would be easy.

Yet, just as with coordination, there remain some strong doubts whether using argument sharing as a conceptual model for apparently shared fillers is truly appropriate. For example, *I hate to smoke* means that I dislike cigarette smoke, while *I hate that I smoke* means that I actually like cigarette smoke (but dislike my own weakness of not being able to give it up). The standard method of formalizing argument sharing is by means of shared variables: if we treat transitive verbs as two-place predicates, `try(Someone, Something)`, `hate(Someone, Something)`, and intransitives as one-place predicates, `run(Someone)`, or as two-place predicates with implicit defaults, `smoke(Someone,Cigarettes)`, we can write *John tried to run* as `x=John, try(x, run(x))` and *I hate to smoke* as `x=I, hate(x, smoke(x,Cigarettes))`, but this representation is clearly more appropriate for *I hate it that I smoke*.

How, then, is *I hate to smoke* to be represented? One possibility is to treat 'hating-of-smoking' as a compound verb that has only one valence and relegate the process of forming such verbal complexes to the lexicon, where other valency-changing processes, such as forming passives or causatives, are also known to operate. In syntax, then, we can retain a pure form of valence theory that assigns a valence of $+1$ to every NP, including coordinated forms, and a valence of $-n$ to n-place predicates. However, the situation is further complicated by free adverbials and other adjunct NPs that do not fill any slot associated to the verbal arguments – this is what makes typing the slots and fillers essential.

In the commutative case, we can restate matters using additive rather than multiplicative inverses and direct sums rather than direct products. We will say that a predicate has a **valence** vector $(x_1, \ldots, x_k)$, where the x_i are 0 or -1 and the basis dimensions correspond to grammatical functions or deep cases – for the sake of concreteness, we will use Agent, Goal, Recipient, Instrument, Locative, and Source in this order. The valence vector of *rent* will thus be $(-1, -1, 0, -1, -1, -1)$, and for our example (5.11), *John rented the room from a slumlord for a barbershop for the first three months for fifteen hundred dollars*, we have five NPs that can fill the slots: *John* as the Agent, with valence vector $(1, 0, 0, 0, 0, 0)$; *the room* as the Goal, with valence vector $(0, 1, 0, 0, 0, 0)$; *a slumlord* as the Source, with valence vector $(0, 0, 0, 0, 0, 1)$; *the first three months* as the Location, with valence vector $(0, 0, 0, 0, 1, 0)$; and *fifteen hundred dollars* as the Instrument, with valence vector $(0, 0, 0, 1, 0, 0)$. The purpose clause, *for a barbershop*, is floating freely (the Goal is the immediate result of the renting action in view, namely *the room*) and thus has valence vector $(0, 0, 0, 0, 0, 0)$. As long as the valence vectors sum to 0, we accept the sentence, seemingly irrespective of phrase order; the reason why **John the room rented* is unacceptable is that in English only the NP in the immediate preverbal position can be the Agent and only the NP in immediate postverbal position can be the Goal. In languages where deep cases are morphologically marked (by surface cases), the expectation is that the clauses can be freely permuted as long as their morphological marking is preserved.

5.3 Semantics-driven theories

In the most extreme form of semantics-driven theories of syntax, the relation that connects the form and the meaning of sentences is one of causation: a sentence has a given form *because* this is the best way to express some idea or state of affairs. Given the bewildering variety of ways different languages can express the same thought, this appears a hopelessly naive approach, yet a great deal of the methods and motivation are shared between the semantics-driven and the more 'pure' theories of syntax discussed so far. To solve the problem that different languages use different constructions to express the same idea, all that is needed is a parametric theory of syntax and a slight relaxation of direct causality. Instead of saying that a sentence has a given form because this is the best way to express an idea, we now say that it has this form because this is the best *given* the language-specific setting of the parameters.

To give an example, consider transitive sentences of the (5.8) type. Semantics-driven theories maintain that the structure of thought is universal: there is an act of killing, with *the farmer* as the agent and *the duckling* as the patient. Following Greenberg (1963), we divide languages in six types corresponding into the six possible permutations of Subject, Object, and Verb. English is an SOV language, so the sentence comes out as above. In Japanese, an SOV language, we get *Hyakusyoo wa ahiru o korosita 'As for farmer, duck killed'*, while in Irish, a VSO language, we get *Mhairiagh an feirmeoir an lacha*, and so forth. Given that word order in different languages can take any value (according to Hawkins (1983), 45% of languages are SOV, 42% SVO, 9% VSO, 3% VOS, 0.9% OVS, and 0.1% are OSV), proponents of the Sapir-Whorf hypothesis must explain why the majority of languages do not follow the natural order (whichever that may be).

The research program of eliminating nonparametric variation from syntax is common to many theories, including some of the most pure combinatorical theories, such as classical transformational grammar and modern minimalist syntax. The standard form (see Chomsky and Lasnik 1993) is often called *principles and parameters* theory, assuming a common core of grammatical principles that operate in all languages, and a finite set of binary parameters that govern them (so that altogether there are only finitely many core systems). This is an extremely attractive model, but one that faces many serious technical difficulties that come to light as soon as we attempt to flesh out the typological system. First of all, there are many languages that are *split* (display a mixture of the pure types). For instance, German, while widely regarded as an SOV or V2 (verb second) language, displays a pure SVO construction and has several different word orders both in main and in subordinate clauses. Traditional descriptions of German word order, such as Drach (1937), therefore employ a linear structure composed of separate structural positions or *topological fields*, a notion that has little typological generality. Even if we could separate out the parochial from the universal factors in such cases, there is a larger problem: the terms that we use in establishing our typology may have no traction over the actual variety of languages.

The marked word orders VOS, OVS, and OSV account for less than 5% of the world's languages. However, about one language in ten has no identifiable subject category at all according to Schachter's (1976) criteria: a prime example is Acehnese [ACE] (see Durie 1987). For many languages that fall outside the Greenberg system entirely, a different typological distinction between *ergative* and *accusative* languages is invoked. As the basic pattern is very strange for those whose only exposure has been to the 90% of languages that are accusative, we give an example here from

Warlpiri [WBP] (Hale 1983). The sentences generally considered simplest are the intransitives. These involve just a single argument:

The baby cried
NP.NOM V.PAST (5.14)

and in the familiar *accusative* languages that have overt case marking, this argument, the subject, appears with the nominative case. Transitives, the next simplest case, require two arguments. We repeat example (5.8) here with the relevant pseudogloss:

The farmer killed the duckling
NP.NOM V.PAST NP.ACC (5.15)

That the glosses in (5.14) and (5.15) are not entirely fictitious even for English is clear from the use of oblique pronouns *him/her* in the object position as opposed to the nominative *he/she* in the subject position. The Warlpiri pattern is rather different:

Kurdu ka wangka-mi
NP.ABS AUX speak.NONPAST (5.16)
The child is crying

Ngarrka-nguku ka wawirri panti-rni
man.ERG AUX kangaroo.ABS spear.NONPAST (5.17)
The man is spearing the kangaroo

What is striking about this pattern (which is the basic pattern of languages called *ergative* in typology) is that the default case marking (called *absolutive* and realized by zero morphology in many ergative languages, just as nominative is realized by zero in many accusative languages) appears with what from the accusative perspective we would consider the object of the transitive construction, and it is the subject that receives the overt case marking. As this example shows, it is far from trivial to put to use grammatical primitives such as *subject* or *object* even for their stated purpose of grammatical description – ergative languages group together the subjects of intransitives with the objects of transitives.

Example 5.3.1 The basic accusative pattern. In defining the accusative and ergative patterns, we put to use the algebraic apparatus we started to develop in Section 5.1.1. For both cases, we will have nominal elements in the category n *(John, the baby)*, purely intransitive verbs *(sleeps, walks)*, purely transitive verbs *(loves/kills)*, and verbs that have both kinds of subcategorization *(eats, sees)*, which we place in categories i, t, and d, respectively. To simplify matters, we shall ignore person/number agreement but obviously not case marking. For the relevant case suffixes, we will use N (Nominative) A (Accusative), B (aBsolutive), and E (Ergative), and since these always attach to nominal elements we will treat them as subcategories of n, writing simply A instead of the morphemic gloss $n.A$. The **accusative** pattern is given by a finite language X that will have the strings corresponding to intransitive and transitive sentences, with the appropriate case markings Ni, NtA, Nd, NdA and all their permutations (free word order).

The syntactic monoid C_X has two distinguished elements: the empty string (whose equivalence class we denote by I since it is the multiplicative identity) and a large grab-bag equivalence class U for all strings with more than one verbal or more than two nominal elements. Members of this class are all cases of *unrecoverable ungrammaticality* in the sense that no further element can be added to create a grammatical sentence. Other strings, such as NA, are also ungrammatical, but in a recoverable way: adding d or t to the string will create a grammatical sentence, and in fact it simplifies matters to arrange all equivalence classes in layers running from the innermost (I) to the outermost (U) in accordance with how many further elements it takes to make them grammatical. In the innermost layer, we find two classes, represented by t and A, respectively (these require two further elements to complete a sentence), in the next layer we find five classes, represented by Nt, i, NA, N, and d, respectively (these require only one further element), and in the outermost layer we find two classes represented by Ni and Nd, respectively (both require zero additional elements, but they are not in the same class since for one A can still be added). Altogether, the four basic grammatical strings in the accusative pattern (or five, if we declare the empty string grammatical) and their permutations yield a total of eleven syntactic congruence classes i.e. a syntactic monoid with eleven elements.

Discussion The progression from inner to outer layers is a way of recapitulating the idea of valency discussed in Section 5.2.1, and serves to show an essential property of the system, namely that no simple sentence ever has a (semigroup-theoretic) inverse: once valencies are filled in, there is no return. Complex sentences (with more than one verb) may still show periodic behavior (by filling in a verbal dependent we may end up with more open valences than when we started out), but the syntactic semigroup built on simple sentences will always be aperiodic. Since this result is clearly independent of the simplifying assumption of free word order made above, we state it as a separate postulate.

Postulate 5.3.1 In any natural language, the syntactic monoid associated to the language of simple sentences is aperiodic.

In Chapter 2, we already emphasized the importance of Chomsky's (1965) observation that natural languages are noncounting in the informal sense that they do not rely on arithmetic notions. Postulate 5.3.1 provides a more formal statement to the same effect, and in Section 5.5 we shall offer a formal definition of *noncounting* that is, perhaps, easier to falsify empirically than Postulate 5.3.1, but as Yli-Jyrä (2003) notes, even a large-scale wide-coverage grammar of English (Voutilainen 1994) can be modified to avoid the Kleene * operator altogether. Note, however, that the statement is clearly relevant only for syntax since phonology evidently shows periodic patterns, e.g. in the placement of stress (typically binary, sometimes ternary), as discussed in Section 4.1. This runs counter to the generally unspoken but nevertheless very real prejudice of early generative grammar that phonology is easier than syntax.

Returning to the ergative pattern for a moment, the language Y will have *Bi, EtB, Bd, EdB*, and their permutations. It is a trivial exercise to verify that the monoid C_Y is isomorphic to C_X. By mapping N to B and A to E, the accusative set of patterns is mapped on the ergative set in a concatenation-preserving manner. However, this is an isomorphism that fails to preserve meaning: *The man is spearing the kangaroo* would become *The kangaroo is spearing the man*, and it is easy to see that in fact no meaning-preserving isomorphism exists between the two patterns. We return to this matter in Chapter 6, where we investigate not just the stringsets but rather more complex structures that contain both the words and their meanings.

The study of the ergative pattern makes it abundantly clear that the overall plan of using a few examples and counter examples to help navigate the typological decision tree is very hard to put into practice. Compared with stress typology (see Section 4.1.4), the situation is rather dismal. The empirical correlates to the primitive notions are unclear, and the system, however organized, is riddled with split cases. For example Dyirbal [DBL] has an accusative system in the first and second persons, ergative in the third person (Schmidt 1985), and many Indo-Iranian languages are ergative in perfect and accusative in imperfect aspects and so on.

This gives rise to the suspicion, shared by many in the field of artificial intelligence (AI), that mainstream linguistics goes about the whole matter the wrong way, invoking too many hard to define and hard to apply grammatical intermediaries in what should be a straightforward mapping from thought to expression. The goal should be simply to specify the *language of thought* and explain syntactic complexities, to the extent we care about them, by means of parochial imperfections in natural languages. This line of thought goes back at least to the great rationalist thinkers of the 17th century, in particular Descartes, Pascal, and Leibniz. Early AI research such as Quillian's (1969) Teachable Language Comprehender and Anderson and Bower's (1973) Human Associative Memory and the subsequent ACT-R took quite seriously the task of accounting for human long-term memory, including performance effects such as reaction times.

In modern research, the mental aspects of the language of thought are generally downplayed, with many researchers being downright hostile to the idea of looking for brain activity correlates to formulaic descriptions of thoughts or ideas. Rather, to eliminate any reference to the mental aspects, emphasis is placed on what the thoughts or ideas are about, in particular on real-world objects (entities) and assertions about them.

In this view, the task of *knowledge representation* (KR) begins with hierarchically organizing the entities that are typically expressed by common nouns. The central organizational method is Aristotelian, with increasingly special species subsumed under increasingly general genera. Economy is achieved by a system of *default inheritance* whereby lower nodes on the hierarchy inherit the properties of the higher nodes unless specified otherwise. Thus, we need not list the property of breathing with Baby Shamu since she is a killer whale, killer whales are dolphins, dolphins are mammals, mammals are animals, and animals breathe. However, the inference that fishes breathe has to be blocked by explicitly stating that fishes do *not* breathe air, that indeed breathing water through gills is part of their definition. (Going further, the default fish rule needs to be suspended for lungfish and other fish species that can in fact breathe air.) The creation of systems of logic that can sustain nonmonotonic inferences (overriding default conclusions in special cases) is an important thread in modern AI and philosophical logic (see Ginsberg 1986a, Antonelli 1999).

Linguists have long used this technique to hierarchically classify lexical entries, in particular verbs. At the top of the hierarchy, we find the **telic/atelic** distinction that roughly corresponds to the count/mass distinction among nouns. Telic verb phrases denote events that have a definite endpoint, while atelic VPs lack this; compare *John cleaned the dishes for an hour* to **John recognized Bill for an hour*. Next, atelic events are divided into *states* and *activities*, telic events into *achievements* and *accomplishments*; see Vendler (1967), Dowty (1979), and Verkuyl (1993). The situation is greatly complicated by the existence of other largely orthogonal classification schemes such as *stage-level* and *individual-level* predicates (for a modern summary, see Kratzer 1995) or other semantically motivated groups such as *verbs of motion* or *psych verbs*. Levin (1993) offers a rich selection of grouping criteria but refrains from hierarchically organizing the data and thus avoids the need to specify exactly what aspects of the competing cross-classification schemes inherit and to what extent the different flavors can be mixed.

Just as common nouns correspond to entities, adjectives correspond to properties. Unlike entities, which are taken to be an atomic type E, properties are taken as functions from entities to truth values T. Under this analysis, the meaning of an adjective a is simply the set of entities N such that $a(n)$ is true, and n enjoys property a just in case $n \in N$. A key aspect of this model is the division of properties that an entity has into *accidental* and *essential* properties. Continuing with the example above, not only do fishes breathe water but this is essential to their being fishes. The fact that they generally have iridescent scales is accidental: we can imagine fishes without this property, and indeed we find many, such as sharks or eels, that have no such scales. What is accidental in one kind of entity may be essential in another, and to keep track of which is which Minsky (1975) introduced the notion of *frames*. These are not to be confused with the case frames introduced in Section 5.2 above – KR frames apply to nouns, case frames to verbs.

For example, the *dog* frame will contain slots for *name* and *owner*, which are essential to the definition of dogs as cultural constructs in modern urban life, even though they would not be relevant for a biological definition. That it is indeed the cultural, rather than the biological, definition of dogs that is relevant to concerns of natural language understanding can be seen from the following examples: *Rover has diarrhea. The owner has a hard time complying with cleanup regulations/*The coenzyme Q10 overdose is evident.* Once *Rover* has been introduced to the discourse, referring to *the owner* proceeds smoothly because the former is a definite (singular) entity and as such implies a definite owner. For people outside the medical and veterinary professions, the relationship between dogs, diarrhea, and coenzyme Q10 is anything but evident, and introducing *coenzyme Q10 overdose* with the definite article *the* is infelicitous.

For another example, consider the dictionary definition of *cup* as *a usually open bowl-shaped drinking vessel with or without a handle*. What does this really mean? Clearly, every individual object is with or without a handle, and it is just as true of the genus *fish* that it comes with or without a handle as it is of the genus *cup*. Yet there is something right about the definition: *I found a cup/*fish but the handle was broken* shows the same effortless move from *cup* (but not from *fish*) to *handle* that we had from *dog* to *owner*. So the *cup* frame must contain a slot for *handle* meaning that having a handle, or not, is an essential property of cups.

Frames extend more or less naturally to events. To avoid confusion with case frames, we call the frame representation of event objects *scripts*, as has been standard in KR since the work of Schank and Abelson (1977). The original intention was to use scripts as repositories of commonsense procedural knowledge: what to do in a restaurant, what happens during a marriage ceremony, etc. Scripts have actors fulfilling specific roles, e.g. that of the waiter or the best man, and decompose the prototypical action in a series of more elementary sub-scripts such as 'presenting the menu' or 'giving the bride away'. There are some linguistically better motivated models, in particular *discourse representation theory*, that rely on lexically stored commonsense knowledge, but their scope is more modest, being concerned primarily with the introduction of new entities *(the owner, the best man)* in the discourse. Also, there are more systematic studies of *ritual*, in particular in the Indian tradition (Staal 1982,1989), but their cross-fertilization with the Western KR tradition has been minimal so far.

The AI program, then, offers specific solutions to many issues surrounding the representation of nouns, adjectives, and verbs, with both active and stative verbs freely used as elementary subunits in scripts. In fact, these subunits are not viewed as atomic: *conceptual dependency* (Schank 1972) is a representational theory that extends ideas familiar from dependency grammar down to the level of mental language. To apply the MIL formalism of Section 5.2 to Conceptual Dependency, we take the primitive objects to be the atomic symbols PP, PA, and AA (which in CD correspond roughly to nouns, adjectives, and adverbs), ATRANS, PTRANS, MTRANS, GRASP, PROPEL, MOVE, INGEST, EXPEL, ATTEND, SPEAK, MBUILD, and DO (which correspond roughly to verbs), and choose the P_i as CAUSE, BI-CAUSE, MOBJECT, INSTRUMENT, ENABLES, RESULTS, INITIATES, and REASON (which correspond roughly to cases).

Exercise 5.2 Consider all the English example sentences in Chapter 6 and write the equivalent graphs and formulas.

This skeletal picture would need to be supplemented by a host of additional axioms to recapitulate the exact combinatorical possibilities of CD, e.g. the notion that objects need to be AT-LOC before they can PTRANS out of it (Schank 1973). We will not pursue CD in this detail because the representations preferred in linguistics tend to use a slightly different set of primitives; e.g. for *John gave the book to Bill* we could have DO(John,CAUSE(HAVE(Bill,book))) as the underlying semantic structure (Jackendoff 1972). The explicit use of implicit illocutionary primitives such as DO in this example is particularly characteristic of the *generative semantics* school (Harris 1995).

An important version of the same broad program has been pursued by Wierzbicka (1972, 1980, 1985, 1992) and the NSM school (Goddard 2002). The set of primitives is broader, including pronouns *I*, *YOU*, *SOMEONE*, *SOMETHING/ THING;* determiners *THIS*, THE SAME, OTHER; quantifiers ONE, TWO, SOME, ALL, MANY/MUCH; adjectives *GOOD*, BAD, BIG, SMALL, TRUE; adadjectives

VERY, MORE; verbs *THINK*, *KNOW*, *WANT*, *FEEL*, SEE, DO, HEAR, *SAY*, HAPPEN, MOVE, TOUCH, LIVE, DIE; adverbials *WHERE/PLACE*, HERE, ABOVE, BELOW, FAR, NEAR, SIDE, INSIDE, TOUCHING, WHEN/ TIME, NOW, BEFORE, AFTER, A LONG TIME, A SHORT TIME, FOR SOME TIME; connectives *NOT*, MAYBE, CAN, BECAUSE, IF; relations KIND OF, PART OF, THERE IS/EXIST, HAVE, LIKE; and a handful of nouns: BODY, WORDS, MOMENT, PEOPLE (elements in italics were included in the early work, the rest were added since the 1990s). Unlike many of the early AI researchers whose work aimed at immediate algorithmization, Wierzbicka and the NSM school, with a commendable lack of pretense, eschew formalization and operate entirely with natural language paraphrases such as the following definition (Wierzbicka 1992:36) of *soul*:

> one of two parts of a person
> one cannot see it
> it is part of another world
> good beings are part of that world
> things are not part of that world
> because of this part a person can be a good person

To recapitulate this analysis in a more formal framework, we would need to introduce two worlds, one with visible things and one without, a conceptual model of persons with parts in both worlds, goodness and visibility (or the lack thereof) as essential properties of the subclass of persons and the superclass of beings, and so forth. This kind of conceptual modeling of the *folk theory* or *naive theory* behind conceptual entities fits very well in the style of logical analysis undertaken in AI (Hayes 1978), except perhaps for the issue of uniqueness: in AI it is commonly assumed that there will be a unique correct solution reflecting the fundamental nature of reality, while the NSM school assumes that such definitions may show considerable variation across languages and cultures.

An even larger set of 2851 primitives, the Longman Defining Vocabulary (LDV), is used throughout the *Longman Dictionary of Contemporary English* (LDOCE). For the reductionist, the NSM list already offers some tempting targets: do we really need LIVE and DIE as primitives given the availability of NOT? Do we need TOUCH and TOUCHING? Because the LDV list contains a large number of cultural constructs, often trivially definable in terms of one another (e.g. Monday, Tuesday, . . ., Sunday are all listed), it is clearly not a candidate list for the primitive entities in a presumably genetically transmitted universal internal language of thought. Yet LDOCE performs the bulk of the work we expect from a reductionist program: it covers over a hundred thousand word and phrase senses, and clearly every word of English ever encountered is definable in terms of those defined there. Thus, the more radical reductionist programs such as CD or NSM only need to cover the 2851 LDV entries; the rest of English will follow from these (assuming of course that the reductions will not implicitly rely on background knowledge).

Exercise 5.3 Pick ten words randomly from LDV and define them in terms of the CD or NSM primitives.

Given the large number of elusive but nevertheless attractive generalizations that link semantical notions such as volition or telicity to syntactic notions such as case or subjecthood, one may wonder why most linguists consider the task of explaining syntax from semantics hopeless. One particularly strong reason is that natural language has an immense number of *constructions* that are fixed both in form and meaning, but the relationship between the two is arbitrary. We already discussed in Chapter 3 the arbitrariness of signs: the meaning of words is not predictable from the sounds of which they are composed. As it turns out, the meaning of syntactic constructions is also not predictable from the words of which they are composed. Consider arithmetic proportions such as $3 : 5 = 6 : 10$. In English, we say *3 is to 5 as 6 is to 10*, and the construction extends well beyond arithmetic. We can say *London is to England as Berlin is to Germany*. Notice that the key predicate *is to* does not exist in isolation; the only way a phrase like *Joe is to Bill* can appear is as part of the *X is to Y as Z is to W* construction. There appears to be no rational calculus whereby the overall meaning $X : Y = Z : W$ could be derived from the meaning of the function words *as*, *is*, *to*, especially as there must be other factors that decide whether we solve *London is to England as Z is to the US* by *New York* (analogy based on the biggest city) or *Washington* (analogy based on capital). The number of such constructions is so large and their range is so varied (see the entry *snowclones* in `http://www.languagelog.org`) that some theories of grammar, in particular *construction grammar* (Fillmore and Kay 1997, Kay 2002), take these as the fundamental building blocks of syntax, relegating more abstract rules such as (5.7) to the status of curiosities.

Faced with the immense variety of constructions, most linguists today subscribe to the *autonomy of syntax* thesis that both the combinatorical properties of words and the grammatical descriptions of sentences are independent of their meaning. In this view, it is necessary to make the relationship between form and meaning the subject of a separate field of inquiry. This field, what most linguists would call semantics proper, will be discussed in Chapter 6. Here we are concerned with the issue of developing a formal theory of semantics-driven syntax, leaving it to the future to decide how far such a program can actually get us. As is clear from the foregoing, the formal theory begins with a set of conceptual primitives such as LIVE and MOVE, which are used in three settings: first, there are some *axioms* connecting these to each other, e.g. that good things are not the same as bad things; second, there are some dictionary definitions or *meaning postulates* that connect the vast majority of lexical items to those few we consider primitive; and third, some kind of *inner syntax* that regulates how conceptual representations, both primitive and derived, combine with one another.

In most versions of the theory, in particular in the AI/KR approach, the primitives come equipped with a frame not so different from the grammarian's case frame: primitives are modeled as functions with a definite signature, both in terms of having a fixed arity (number of arguments) and in terms of having type restrictions imposed both on their input arguments and on their output. The inner syntax does little more than type-checking: well-formed conceptual representations correspond to well-typed programs evaluating to a few distinguished types such as T (truth value) for sentences and E (entity) for noun phrases. To obtain predictions about the combinatorical possibilities of words, it is necessary to couple this inner syntax to some statements about word order (e.g. whether an adjective precedes or follows the noun it modifies) and to morphology (e.g. whether a property is signaled by word-internal processes or by adding a separate word). The coupling is language-specific, while the inner syntax is assumed to be universal.

While the AI/KR approach is using function arguments for the slot/filler mechanism, the NSM work is more suggestive of categorial grammar in this regard. For example, a primitive such as VERY could be treated as a function that takes an adjective and returns an adjective or as an adadjective that combines with a following adjective to yield another adjective. These two approaches are not incompatible: in full generality, the former corresponds to lambda calculus and the latter to combinatory logic. However, both of these are Turing-equivalent, while there is no reason to suppose that syntax, either alone or in combination with semantics, goes beyond the context-sensitive (linear bounded) domain.

Exercise 5.4 Write a context-sensitive grammar describing the set of first-order formulas where no quantifier is dangling (every quantifier binds at least one variable).
Exercise 5.5* Write an indexed grammar describing the same set.

In terms of the Chomsky hierarchy, context-sensitive grammars provide an upper bound on the complexity of natural languages, and many take examples like (5.13) as indicative of a lower bound, namely that natural languages cannot be properly described by context-free grammars. Finding *mildly* context-sensitive grammars that cover all such examples without sacrificing polynomial parsability has been a focal point of research in mathematical linguistics since the 1980s, with special attention on linear and partially linear versions of indexed grammars – for a good summary, see Kracht (2003). The larger issue of whether structured sets of examples such as (5.13) can actually provide lower bounds remains unsettled. The first such set of examples, used in Chomsky (1957) to demonstrate that English is not finite state (regular), was attacked almost immediately (Yngve 1961) on the grounds that as the length of such examples increases so does the uncertainty about their grammaticality. We return to this issue in Section 5.4 from the perspective of weighted languages.

5.4 Weighted theories

Both grammatical and semantics-driven theories rely on devices, such as case, grammatical function, thematic role, or frame slot, that are in many constructions obscured from the view of the language learner and must be inferred by indirect means. To the extent that combinatorical theories rely on deletion (e.g. the use of *traces* in modern transformational theory), they are open to the same charge of multiplying entities beyond necessity. Since the proper choice of such partially observable devices is anything but clear, it makes a great deal of sense to attempt to bring as much observable evidence to bear as possible. One important range of facts that is becoming increasingly accessible with the advance of computers and the growth of on-line material is the frequency of various word strings: though equally grammatical, *Hippopotami are graceful* has one Google hit, while *What's for lunch?* has over 500,000. The discrepancy is large enough to lend some plausibility to the assumption that a child learning English will likely encounter the second, but not the first, early on. Given the paucity of *NP's for lunch*, a theory of language acquisition that relies on direct memorization (example-based learning) for the contracted copular *'s* in questions that have the *wh*-element in situ is far more credible than one that would crucially rely on the availability of examples like *?Fruit's for lunch* to the language learner.

To the extent that syntactic theories are incapable of accommodating such facts, the combinatorical statement of the problem as a membership problem is incomplete. We address this defect by introducing **weighted languages**, defined as mappings f from strings $w \in \Sigma^*$ to values in a semiring R. $f(w)$ is called the **weight** of w. Our primary example of R will be the set of nonnegative reals $\mathbb{R}^+$ endowed with the usual operations, but the overall framework carries as special cases both standard formal language theory (with R taken as $\mathbb{B}$, the Boolean semiring with two elements) and theories that rely on degrees of grammaticality (Chomsky 1967), as well as theories that use weights in $\mathbb{N}$ to count the number of ways a string can be derived. When the weights taken over the set of all strings sum to 1, we will talk of **probabilistic languages** and write P instead of f.

By *language modeling* we mean the development of models with the goal of approximating the pattern of frequencies observed in natural language. This includes not just probabilistic theories but all the theories discussed so far, as these can be interpreted as offering a crude 0-1 approximation of the actually observable pattern. To be sure, most theories of syntax were not designed with this kind of explanatory burden in mind, and many grammarians actually disdain frequency counts, be they from Google or from better organized corpora. But the overall question of how successfully one (weighted) language *approximates* another remains valid for the standard (unweighted) case, and the mathematical theory of *density* developed in Section 5.4.1 covers both. As we shall see, the central case is when density equals zero, and in Section 5.4.2 we describe some finer measures of approximation that are useful within the zero density domain. In Section 5.4.3, we introduce the general notion of weighted rules and discuss their interpretation in sociolinguistics as *variable rules*. In Section 5.5, we turn to weighted regular languages and discuss the main devices for generating them, weighted finite state automata/transducers and hidden Markov models. The larger issues of bringing external data to bear on syntax, be it from paraphrase or translation, from language acquisition and learnability, from the study of dialects and historical phases of the language, or from performance considerations such as parsing speed or memory issues, will all have to be rethought in the probabilistic setting, a matter we shall turn to in Section 5.6.

5.4.1 Approximation

Here we first investigate what it means in general for one (weighted) language to approximate another. Given $f : \Sigma^* \to \mathbb{R}^+$ and a threshold $\varepsilon \geq 0$, the *niveau sets* $f_\varepsilon^+ = \{w \in \Sigma^* | f(w) > \varepsilon\}$, $f_\varepsilon^- = \{w \in \Sigma^* | f(w) < \varepsilon\}$, $f_\varepsilon^0 = \{w \in \Sigma^* | f(w) = \varepsilon\}$

are ordinary (nonweighted) formal languages over Σ. In engineering applications, it is convenient to restrict the mapping to strictly positive values. Even though a naive frequentist would assign zero to any string w that has not been observed in some very large sample, it is hard to entirely guarantee that w will never be seen (e.g. as a result of a typo), and it is best not to let the model be driven to a singularity by such random noise. As we shall see in Chapter 9, a great deal of engineering effort is spent on deriving reasonable nonzero estimates for zero observed frequency cases. Such estimates, depending on the status of w, can differ from one another by many orders of magnitude. For example, Saul and Pereira (1997) estimate the ratio of the probabilities P(colorless green ideas sleep furiously)/P(furiously sleep ideas green colorless) to be about $2 \cdot 10^5$. This suggests that with a good probabilistic model of English we may have a broad range from which to choose a threshold ε such that f_ε^+ (f_ε^-) approximates the set of grammatical (ungrammatical) strings.

But if P is a probability measure, then P_ε^+ will be finite for any ε. Since all niveau sets will be either finite or cofinite, all niveau sets are regular, rendering our primary navigation aid, the Chomsky hierarchy, rather useless for probabilistic languages. This is not to say that there is no way to generalize finite automata, CFGs, TAGs, or even Turing machines to the probabilistic case (to the contrary, such generalizations are readily available) but rather to say that studying their niveau sets, which is generally the focal point of the analysis of probabilistic systems, suggests that the regular case will be the only one that matters. Since the set of grammatical strings could in principle have nonregular characteristics, while the niveau sets we use for approximation are of necessity regular, our driving example will be the approximation of the Dyck language D_1 over a two-letter alphabet. As D_1 has infinite index (for $i \neq j, a^i$ and a^j always have different distributions), it cannot be described by a finite automaton, so no regular approximation can ever be perfect. The language D_1^1 of matched parentheses of depth one, as given by the CFG $S \rightarrow aTb|SS|\lambda, T \rightarrow ab|abT$, can be easily described without this CFG by a finite automaton, and so could be the language D_1^2 of matched parentheses of depth at most two, and so forth. In general, we define a **bounded counter** of depth k as a finite automaton having states $\{0, 1, \ldots, k-1\}$ and with transitions under $a(b)$ always increasing (decreasing) state number, except in state $k-1$ (resp. 0) where a (resp. b) keeps the automaton looping over the same state. It is intuitively clear that with increasing k the D_1^k get increasingly close to D_1: our concern here is to capture this intuition in a formal system.

As in classical analysis, the general problem of approximating an arbitrary weighted language f by a series of weighted languages f_k reduces to the special case of approximating zero by a series. We will say that the f_k **tend to** f (denoted $f_k \rightarrow f$) if the symmetric differences $(f \setminus f_k) \cup (f_k \setminus f) \rightarrow 0$ (here 0 is the language in which every string has weight 0). We discuss a variety of measures μ for quantitative comparison. These will all assign numerical values between zero and plus infinity, and if $U \subseteq V$ or $f \leq g$ we will have $\mu(U) \leq \mu(V)$ or $\mu(f) \leq \mu(g)$. Some of them are true measures in the sense of measure theory; others are just useful figures of merit. We begin with weighted languages over a one-letter alphabet and see how a simple quantitative measure, *density*, can be properly generalized for

k-letter alphabets. Eilenberg (1974:225) defines the **density** of a language L over a one-letter alphabet as

$$\lim_{n\to\infty} \frac{|\{\alpha \in L \| \alpha| < n\}|}{n}$$

if this limit exists. This definition can be generalized for languages over a k-letter alphabet Σ in a straightforward manner: if we arrange the elements of Σ^* in a sequence ϕ and collect the first n members of ϕ in the sets Σ_n,

$$\lim_{n\to\infty} \frac{|L \cap \Sigma_n|}{n} = \rho_\phi(L) \tag{5.18}$$

can be interpreted as the density of L when it exists. Since this definition is not independent of the choice of the ordering ϕ, we need to select a canonical ordering. We will call an ordering ψ **length-compatible** if $|\psi(n)| \leq |\psi(m)|$ follows from $n < m$. It is easily seen that for arbitrary alphabet Σ and language L, if ϕ is a length-compatible ordering of Σ^* and the limit in (5.18) exists, then it exists and has the same value for any other length-compatible ordering ψ. In such cases, we can in fact restrict attention to the subsequence of (5.18) given by $\Sigma^0, \Sigma^1, \Sigma^2, \ldots$. If we denote the number of strings of length n in L by r_n, **natural density** ν can be defined by

$$\nu(L) = \lim_{n\to\infty} \frac{\sum_{i=0}^{n} r_i}{\sum_{i=0}^{n} k^i} \tag{5.19}$$

To define density by (5.19) over k-letter alphabets for $k > 1$ would have considerable drawbacks since this expression fails to converge for some simple languages, such as the one containing all and only strings of even length (Berstel 1973). To avoid these problems, we introduce the generating function $d(z) = \sum_{n=0}^{\infty} r_n z^n$ and define the **Abel density** ρ by

$$\rho(L) = \lim_{z\to 1} (1-z)d(z/k) \tag{5.20}$$

if this limit exists. A classical theorem of Hardy and Littlewood asserts that whenever the limit (5.19) exists, (5.20) will also exist and have the same value, so our definition is conservative. (We use the name Abel density because we replaced the Cesàro summation implicit in Berstel's and Eilenberg's definition with Abel summation.)

For weighted languages, the number of strings r_n is replaced by summed weights $R_n = \sum_{|w|=n} f(w)$ of the strings of length n, otherwise the definition in (5.20) can be left intact. As long as the individual values $f(w)$ cannot exceed 1, the Abel density will never be less than zero or more than one irrespective of whether the total sum of weights converges or not. As Berstel notes (1973:346), natural density will not always exist for regular languages, even for relatively simple ones such as the language of even length strings over a two-letter alphabet. Abel density does not suffer from this problem, as the following theorem shows.

Theorem 5.4.1 Let L be a regular language over some k-letter alphabet Σ. The Abel density $\rho(L)$ defined in (5.20) always exists and is the same as the natural density whenever the latter exists. The Abel density of a regular language is always a rational number between 0 and 1.

Proof Since $r_n \leq k^n$, $d(z) \leq 1/(1-kz)$ so $\rho(L) \leq 1$ will always hold. The transition matrix A associated with the finite deterministic automaton accepting L has column sums k, so $B = A/k$ is stochastic. Define $H(z)$ as $(1-z)(E-zB)^{-1}$. The limiting matrix $H = \lim_{z\to 1} H(z)$ always exists, and the density of L is simply $\mathbf{v}H\mathbf{e}_i$, where the jth component of $\mathbf{v}$ is 1 of the jth state is an accepting state (and 0 otherwise), and the initial state is the ith. Since $H(z)$ is a rational function of z and the rational coefficients of B, and its values are computed at the rational point $z = 1$, every coefficient of H is rational and so is the density.

This is not to say that nonregular languages will always have a density. For example, the context-sensitive language $\{a^i | 4^n \leq i < 2 \cdot 4^n, n \geq 0\}$ can be shown not to have Abel density over the one-letter alphabet $\{a\}$. When it exists, Abel density is always additive because of the absolute convergence of the power series in $z = 1$.

The proof makes clear that shifting the initial state to i' will mean only that we have to compute $\mathbf{v}H\mathbf{e}_{i'}$ with the same limiting matrix H, so density is a bilinear function of the (weighted) choice of initial and final states. Because of this, density is more naturally associated to **semiautomata**, also called **state machines**, which are defined as FSA but without specifying initial or accepting states, than to FSA proper. The density vector $H\mathbf{e}_i$ can be easily computed if the graph of the finite deterministic automaton accepting L is strongly connected. In this case the Perron-Frobenius theorem can be applied to show that the eigenvalue k of the transition matrix has multiplicity 1, and the density vector is simply the eigenvector corresponding to k normed so that the sum of the components is 1. If this condition does not hold, the states of the automaton have to be partitioned into strongly connected equivalence classes. Such a class is **final** if no other class can be reached from it, otherwise it is **transient**.

Theorem 5.4.2 The segment corresponding to a final class in the overall density vector is a scalar multiple of the density vector computed for the class in question. Those components of the density vector that correspond to states in some final class are strictly positive, and those that correspond to states in the transient class are 0.

Proof By a suitable rearrangement of the rows and columns of the transition matrix A, $B = A/k$ can be decomposed into blocks D_i, which appear in the diagonal, a block C, which corresponds to transient states and occupies the right lowermost position in the diagonal of blocks, and blocks S_i appearing in the rows of the D_i and the columns of C. The column norm of C is less than 1, so $E - C$ can be inverted, and its contribution to the limiting matrix is 0. The column sum vectors of S_i can be expressed as linear combinations of the row vectors of $E - C$, and the scalar factors in the theorem are simply the nth coefficients in these expressions, where n is the number of the initial state. Moreover, since $(E-C)^{-1} = \sum_{i=1}^{\infty} C^i$ holds, all these scalars will be strictly positive. By the Perron-Frobenius theorem, the density vectors corresponding to the (irreducible) D_i are strictly positive.

We will say that a language L over Σ is **blocked off** by a string β if $L\beta\Sigma^* \cap L = \emptyset$. L is **vulnerable** if it can be blocked off by finitely many strings; i.e. iff

$$\exists \beta_1, \ldots, \beta_s \forall \alpha \in L \exists \beta \in \{\beta_1, \ldots, \beta_s\} \forall \gamma \in \Sigma^* \alpha\beta\gamma \notin L$$

Theorem 5.4.3 For a language L accepted by some finite deterministic automaton A, the following are equivalent:

(i) $\rho(L) = 0$.
(ii) The accepting states of A are transient.
(iii) L is vulnerable.

Proof (iii) $\Rightarrow$ (i). If $\rho(L) > 0$, A has accepting states in some final class by Theorem 5.4.2. If $\alpha \in L$ brings A in such a state, then no $\beta \in V^*$ can take A out of this class, and by strong connectedness there is a $\gamma \in V^*$ that takes it back to the accepting state, i.e. $\alpha\beta\gamma \in L$. Thus, α cannot be blocked off.

(i) $\Rightarrow$ (ii). This is a direct consequence of Theorem 5.4.2.

(ii) $\Rightarrow$ (iii). If the accepting states of A are transient, then for every such state i there exists a string β_i that takes the automaton in some state in a final class. Since such classes cannot be left and contain no accepting states, the strings β_i block off the language.

Theorem 5.4.4 Probabilistic languages have zero density.

Proof We divide the language into three disjoint sets of strings depending on whether $f(w) = 0$, $0 < f(w) < \varepsilon$, or $f(w) \geq \varepsilon$ holds. The first of these, f_0^0, obviously does not contribute to density. The third, f_ε^{0+}, can have at most $1/\varepsilon$ members, and will thus have $R_n = 0$ for n greater than the longest of these. In other words, the generating function for these is a polynomial, of necessity bounded in the neighborhood of 1, so the limit in (5.20) is 0. Thus only words with weight $0 < f(w) < \varepsilon$ can contribute to density, but their contribution is $\leq \varepsilon$, and we are free to choose ε as small as we wish.

Theorem 5.4.5 Hidden Markov weighted languages have zero density.

The proof of this theorem is deferred to Section 5.5 where hidden Markov models are defined – we stated the result here because it provides additional motivation for the study of zero density languages, to which we turn now.

5.4.2 Zero density

In Theorem 5.4.3, blocking off corresponds to the intuitive notion that certain errors are nonrecoverable: once we have said something that arrests the grammatical development of the sentence (generally any few words of nonsense will do), no amount of work will suffice to get back to the language *within* the same sentence. One needs an explicit pause and restart. There are some sentence-initial strings that are harder to block, e.g. whenever we gear up for explicit quotation; *And then, believe it or not, he said* is easily followed by any nonsense string and can still be terminated grammatically by *whatever that means*. But such sentence-initial strings can still be blocked off by better crafted βs; e.g. those that explicitly close the quotation and subsequently introduce unrecoverable ungrammaticality.

It is of course highly debatable whether natural languages can be construed as regular stringsets, but to the extent they can, Theorem 5.4.3 applies and zero density follows. Clearly any length-delimited subset, e.g. English sentences with less than 300 words (which contains the bulk of the data), will be regular, so to escape Theorem 5.4.3 one would need a strong (infinite cardinality, nonzero density) set of counterexamples to demonstrate that the entire stringset is not zero density. No such set has ever been proposed. Even if we accept as grammatical, without reservation, for arbitrary length, the kind of center-embedded or crossed constructions that have been proposed in the literature, these carry very strong conditions on the category of elements they can contain. For example, the Dutch crossed constructions must begin with a specified formative *dat* followed by NPs followed by infinitival VPs – this alone is sufficient to guarantee that they have zero density.

Since the most important weighted languages, probabilistic languages, and natural languages all have zero density, it is of great importance to introduce finer quantitative measures. We will consider three alternatives: *Bernoulli density, combinatorial density*, and *saturation*. Aside from finite lists, the simplest class of weighted languages is *Bernoulli languages* over k letters $a_1, \cdots, a_k$ that have positive probabilities $p_1, \cdots, p_k$ summing to one: the weight f of a string $a_{i_1} a_{i_2} \cdots a_{i_r}$ is defined as $p_{i_1} p_{i_2} \cdots p_{i_r}$. Note that this is a probabilistic *process* but not a probabilistic language. For any $n > 0$ the probabilities of the strings of length exactly n sum to 1, so the overall probabilities diverge (a trivial problem that we will fix in Example 5.5.3 by penalizing nontermination at each stage). Beauquier and Thimonier (1986) take Bernoulli languages as their starting point and define the **Bernoulli density** δ of a language by the weights of its prefixes (minimal left factors):

$$\delta(L) = \sum_{\alpha \in Pref(L)} f(\alpha) \tag{5.21}$$

where *Pref*(L) contains all those strings in L that have no left factors in L. In the equiprobable case, for languages where every word is a prefix, this coincides with the **combinatorial density** $\kappa(L) = \sum_{n=1}^{\infty} r_n/k^n$. Finally, the **saturation** $\sigma(L)$ of a language L over a k-letter alphabet is given by the reciprocal of the convergence radius of $d(z/k)$ – this again generalizes trivially to arbitrary weighted languages.

Although clearly inspired by Bernoulli languages, Bernoulli density as a formal construct is meaningful for any weighted or probabilistic language, though it may be infinitely large. Combinatorial density is restricted to weighted languages and languages with Abel density 0. For $k > 1$ if $\rho(L) > 0$, the terms in κ will converge to this value so combinatorial density itself will diverge. As for saturation, we have the following theorem.

Theorem 5.4.6 If $\rho(L) > 0$, then $\sigma(L) = 1$. If $\rho(L) = 0$ and L is regular, then $\sigma(L) < 1$. If $L \subset L'$, then $\sigma(L) \leq \sigma(L')$. $\sigma(L) = 0$ iff L is finite.

Proof If $\lim_{z \to 1}(1-z)d(z/k) > 0$, then $d(z/k)$ tends to infinity in $z = 1$, and since it is convergent inside the unit circle, σ must be 1. If L is regular, $d(z/k)$ is rational (since it is the result of matrix inversion). Therefore if it is bounded in $z = 1$, it has to be convergent on a disk properly containing the unit circle. If $L_1 \subseteq L_2$, then $d_1(z/k) \leq d_2(z/k)$, and since the Taylor coefficients are nonnegative, it is sufficient to look for singularities on the positive half-line. There $d_1(z/k)$ must be convergent if $d_2(z/k)$ is convergent, so $\sigma(L_1) \leq \sigma(L_2)$. Finally, if $d(z/k)$ is convergent on the whole plane, then $f(1) = \sum_{n=0}^{\infty} r_n < \infty$, so L must be finite.

Discussion Which of these three measures is more advantageous depends on the situation. Neither Bernoulli nor combinatorial density is invariant under multiplication with an arbitrary string, but for Abel density and for saturation we have $\rho(L) = \rho(\alpha L) = \rho(L\alpha)$ and $\sigma(L) = \sigma(\alpha L) = \sigma(L\alpha)$ for any string α and for any L, not just those with zero density. While (5.20) does not always yield a numerical value, Bernoulli density always exists. Although this suggests that δ would be a better candidate for a basic measure in quantitative comparisons than ρ, there is an important consideration that points in the other direction: while Bernoulli density is only an exterior measure, additive only for languages closed under right multiplication, Abel density is always additive. If L is not closed, there is an $\alpha \in L$ and a $\beta \in V^*$ such that $\alpha\beta \notin L$. Either α is a prefix or it contains a left factor $\alpha_0 \in L$ that is. Consider the two-member language $X = \{\alpha_0, \alpha\beta\}$:

$$\delta(X) = p(\alpha_0) \neq p(\alpha_0) + p(\alpha\beta) = \delta(X \cap L) + \delta(X \setminus L)$$

Thus, by Caratheodory's theorem, L cannot be measurable. Note also that for languages closed under right multiplication $r_{n+1} \geq kr_n$, so the coefficients in $d(z/k) = \sum_{n=0}^{\infty} r_n z^n/k^n$ are nondecreasing. Therefore the coefficients of the Taylor expansion of $(1-z)d(z/k)$ are nonnegative, and the Abel density ρ also exists.

Now we are in a better position to return to our motivating example, the approximation of the Dyck language D_1 by the languages D_1^k that contain only matching parentheses of depth k or less. The number of strings of length $2n$ in D_1 is given by $r_{2n} = \binom{2n}{n}/(n+1)$, so $d(z) = (1 - \sqrt{1-4z^2})/2z^2$ and thus $\rho(D_1) = 0$. This makes all three finer measures discussed so far usable: taking the left and the right parenthesis equiprobable, $\delta(D_1) = 1/2^2 + 1/2^4 + 1/2^6 + \ldots = 1/3$, $\kappa(D_1) \approx 0.968513$, and $\sigma(D_1) = 1$. Using Bernoulli density, it is clear that the D_1^k approximate D_1: the shortest string in $D_1 \setminus D_1^k$ has length $2k+2$ and $\delta(D_1 \setminus D_1^k) = 3/4^k$, which tends to zero as k tends to infinity. Using saturation, it is equally clear that the D_1^k do not approximate D_1: no matter how large k we choose, the convergence radius of the differences remains 1.

Exercise 5.6 Is D_1^k defined by a bounded counter of depth k? Compute $\kappa(D_1 \setminus D_1^k)$.

One class of weighted languages that deserves special attention is when the weights are set to be equal to the number of different derivational histories of the string. In the case of CFGs, we can obtain the generating function $d(z)$ that counts the number of occurrences by solving a set of algebraic equations that can be directly read off of the rules of the grammar (Chomsky and Schützenberger 1963). For example, the grammar $S \to aSb|SS|\lambda$ generates D_1, and the generating function associated with the grammar will satisfy the functional equation $d(z) = z^2 d(z) + d^2(z) + 1$. Thus $d(z)$ will have its first singularity in $\sqrt{3} > 1$, so the language is supersaturated: its saturatedness $2\sqrt{3}$ can be interpreted as the degree of its ambiguity.

If strings of length n are generated by some CFG approximately a_n times, then $a_{n+m} \approx a_n a_m$ because context-freeness makes disjoint subtrees in the generation tree independent. Therefore, $a_n \approx c^n$ and the base c is a good measure of ambiguity. By the Cauchy-Hadamard theorem, $\sigma = \limsup \sqrt[n]{a_n} = c$. Note also that in the unambiguous case, $\log(\sigma) = \limsup \log(a_n)/n$ can be interpreted as the *channel capacity* of the grammar (Kuich 1970). In the unambiguous case as well as in the case of context-free languages where weights are given by the degree of ambiguity, the generating functions corresponding to the nonterminals satisfy a system of algebraic equations, and therefore $d(z)$ will have its first singularity in an algebraic point. Therefore, in such cases, saturation and Abel density are algebraic numbers.

Although from the applied perspective issues such as approximating the Dyck language are meaningful only if probabilities, rather than derivational multiplicities, are used as weights, the following definitions are provided in their full generality. Given two weighted languages f and g over the same alphabet T and a precision $\varepsilon > 0$, we define the **underestimation error** $U(\varepsilon)$ of g with respect to f by

$$U(\varepsilon) = \sum_{\substack{\alpha \in T^* \\ g(\alpha) < f(\alpha) - \varepsilon}} f(\alpha) - g(\alpha) \tag{5.22}$$

and the **overestimation error** by

$$T(\varepsilon) = \sum_{\substack{\alpha \in T^* \\ g(\alpha) > f(\alpha) + \varepsilon}} g(\alpha) - f(\alpha) \tag{5.23}$$

If we start from an unweighted language and use the values of the characteristic function as weights, (5.22) and (5.23) are often divergent, but for probabilistic languages they always converge and tell us a great deal about the structure of the approximation. To apply them to the Dyck case, where different measures of approximation so far have led to different conclusions, we need to endow both D_1 and D_1^k with a probability function. We could start by setting all Dyck strings of length $2n$ equiprobable and prescribing a reasonable distribution, such as lognormal, on overall length. Yet one would inevitably feel that this is more in the nature of a problem book exercise than something definitive about the way language works – what we need is empirical data.

Here we shall briefly consider parenthetical constructions in the Linux kernel. This has the advantage of removing performance limitations: on the 'hearer' side, the compiler, unlike humans, is ready to handle parentheticals of large depth, and on the 'speaker' side, the authors write the code with a great deal of attention and using long-term memory aids rather than relying on short-term memory alone. Purists may object that kernel hacking relies on skills very distant from human language production and comprehension, but in written, let alone spoken, language we rarely find embedding of depth 5, and depth 7 is completely unattested, while in the kernel, in just three hundred thousand expressions, we find over a thousand with depth 7 and dozens with depth 10 (topping out at depth 13). Restricting our attention to English prose would only serve to trivialize the issue.

A simple but attractive model of D_1^k has one state for each depth starting at 0: at each level, we either open another parenthesis with probability p_i or close one with probability $1 - p_{i+1}$. At the bottom, we have $p_0 = 1$ and at the top we assume $p_k = 0$, so as to make the automaton finite. Since the model only has $k-1$ free parameters and is perfectly symmetrical left to right, it will fail to account for many observable regularities; e.g. that back-loaded constructions like (()(())) are more frequent than their front-loaded counterparts like ((())()) by about half.

Exercise 5.7† Take a large body of code, remove program text, format strings, and comments, and replace all types of left and right parentheses by (and), respectively. Fit a model with parameters $p_1, \ldots p_{k-1}$ as above, and compute under- and overestimation errors for different values of $\varepsilon \approx 1/\sqrt{N}$, where N is your corpus size. For $\varepsilon > 0$ are the under- and overestimation errors always coupled in the sense that changing one will necessarily change the other as well? Does the model improve from increased k? Can better models be found with the same number of free parameters?

5.4.3 Weighted rules

The idea that we should investigate weighted languages using weighted grammars is a natural one, and probabilistic generalizations of the entire Chomsky hierarchy of grammars were presented early on (for finite automata, see Rabin 1963). In these systems, we assign some value between 0 and 1 to each production, and require these values to sum to one for all productions sharing the same left-hand side. Here we depart significantly from the historical line of development because probabilistic grammars at or above the CFG level have shown very little ability to characterize the distributions that occur in practice. The probabilistic generalizations of finite automata that are of practical and theoretical significance, finite (k-)transducers and (hidden) Markov processes, will be discussed in Section 5.5. Here we turn directly to context-sensitive rules because the practice of associating probabilities to such rules has long been standard practice within *sociolinguistics*, the study of language variation.

In his study of the speech patterns in New York City, Labov (1966) noted that contraction of *is* to *'s* as in *John's going* is almost universal among both white and African-American speakers when the subject is a pronoun *(He's going)*. When a full noun ending in a vowel precedes, contraction is more likely than when a full noun ending in a consonant precedes, p(*Martha's going*) $>$ p(*Robert's going*). When the contracting *is* is copulative, as in *John's a good man*, or locative, as in *John's in the bathroom*, contraction is less likely than when it appears before an ordinary VP as in *John's going*, and contraction is most likely preceding the future auxiliary, as in *John's gonna go*.

Arranging the factors in three rows by four columns, with i running over the values *pronoun, V-final, C-final* and j running over the values *copula, locative, ordinary, gonna*, the probabilities (observed frequencies) of *is*-contraction form a $3 \cdot 4$ matrix whose values p_{ij} decrease with i and increase with j. The original **additive variable rule model** (Cedergren and Sankoff 1974) simply assumed that there are additive constants γ_1, γ_2 and $\delta_1, \delta_2, \delta_3$ such that $p_{ij} = p_0 + \sum_{k=1}^{i} \gamma_k + \sum_{l=1}^{j} \delta_l$.

Obviously, there is no guarantee that the observed pattern of frequencies can be fully replicated. The model assumed finding thc parameters by maximum likelihood fit.

When the observed p_{ij} are small (close to 0), often a good fit can be found. When the p_{ij} are large (close to 1), we can take advantage of the fact that the same outcome, e.g. deleting a vowel or voicing a consonant, can be equally well analyzed by the obverse rule (addition of a vowel, devoicing a consonant), for which the application probability will come out small. For the case where the p_{ij} are close to 1/2, Cedergren and Sankoff (1974) also introduced a **multiplicative variable rule model** where the sums are replaced by products or, what is the same, an additive model of log probabilities is used. However, it is unclear on what basis we could choose between the additive and the multiplicative models, and when the probabilities are outside the critical ranges neither gives satisfactory results.

As readers familiar with logistic regression will already know, most of these difficulties disappear when probabilities p are replaced by **odds** $p/(1-p)$. We are interested in the effect some factors $F_1, \dots, F_k$ have on the odds of some event H like *is*-contraction. We think of the F_i as possibly causative factors over which we may have control in an experiment. For example, if the experimenter supplies the proper name at the beginning of the sentence, she may decide whether to pick one that ends in a consonant or a vowel. In order to account for factors that lie outside the experimenter's control (or even awareness), we add a background factor F_0. By Bayes' rule, we have $P(H|F_0F_1 \dots F_k) = P(H|F_0)P(F_1 \dots F_k|HF_0)/P(F_1 \dots F_k|F_0)$ and similarly $P(\overline{H}|F_0F_1 \dots F_k) = P(\overline{H}|F_0)P(F_1 \dots F_k|\overline{H}F_0)/P(F_1 \dots F_k|F_0)$ for the complementary event $\overline{H}$. Dividing the two, we obtain the odds $P(H)/P(\overline{H})$ as

$$O(H) = \frac{P(H|F_0)P(F_1 \dots F_k|HF_0)}{P(\overline{H}|F_0)P(F_1 \dots F_k|\overline{H}F_0)} \tag{5.24}$$

because the terms $P(F_1 \dots F_k|F_0)$ simplify. Taking logarithms in (5.24), we see that the log odds of H given the factors $F_0 \dots F_k$ are now obtained as the log odds of H given the background plus $\log(P(F_1 \dots F_k|HF_0)/P(F_1 \dots F_k|\overline{H}F_0))$. When the $F_1, \dots, F_k$ are independent of F_0, which lies outside the control of the experimenter, and of each other (which is expected given that the experimenter can manipulate each of them separately), we can apply the product rule of conditional probabilities and obtain

$$e(H|F_0 \dots F_k) = e(H|F_0) + \sum \log \frac{P(F_i|HF_0)}{P(F_i|\overline{H}F_0)} \tag{5.25}$$

where we use $e(A|B)$ to denote the log odds of A given B. While (5.25) is exact only as long as the independence assumption is met, we take it to be indicative of the general class of models even when independence is not fully assured. Log odds define a surface over the space spanned by the F_i, and (5.25) is a gradient expansion with constant term $e(H|F_0)$ plus some functions (logs of conditional probability ratios) that we may as well assume to be linear since they are known to us only at two points, when the trigger F_i is present and when it is absent.

In the terminology of modern machine learning, the independent (experimenter-controlled) variables or triggers F_i are called *features* or *predicates*.[2] Generally, features are categorial (present or absent, on or off), and we do not assume that they are logically or empirically independent of each other. For example, the feature 'preceding word is vowel-final' is always on when the subject is *he* or *she*, yet the presence of *he/she* in that position is such a salient trigger of *is*-contraction that we may want to treat it as a feature on its own. From (5.25) we take the lesson that the log odds of the dependent variable are to be sought in the form $e(H) = \lambda_0 + \sum \lambda_i F_i$, but note that in sociolinguistics the primary goal is *data modeling*, describing the observed probabilities with the least number of parameters, while in machine learning the goal is *predicting* the value of the dependent variable given some features F_i (these two goals are not necessarily different).

5.5 The regular domain

Given a system of grammatical description, we would like to characterize both the set of grammars that can be written in this system and the set of languages that can be described by such grammars. In the combinatorical view, a language is identified as a stringset, and **weak generative capacity** refers to the set of formal languages that can be generated by the permissible grammars, while **strong generative capacity** is defined as the structure-generating capacity of the system (Chomsky and Halle 1965a). In some cases, it is possible to investigate issues of strong generative capacity by linearization, but, on the whole, less combinatorical views of syntax are not served well by these definitions. Strong generative capacity needs significant reworking before it can be put to use over the wider range of grammatical theories considered here (see Rogers 1998, Miller 1999).

Weak generative capacity is practically meaningless when it comes to probabilistic versions of the theory: if defined through niveau sets, only the full stringset Σ^* can be obtained in the limit. One could in principle use f_0^0 to encode an arbitrary stringset, but this runs counter to the probabilistic interpretation, where small differences in weight are considered empirically undetectable. If defined so as to include the actual numerical values, the study of weak generative capacity quickly turns into a study of algebraic independence and transcendence degree: we discuss the reasons for this in Section 5.5.1, where we treat weighted FSA and transducers. In Section 5.5.2, we turn to hidden Markov models, which play a critical role in the applications.

5.5.1 Weighted finite state automata

Probabilistic finite state automata (PFSAs) are defined as FSAs with the additional requirement that the weights associated to all transitions that can be taken

[2] This terminology is completely foreign to sociolinguistics, where 'feature' always means distinctive feature in the sense of Section 3.2. Since in syntax and semantics 'predicate' is always used in the logical sense, there is no easy way out.

from a given state on scanning a symbol sum to one. Note that we do not require the probabilities of all transitions that leave a state to sum to one. In fact, if the alphabet has n symbols the total weight of such transitions will be n. For a one-letter alphabet already, there is the following theorem.

Theorem 5.5.1 (Ellis 1969) There exist probabilistic languages $f : \{a\}^* \rightarrow \mathbb{R}^+$ that cannot be characterized by any PFSA.

The original proof by Ellis explicitly constructs an infinite set of weights as $1/\sqrt{p_i}$, where p_i is the smallest prime larger than 4^i and proceeds by counting degrees of field extensions. In response, Suppes (1970) argued that

> From the empirically oriented standpoint ... Ellis' example, while perfectly correct mathematically, is conceptually unsatisfactory, because any finite sample of L drawn according to the density p could be described also by a density taking only rational values. Put another way, algebraic examples of Ellis' sort do not settle the representation problem when it is given a clearly statistical formulation. Here is one such formulation. ...
>
> *Let L be a language of type i with probability density p. Does there always exist a probabilistic grammar G (of type i) that generates a density p' on L such that for every sample s of L of size less than N and with density p_s the null hypothesis that s is drawn from (L, p'_s) would not be rejected?*
>
> I have deliberately imposed a limit N on the size of the sample in order to directly block asymptotic arguments that yield negative results.

Suppes conjectured that the problem, stated thus, has an affirmative solution. To approach the issue formally, we first define a **weighted transducer** as a mapping $f : \Sigma^* \times \Gamma^* \rightarrow R$, where Σ and Γ are finite alphabets and R is a semiring. Note that in the general case we do not require the mapping to be homomorphic in the sense that if $\sigma, \sigma' \in \Sigma^*$ and $\gamma, \gamma' \in \Gamma^*$, then $f(\sigma\sigma', \gamma\gamma') = f(\sigma, \gamma) \cdot f(\sigma', \gamma')$ (here $\cdot$ is the product operation of R).

To recover our earlier definition of weighted languages as a special case, we set $\Sigma = \Gamma$ and take the transduction between Σ^* and Σ^* to be the identity mapping. (Again, there is no requirement that weights multiply and $f(\sigma\sigma') = f(\sigma) \cdot f(\sigma')$ – were we to impose such a requirement, the only weighted languages would be Bernoulli languages.) This more complicated definition has the advantage that the definition of composing weighted transductions given below will extend smoothly to the case of transducing a weighted language by a weighted transduction. If $f : \Sigma^* \times \Gamma^* \rightarrow R$ and $g : \Gamma^* \times \Delta^* \rightarrow R$ are two weighted transductions, their **composition** $f \circ g$ is defined to assign (σ, δ) the weight $\sum_{\gamma \in \Gamma^*} f(\sigma, \gamma) \cdot g(\gamma, \delta)$, where $\cdot$ refers to the multiplication operation and $\sum$ to the addition operation of R. (While the definition is meaningful for all kinds of weighted transductions, the motivating case is that of probabilities.)

Of central interest are **weighted finite state transducers (WFST)** given by a finite set Σ of states, and a finite set of **weighted transitions** $(b, a, l_1, \cdots, l_n, r)$ where b and a are the states *b*efore and *a*fter the move, the l_j are letters scanned on the jth tape during the move or λ, and r is a weight value. When for every b

and $l_1, \cdots, l_n$ there is at most one $a \in S$ such that $(b, a, l_1, \cdots, l_n, r) \in T$ with $r \neq 0$, the transducer is **deterministic**, and when λ never appears in any transition it is called **length-preserving**. There are two natural ways to proceed from weighted transductions to weighted automata: either we can zero out the input alphabet and obtain a pure generating device, or we can zero out the output alphabet and obtain a pure accepting device. Either way, to fully characterize a **weighted FSA** over a one-letter alphabet $\{a\}$ requires only a set of states $\Sigma = \{s_1, \ldots, s_n\}$, for each state i a set of values $t_{i,j}$ that characterizes the probabilities of moving from state s_i to s_j upon consuming (emitting) an alphabetic symbol, and a set of values $l_{i,j}$ that characterizes the probabilities of moving from state s_i to s_j by lambda-move (i.e. without consuming (emitting) an alphabetic symbol). For the sake of concreteness, in what follows we will treat PFSA as generating devices – the results presented here remain true for acceptors as well. To simplify the notation, we add a start state s_0 that only has λ-transitions to s_i for $i > 0$ and replace all blocked transitions by transitions leading to a sink state s_{n+1} that has all (emitting and nonemitting) transitions looping back to it and has weight 0 in the vector $\mathbf{w}$ that encodes the mixture of accepting states. This way, we can assume that in every state s_i and at every time tick the automaton A will, with probability 1, move on to another state s_j and emit (or not emit) a symbol during transition with probability $t_{i,j}(l_{i,j})$.

The probability of A emitting a^k is the sum of the probabilities over all paths that emit a k times. Let us introduce a zero symbol z and the automaton A' that emits z wherever A made a λ-transition. This way, the probability of A emitting a, $P(a|A)$ is the same as $\sum_{k,l \geq 0} P(z^k a z^l | A')$, similarly $P(a^2|A) = \sum_{k,l,m \geq 0} P(z^k a z^l a z^m | A')$, and so forth. To compute the probability over a fixed path $s_{i_1}, s_{i_2}, \ldots, s_{i_n}$, we simply multiply the r_i associated to the transitions. If the transition matrix is $T + L$, where $t_{i,j}$ is the probability of the emitting and $l_{i,j}$ the probability of the nonemitting (lambda) transition from s_i to s_j, the probability of going from s_i to s_j in exactly k steps is given by the (i, j)th element of $(T + L)^k$. It is convenient to collect all this information in a formal power series $p(a, z)$ with coefficients in R and noncommuting variables a and z: in matrix notation, $p(a, z) = \sum_{k \geq 0} (aT + zL)^k$. Given a fixed start state s_0 and some weighted combination $\mathbf{w}$ of accepting states, the probability of a string $x_1, \ldots, x_n \in \{a, z\}^n$ being generated by A' is obtained as the inner product of the zeroth row of $(T + L)^n$ with the acceptance vector $\mathbf{w}$ just as in Theorem 5.4.1. To obtain the probability of, say, a according to A, we need to consider $\sum_{k,l \geq 0} L^k T L^l$, to obtain $P(a^2|A)$ we need to consider $\sum_{k,l,m \geq 0} L^k T L^l T L^m$, and so forth. By collecting terms in expressions like these, it is clear that we are interested in matrix sums of the form $I + L + L^2 + L^3 + \ldots$.

Notice that the spectral radius of L is less than 1. Since $T + L$ is stochastic, the rows of L sum to at most 1, and strictly less for all rows that belong to states that have outgoing emitting transitions with positive weight. The row norm of the submatrix R obtained by deleting the zeroth rows and columns will be strictly less than one as long as pure rest states (those with no emission ever) are eliminated, a trivial task. Therefore the eigenvalues of L are the eigenvalues of R and zero, and as the eigenvalues of R are all less than one, the eigenvalues of L are less than one, and thus the matrix series $I + L + L^2 + L^3 + \ldots$ converges to $(I - L)^{-1}$. This gives us a simple formula for $P(a|A) = \mathbf{e}L(I-L)^{-1}T(I-L)^{-1}\mathbf{w}$ (**e** is the vector that is 1 on the zeroth component and zero elsewhere, and **w** is the vector encoding the weighting of the final states – the initial L is present because the zeroth state by convention has no emitting transitions), $P(a^2|A) = \mathbf{e}L(I-L)^{-1}T(I-L)^{-1}T(I-L)^{-1}\mathbf{w}$, and in general

$$P(a^k|A) = \mathbf{e}L((I-L)^{-1}T)^k(I-L)^{-1}\mathbf{w} \tag{5.26}$$

Since the only parts of (5.26) dependent on k are the kth powers of a fixed matrix $(I-L)^{-1}T$, the growth of $P(a^k|A)$ is expressible as a rational combination of kth powers of constants $\lambda_1, \lambda_2, \ldots, \lambda_n$ (the eigenvalues of $(I-L)^{-1}T$) with the fixed probabilities $t_{i,j}$ and $l_{i,j}$. Therefore, the ratios of probabilities $P(a^k|A)$ and $P(a^{k+i}|A)$ will tend to fixed values for all fixed i. This proves the following characterization of PFSA languages over a one-letter alphabet:

Theorem 5.5.2 Any PFSA language $p : \{a\}^* \to \mathbb{R}^+$ is ultimately periodic in the sense that there exists a fixed k and l such that for all $0 \le i < k$ either all weights $p(a^{i+rk})$ are zero once $i + rk > l$ or none of the weights $p(a^{i+rk})$ are zero for any r such that $i + rk > l$ and all weight ratios $p(a^{i+rk+k})/p(a^{i+rk})$ tend to a fixed value $\lambda_1^k < 1$.

Discussion Since both $(I-L)^{-1}$ and T are nonnegative, so is $(I-L)^{-1}T$. If this is irreducible (every state is reachable from every state), the Perron-Frobenius theorem guarantees the existence of a unique greatest eigenvalue λ_1. If it is reducible, each transitive component corresponds to an irreducible block on the main diagonal, and will have its unique largest eigenvalue. When there is a unique largest one among these, it will dominate the high powers of the matrix, but if different blocks have the same largest eigenvalue λ_1, this will appear with multiplicity in the overall matrix and entries over the main diagonal can contribute $O(k^c\lambda_1^k)$ where c is the number of components. However, such a linear factor to the exponentially decreasing main term will be removed by taking ratios.

Example 5.5.1 Let $p_0 = 2^{-1}, p_1 = p_2 = p_3 = p_4 = 2^{-2}/2^{2^1}, p_5 = \ldots = p_{20} = 2^{-3}/2^{2^2}$, and in general divide the probability mass 2^{-n} among the next 2^{2^n} strings. By Theorem 5.5.2, this distribution will differ from any distribution that is obtained from a PFSA by inspecting frequencies in finite samples, even though all probabilities are rational, fulfilling Suppes' dictum.

Discussion Theorem 5.5.2 provides, and Example 5.5.1 exploits, exactly the kind of asymptotic characterization that Suppes wanted to avoid by limiting attention to samples of a fixed size $< N$. In hindsight, it is easy to see where the strict empiricism embodied in Suppes' conjecture misses the mark: with the availability of corpora (samples) with $N > 10^{10}$, it is evident that our primary goal is not to characterize the underlying distribution to ten significant digits but rather to characterize the tail, where probabilities of 10^{-40} or many orders of magnitude below are quite common. Recall from Section 4.4 that perfectly ordinary words often have text frequencies below 10^{-6} or even 10^{-9}, so sentences like *In our battalions, dandruffy uniforms will never be tolerated* will have probability well below 10^{-40}. Even if we had corpora with $N > 10^{40}$, the goal of reproducing the exact measured frequencies would be secondary: the primary goal is to make reasonable predictions about unattested events *without* memorizing the details of the corpus.

The entire corpus available to language learners in the course of, say, the first twenty years of their lives, is much less than 10^{10} words, yet they have a clear sense that some hitherto unseen strings such as *furiously sleep ideas green colorless* are much less likely than other, presumably also unseen, strings such as *colorless green ideas sleep furiously*. By now it is clear that our interest is precisely with comparing low-probability events, and the central measure of success is whether, by making good enough predictions about the high-probability examples that are observable in the sample, we obtain enough generality to cover the low-probability cases, which will in general not be observable. In an automaton with 10^6 states (quite feasible with today's technology) and 10^2 letters (well below the size of commonly used tagsets), we would have over 10^{14} free parameters, a huge number that could only lead to overfitting, were we to follow Suppes' dictum and restrict ourselves to precisely matching samples of size 10^{12}. The key issue, as we shall see in Chapter 8 and beyond, is to reduce the number of parameters, e.g. by appropriately *tying* together as many as possible.

By homomorphically mapping all tokens on the same token a, we obtain the **length distribution** of the sample. If the initial weighted language is regular, its homomorphic image will be also, which by Theorem 5.5.2 means it must be ultimately periodic. As the proof above makes clear, for any N we are at total liberty to prescribe a probability distribution given by a nonincreasing sequence of the first N values (with judicious use of λ-moves, nonmonotonic orders can also be simulated), and in this sense the conjecture proposed by Suppes trivially holds – in fact it holds in the stronger form that for any language of type $i < 3$ we can also fit a PFSA (type 3 grammar) to the first N terms of a distribution to any required precision. A more interesting question would be to look at the intrinsic complexity of the parameter space: how many parameters do we really need to describe the first N length observations. If the probabilities are $p_1, \ldots, p_r$, we expect Np_r copies of a^r so the last r to show up in the sample will be $p_r \approx 1/N$. In the regular case, p_r is asymptotically p^r for some $p < 1$, so we expect $r \log p \approx -\log N$ or, since $\log p$ is a constant (negative) factor, $r \approx \log N$. In other words, for corpora with 10^{10} strings, a realistic task is to fit a PFSA with about 23 states ($2 * 23^2 \approx 10^3$ free parameters) to approximate its length distribution by regular means.

Exercise 5.8† Obtain two gigaword or larger corpora from the Linguistic Data Consortium or by crawling the web, and parse them into sentences along the lines of Mikheev (2002). What is the length distribution of the first corpus? How many parameters are needed for a PFSA that provides a good fit? How well does this PFSA describe the length distribution of the second corpus?

To get a finer picture than what can be provided by the length distribution, and also to extend the picture from the regular to the context-free case, let us briefly consider t-letter alphabets, where t is the size of the category system, by mapping each observed sequence of words on the sequence of corresponding preterminals. (As a practical matter, for corpora of the size discussed here, disambiguating the category can no longer be done manually: instead of a perfect data set, we have one that may already carry considerable POS tagging error.) At first, we will ignore word order, and replace strings over a t-letter alphabet by **count vectors** of dimension t whose jth component c_j counts how often the jth symbol appeared in the string. A classic theorem of Parikh (see e.g. Salomaa 1973 Ch. II/7) asserts that for every CFL there is a regular language with the same count vectors. However, this theorem does not smoothly extend to weighted languages where the weight is given by multiplicity (number of derivations):

Example 5.5.2 Binary trees. The grammar $S \rightarrow SS|a$ generates all sequences a^k as many times as there are binary trees with k leaves, i.e. $\binom{2n}{n}\frac{1}{n+1}$ times.

If we assign probability p to the branching and q to the nonbranching rule, the probability assigned by an individual derivation of a^k will be $p^{k-1}q^k$, and it follows from Stirling's formula that the total probability (taken over all derivations) will also contain a term proportional to $k^{-3/2}$. So log probabilities would show linearity for individual derivations but not for the totality of derivations, and therefore by Theorem 5.5.2 the weighted CFL of binary trees cannot be generated by any PFSA. This is worth stating as the following theorem.

Theorem 5.5.3 PCFGs generate more languages than PFSAs.

Example 5.5.3 Bernoulli languages. These are generated by finite automata of a very reduced sort: only one accepting state, the initial state, which has outgoing arcs looping back to itself for each letter s_i of the alphabet, with the transition assigned log probability q_i. To keep the sum of assigned probabilities convergent, it is necessary only to add a sink state with a lambda-transition leading to it with any small positive probability z (for convenience, we will use $z = 0.5$). In a Bernoulli language, the log probability of a count vector $\mathbf{l} = (l_1, \ldots, l_t)$ will be given by the multinomial formula

$$\log \binom{l_1 + \ldots + l_t}{l_1, \ldots, l_t} + (\mathbf{q}, \mathbf{l}) + \log(1/2) \sum_{i=1}^{r} l_i \tag{5.27}$$

Here the multinomial coefficients again come from the multiple orders in which different strings with the same count vector could be obtained. If we are interested in probabilities assigned to strings (with a fixed order of letters), the first term of (5.27) disappears and only linear terms, namely the scalar product $(\mathbf{q}, \mathbf{l})$ of $\mathbf{q}$ and $\mathbf{l}$, and the length normalization term, remain.

Theorem 5.5.4 PCFGs and PCSGs over a one-letter alphabet do not generate all weighted one-letter languages.

Proof Define the weights as a set with infinite transcendence degree over $\mathbb{Q}$. (That such sets exist follows from the fundamental theorem of algebra and from the existence of irreducible polynomials of arbitrary degree. If $t_1, t_2, t_3, \ldots$ is such a set, so will be $s_1, s_2, s_3, \ldots$ where $s_i = |t_i|/(1 + |t_i|)2^i$, and the latter will also sum to ≤ 1). Now consider the generating functions which are defined by taking the nonterminals as unknowns and the terminal a as a variable in the manner of Chomsky and Schützenberger (1963), except using the probabilities assigned to the rules as weights. For example the grammar $S \rightarrow SS|a$ of binary trees used in Example 5.5.2 yields the functional equation $S = pS^2 + qa$. In the general case, solving the set of equations for the generating function associated to the start symbol S is very hard, but over a one-letter alphabet the only variable introduced by CF or CS rules will of necessity commute with all probabilistically weighted polynomials in the same variable. Since all defining equations are polynomial, the output probabilities are algebraically dependent on the rule probability parameters. Since a CF or CS rule system with n rules can generate at most n algebraically independent values, it follows that $s_1, s_2, s_3, \ldots, s_{n+1}$ cannot all be obtained from the CFG or CSG in question. □

These results may leave the reader with some lingering dissatisfaction on two counts. First, what about Type 0 grammars? Second, there is still the Suppes objection: what does all of this have to do with transcendence degree? With Theorem 5.5.2 at hand, we are in a better position to answer these questions.

For Turing machines, an important reduction was presented in de Leeuw et al. (1956), showing that a Turing machine with access to a random number generator that produces 1s and 0s with some fixed probability p is equivalent to a standard TM without random components as long as p itself is computable. Any string of 0s and 1s can be uniquely mapped on a language over a one-letter alphabet: we set $a_i \in L$ if the ith digit was 1 and $a_i \notin L$ if it was 0. If the string was generated by a random number generator with a fixed probability p for 1s, the density of the associated language will be p. It is easy to program a Turing machine that outputs a string (or language) with no density (see Section 5.4.1 above), but it is impossible to program a TM that outputs (or accepts) a language with noncomputable density. In other words, the key issue is not the randomness of the machinery but rather the complexity of the real numbers that express the probabilities.

At the bottom of the complexity hierarchy, we find the rationals: every rational number between zero and 1 (and only these) can be the density of an unweighted language over a one-letter alphabet generated/accepted by some FSA. Adding weights $p_1, \ldots, p_k$ to an FS grammar or automaton accomplishes very little since we will of necessity remain in their rational closure. However, real numbers are powerful carriers of information: as a moment of thought will show, all algorithms can be encoded in a single real number. When we add just one noncomputable p to the weight structure, we have already stepped out of the TM domain.

At the next level of the hierarchy, we find the algebraic numbers. These correspond to polynomials and thus, via generating functions, to context-free grammars and languages. While the relationship is not very visible over the one-letter alphabet because CFLs have letter-equivalent FSLs, and one needs to consider the structure to see the difference, algebraic numbers (and only these) arise naturally at every stage of the analysis of unweighted CFGs – for the weighted case, the same caveat about introducing arbitrary reals applies.

As for Suppes' objection, all algebraic numbers, and some transcendental ones, are computable. By Liouville's theorem, algebraic numbers of degree n can only be approximated by rationals p/q to order q^{-n} – if a number can be approximated at a faster rate, it must be transcendental. (Of course, it may still be transcendental even if it has no faster approximation.) Arguments based on transcendence can be thought of as convenient shorthand for arguments based on order of growth. The analogy is loose, and we shall not endeavor to make it more precise here, but Theorem 5.5.2 is a clear instance of replacing the heavy machinery of transcendence by a more pedestrian reckoning of growth (in our case, exponential decay).

Exercise 5.9* Prove Theorem 5.5.4 by direct appeal to order of growth.

5.5.2 Hidden Markov models

A striking consequence of Zipf's law (see Section 4.4) is that in the same syntactic category we can find words of radically different empirical frequencies. Indeed, by Corollary 4.4.1, vocabulary is infinite, and as there are only finitely many strict lexical categories (see Section 5.1.1), by the pigeonhole principle there will be at least one category with infinitely many words, and in such categories, arbitrarily large differences in the log frequencies of the high- and the low-frequency items must be present. For example, frequent adjectives such as *red* occur over a million times more often than rarer ones such as *dandruffy* in the gigaword corpora in common use today, and as the corpus size grows we will find even larger discrepancies.

Here we begin introducing hidden Markov models (HMMs) by using the POS tagging problem as our example. Formally, a **discrete hidden Markov model** is composed of a finite set S of *hidden states* (in the example, there is one state for each POS tag), a *transition model* that assigns a probability t_{ij} to the transition from state s_i to state s_j and an *emission model* that assigns, for each state s and for each *output symbol* **w**, a probability $E_s(w)$. A **continuous** HMM does not have a discrete set of output symbols, but rather a continuous set of (multidimensional) output values o, and the emission model is simply a continuous probability distribution $E_s(o)$ for each hidden state s (see also Section 8.2). The *hidden* aspect of the model comes from the fact that the same output, e.g. the word *run*, can be emitted from more than one state. In our example, there is a significant nonzero probability that s_{verb} will emit it but also a (smaller, but not zero) probability that it is emitted by the s_{noun} state. In other words, from seeing the output *run* we cannot determine which state the model is in; this information is (at least to some extent) hidden.

Since there are only finitely many states (in the simple POS tagging example considered here, generally about 200), even the smallest nonzero transition probability is expected to be rather large, on the order of 10^{-3}–10^{-4}. Smaller numbers can be used of course, but in any given model there are only finitely many transitions, so there will be a smallest nonzero transition probability. This is in sharp contrast to emission probabilities, where an infinite set of symbols associated to a state will of necessity give rise to arbitrarily small nonzero values. In POS models, emission values going down to 10^{-9} are quite common, and in the continuous case (which plays a key role in speech and handwriting recognition; see Chapters 8 and 9), emission probabilities routinely underflow the 64 bit floating-point range (which bottoms out at around 10^{-300}) so that log probabilities are used instead. Altogether, it is not just the obvious Markovian aspect that makes HMMs so suitable for language modeling tasks, but the clear segregation of the transition and emission models also plays a critical role.

Although typically used for recognition tasks, HMMs are best viewed as generation devices. In a single *run*, the model starts at a designated state s_0 and, governed by the transition probabilities t_{ij}, moves through a set of hidden states $s_{i_1}, s_{i_2}, \ldots, s_{i_n}$, emitting an output symbol $w_1, w_2, \ldots, w_r$ at each state with probability governed by $E_{s_{i_j}}$. We say that the probability the model assigns to the output string $w_1 w_2 \ldots w_r$ in a given run is the product of the transition and emission probabilities over the run and that the probability assigned to the string *by the model* is the sum of the probabilities assigned in all runs that output $w_1 w_2 \ldots w_r$.

HMMs are, at least in the discrete case, clearly finitistic devices that assign probabilities to strings. Yet they differ from WFSAs and WFSTs in some ways even if we ignore the possibility of emitting an infinite variety of symbols. First, WFSAs and WFSTs emit on transitions, while HMMs emit at states. As with Mealy and Moore machines, this distinction carries no theoretical significance but implies significant practical differences in the design of software libraries. Second, older definitions of HMMs may use an *initial probability distribution* to decide in which state the model starts in a given run. Here, as in the case of WFSTs above, we use a designated (silent) start state instead – this again entails no loss of generality. The main novelty that HMMs bring is thus the segregation of emissions from transitions: as we shall see, this can be thought of as a form of *parameter tying*.

While in theory it would be possible to abandon the hidden aspect of HMMs and replace each (state, emission) pair by a dedicated state with only one (weighted but deterministic) nonsilent emission possibility, this would complicate the model enormously. For an output inventory of n symbols and a state space of k states, we would now need nk states, with $(nk)^2$ parameters (transition probabilities) to learn, as opposed to the $k^2 + nk$ required by the HMM. Since k is a fixed (and rather small) number, as n grows to realistic values, the system would very soon leave the realm of practicality. Much of the early criticism leveled at Markov models (Miller and Chomsky 1963) grew out of this issue. The surprising economy provided by collecting all emissions from a given underlying state was discovered only later, as attention moved from ordinary 'open' Markov chains to HMMs.

Hiding the states from direct inspection was a move very similar to the one taken by Chomsky (1957), who introduced *underlying* structure different from the observable *surface* structure and used the former to characterize the latter. In the simple POS tagging model taken as our driving example here, the mental state of the speaker is characterized as a succession of POS tags, and the observables are the words. In a speech recognition model, the underlying states are the phonemes, and the surface forms are acoustic signals. To recover the underlying states, we need to perform analysis by synthesis, i.e. to compute the most likely state sequence that could have emitted the words in the observed sequence. This is done by the *Viterbi algorithm*, which is based on a data structure called a trellis: a matrix M with k rows and as many columns as there are items in the surface string.

The central idea is that to keep track of the sum of the probabilities generated by every possible run, it is best to group together those runs that end in the same state.

To compute the probability of a string $w_1 w_2 \dots w_r$ in a given run requires $2r$ product operations, and as there are k^r possible runs to consider, a naive algorithm would use $2rk^r$ multiplications and k^r additions. Using the trellis, we first populate the first column of M according to the initial transition probabilities times the emission probabilities from each state – this takes $2k$ product operations. Once column i has been populated, the jth element of column $i+1$ is constructed as $\sum_{l=1}^{k} t_{lj} M_{li} E_j(w_{i+1})$, which requires $2k$ product and k sum operations per entry, for a total of $2k^2$ products and k^2 sums per column. Over a run of length r, this is still only $2rk^2$ products and rk^2 sums; i.e. a number of operations *linear* in r for any fixed k. Once the trellis is filled, we simply pick the maxima in each column to obtain our estimate of the maximum likelihood state sequence. The number of compare operations required is again linear in r.

A key aspect of hidden Markov modeling is that the HMMs are *trainable*: rather then setting the transition and the emission model parameters manually, given a set of *truthed* data where both the underlying state and the outputs are known, there exist highly efficient algorithms to set the model parameters so that the posterior probability of the truthed observations is maximized. We return to this matter in Chapter 8 – here we conclude by proving our Theorem 5.4.5 that weighted languages generated by HMMs have zero density.

Proof We use the same trellis structure as above but simplify the calculation by exploiting the fact that for each state both the emission probabilities and the outgoing transition probabilities sum to one. Starting with the 0th state and summing over all possible outputs $\mathbf{w}$, we see that $\sum_{i=1}^{k} \sum_{j=1}^{n} P(w_j|s_i) t_{0i} = \sum_{i=1}^{k} t_{0i} = 1$; i.e. that the total weight assigned to strings of length 1 by the HMM is 1 (or less if silent states are permitted). Similarly, in filling out the second column, by keeping the first output w_1 fixed, the probability mass assigned to all strings of length two that begin with w_1 is at most 1, and therefore the total weight of strings of length two is again at most 1. This means that the coefficients r_i of the generating function $d(z)$ are all ≤ 1, and therefore $d(z) \leq 1/(1-z)$. By (5.20), we have $\rho(L) = \lim_{z \to 1}(1-z)d(z/n) \leq \lim_{z \to 1}(1-z)/(1-(z/n)) = (1-1)/(1-(1/n)) = 0$.

Discussion The same result is obtained by noting the elementary fact that the HMM assigns a weight of at most r to strings of length at most r, and there are $n + n^2 + \dots + n^r$ such strings. This is not a bad result since natural languages, conceived as regular stringsets, also have density zero by Theorem 5.4.3.

Using the finer measures introduced for zero-density languages in Section 5.4.2 above, the saturation σ of the weighted language generated by an HMM is at most $1/n$, where n is the size of the output alphabet. Note that, for any corpus, the generating function is a polynomial, which has saturation 0 (polynomials converge on the whole plane). To model a corpus of size N, we need, by (4.7), about $n = cN^{1/B}$ words. As N tends to infinity, $1/n$ will tend to zero.

The Bernoulli density δ of HMM languages is ≤ 1 by the first step of the inductive proof of Theorem 5.4.5 given above. As for natural languages, the Bernoulli density is hard to estimate, but clearly not every sentence is a prefix, so we have $\delta(L) < 1$.

Finally, for combinatorial density κ in HMM languages, we have $\kappa \leq \zeta n$, where n is the size of the vocabulary (output alphabet), which again tends to 0 as n tends to infinity. Whether this is the right prediction for natural languages is hard to say. As long as we treat vocabulary size as finite (a rather dubious assumption in light of Corollary 4.4.1), combinatorial density will be positive since at least the words that are suitable for isolated utterances such as *Ouch!* or *Yes* will contribute to κ. Since the Zipf constant B is close to 1, the contribution of two-word or longer utterances to κ will vanish as we increase the corpus size, but it is possible that the probability of one-word utterances remains above some positive lower bound. We are far from being able to settle the issue: the total probability mass of one-word utterances is clearly below 0.001, while the best language models we have today have a total of over- and underestimation errors on the order of 0.2. To put the issue in perspective, note that for a fixed HMM with fixed vocabulary it is quite trivial to adjust the probabilities of emitting a sentence boundary so that interjections and similar material are modeled as one-word sentences with a positive probability – in fact, we expect our models to automatically train to such values without making any manual adjustments toward this goal.

5.6 External evidence

Just as rational numbers provide a sufficient foundation for all numerical work in computer science, regular languages with rational weights are sufficient for all empirical work in linguistics (though in practice irrational numbers in the form of log probabilities are used quite often). Our interest in models of higher intrinsic complexity comes entirely from their capability to assign relevant structures to widely attested constructions, not from their improved weak generative capacity, which, at any rate, comes into play only on increasingly marginal examples. In the standard unweighted setup there are already reasons to prefer the rational models over the more complex ones, and these are summarized briefly here.

One key issue in selecting the right model is the amount of resources it takes to compute the structure it assigns. Without some special effort, even innocent-looking grammatical frameworks such as CG require time exponential in the size of input: as we discussed above, if the average ambiguity of a word is d (in practice $d \approx 1.5$), brute force algorithms require the inspection of d^n combinations to find all structures associated to a string of length n. This sets the outer limits of weak generative capacity at or below mild context-sensitivity. Such formalisms are by definition polynomially parsable (see Kracht 2003 Ch. 5 for further discussion). In particular, polynomial (typically n^6) parsing algorithms are known for all widely used grammar formalisms such as CCGs or TAGs. For the average case, requiring a polynomial bound is not very ambitious, inasmuch as humans obviously perform the task in real time on ordinary sentences, but we have little knowledge of how the human parser is implemented. Also, by carefully constructed *garden path* sentences such as

$$\text{Fat people eat accumulates} \tag{5.28}$$

it is quite possible to trip up the human parser, so a worst-case polynomial bound may still make sense.

Another key issue is learnability: it is clear that humans can learn any human language they are exposed to at an early age. Ideally, we are looking for a class of grammars such that the correct one can be selected based on exposure to positive examples only. This puts much more severe constraints on the weak generative capacity of the system. For example, context-free grammars cannot be learned this way (Gold 1967). The best lower bound on this problem is perhaps Kanazawa's (1996) result that CGs in which the degree of ambiguity for preterminals is limited to some fixed k are still identifiable in the limit from positive data. For probabilistic CFGs, the situation is better: algorithms that converge to a probabilistic CFL are widely used. Since language learning takes several years in humans, the known complexity bounds (e.g. the fact that learning regular expressions is NP-hard; see Angluin 1980, 1982) are harder to interpret as imposing external constraints on the class of human grammars.

The structuralist program of defining a *discovery procedure* whereby the grammarian (as opposed to the child learning the language) can systematically extract a grammar by selectively testing for distributional similarities has been at least partially realized in the work of Clark and Eyraud (2005), who prove that it is possible to discover CFGs for those CFLs that are **substitutable** in the sense that positive context sharing between any two γ and δ (the existence of $\alpha\gamma\beta, \alpha\delta\beta \in L$) implies full distributional equivalence (5.5). To the extent that we have already seen examples such as *slept* and *opened Parliament* that share a positive context without being fully distributionally equivalent (cf. **a bed rarely opened Parliament in*), one may rush to the conclusion that the Kanazawa result cited above is more readily applicable to natural language, especially as we have not seen examples of arbitrarily ambiguous words. But in modern versions of CG arbitrary **type lifting** (automatic assignment of higher types) is often present, and a realistic theory of *grammar induction* remains one of the central unsolved problems of mathematical linguistics.

Part of the interest in the regular domain comes from the fact that there grammar induction is feasible, and part comes from the fact that once we add memory limitations, theories with richer structural descriptions such as CFGs or LTAGs also reduce to finite automata. As we shall see in Section 6.3.2, memory limitations, whether imposed externally on a richer theory or inherent in the system as assumed here, can be used in a direct fashion to characterize the syntactic congruence by means of investigating the number of words that are required to complete a partial sentence. One striking observation is that natural language syntax seems to be enjoying the following **noncounting property** Q_4.

$$ax^4b \in L \Leftrightarrow ax^5b \in L \tag{5.29}$$

In general, we say that if a language satisfies $ax^kb \in L \Leftrightarrow ax^{k+1}b \in L$ it is noncounting with *threshold* k (has property Q_k). Obviously the complement of a noncounting language is also noncounting, with the same threshold. In (5.29) the number 4 cannot be further reduced because of agreement phenomena. Some languages, such as Gunwinggu [GUP], distinguish singular, dual, trial, and plural, so x^3

and x^4 may require different a or b. However, to make the performance limitations truly felt, it can be increased: we could say e.g. that no construction distinguishes between x^{13} and x^{14}. As an example, consider expressions such as *missile, anti missile missile, anti anti missile missile missile* etc. According to Carden (1983), these provide a simple way to demonstrate that word formation is not finite state: to get from the n-degree weapon from $n-1$, we need to prefix *anti* and suffix *missile*, which would yield the language $anti^n$ $missile^{n+1}$. But in reality, the enterprise falls apart at around three: people will accept *anti anti missile, anti anti missile missile*, and other forms about as well as those where the number or *anti*s and *missile*s around the first *missile* is properly balanced.

Observation (5.29) has far-reaching consequences for the structure of the formal apparatus. For an n element set Σ, the set of $\Sigma \rightarrow \Sigma$ functions is a monoid, with the role of multiplication played by function composition and the role of identity by the identity function. Subsets of this set are called **transformation semigroups**, **transformation monoids**, and **transformation groups**, provided they are closed under multiplication, contain the identity, and are also closed under inverse, respectively. A classic theorem of Cayley asserts that any group is isomorphic to a transformation group. The proof extends to monoids and semigroups as well. Given an arbitrary semigroup (monoid) S, we can take its members to be the set to be transformed and define the transformation effected by a given semigroup (monoid) element s (also called the **action** of s on S) as the result of multiplying (from the right) by s. Unlike in groups, where each transformation is a permutation of the elements, in semigroups multiplying two different elements r and r' by s may lead to the same result $rs = r's$. The number of elements in the image of S under the transformation is called the **rank** of the transformation – this equals $|S|$ only if the transformation is injective.

The importance of transformation semigroups in linguistics comes from the fact that each element of the alphabet Σ induces a transformation on the syntactic congruence. Since automata states are in one-to-one relation with classes of the right congruence, it is natural to identify the semigroup generators with the transformations of the state space of the automaton induced by the letters of the alphabet. If all elements of a semigroup S except the identity transform the base set into a singleton set, S is called a **reset semigroup (monoid)**. To see the utility of this notion, recall that every word (more precisely, the lexical category of the word) is a member of the syntactic monoid of the language and acts as a transformation on it. By picking several words at random, one is extremely likely to introduce unrecoverable ungrammaticality, i.e. the combined action of these words is to reset the monoid to a sink state. This is not to say that every word (lexical category) performs a reset, but they all narrow down what can come next.

The fundamental theorem of semigroup decomposition (Krohn and Rhodes 1965) asserts that every semigroup is a divisor of some semigroup that can be built up as a *wreath product* of those semigroups in which every element has extreme low rank (reset semigroups) with those where every element has extreme high rank, namely groups. Fortunately, we do not need to consider Krohn-Rhodes theory in its full generality because Schützenberger (1965) proved that the transformation semigroup associated to a language is **group-free** (has no nontrivial subgroups) if and only if the language is noncounting. Combining these results gives the following.

Theorem 5.6.1 (McNaughton and Papert 1968) The semiautomaton associated with a noncounting language is a divisor of a cascade product of reset monoids U_2.

Discussion Cascade products $A \circ B$, defined for semiautomata, and the analogous wreath products $A \wr B$, which are defined directly on semigroups, are somewhat difficult to grasp because the role of the two terms A and B is asymmetrical: once A is given, B must take a certain form for their cascade or wreath product to be defined. This is unusual, but not unheard of. For example, if A were an n row by k column matrix, the only Bs that could participate in a matrix product AB would be those with k rows. Specifically, if A has state space Q_A and input alphabet Σ, B must have state space Q_B and input alphabet $Q_A \times \Sigma$ – the construction is called a *cascade* because the higher automaton A can transmit its state as part of the input to B, but the lower automaton B does not have the means to inform A of its own state.

One way of thinking about this construction is to reproduce the lower automaton in as many copies as there are states in Q_A and associate one copy to each state of Q_A. During this process, the state space Q_B is preserved exactly, but the transition function of the copies is made dependent on the state to which they are associated. We can take the A to be the prime minister and the Q_B clones as the cabinet. In each state, the prime minister listens only to the cabinet member associated to that state: the higher automaton Q_A takes transition according to its own transition function, and the Q_B copies must all fall in line and move to the same state that was dictated by the transition function of the copy that was distinguished before the transition. More formally, we have the following definition.

Definition 5.6.1 Given two semiautomata A and B with state spaces Q_A and Q_B, input alphabets Σ and $Q_A \times \Sigma$, and transition functions $\delta_A : Q_A \times \Sigma \rightarrow Q_A$ and $\delta_B : Q_B \times Q_A \times \Sigma \rightarrow Q_B$, their **cascade product** $A \circ B$ has as its state space the direct product $Q_A \times Q_B$ and has the input alphabet Σ as its input alphabet. From the product state (q_1, q_2), upon input of $\sigma \in \Sigma$ the cascade product automaton will transition to $(\delta_A(q_1, \sigma), \delta_B(q_2, q_1, \sigma))$. By definition, cascade products $A_1 \circ A_2 \circ \ldots \circ A_n$ are associating $((\ldots(A_1 \circ A_2) \circ A_3) \ldots \circ A_n)$.

The cardinal building block used in Theorem 5.6.1 is the reset monoid U_2, which acts on two elements u and v as a transformation semigroup and has only three elements: the transformation C that maps everything to u (i.e. the constant function whose only value is u), the transformation D that maps everything to v, and the identity transformation I. If we take u and v to be the open and closed parens a and b of the bounded counter of depth 2, it is easy to see that the semigroup of the language associated to this automaton is exactly U_2.

Exercise 5.10 Consider the k-term cascade product $U_2^k = U_2 \circ \ldots \circ U_2$ over a two-letter alphabet $\Sigma = \{a, b\}$. What is the largest n for which U_2^k is a cover for the bounded counter of depth n?

From the linguistic standpoint, this is closest to a single binary feature, which can take the value '+', the value '–', or be left *underspecified*. Note that the fourth possible action P, mapping u on v and v on u, is missing from U_2. Were we to add this (a mapping that would correspond to 'negating' the value of a binary feature), we would still not obtain a group, since the constant functions C and D have no inverse, but we would destroy the group-free property since P is its own inverse and the subset $\{I, P\}$ would be a group. In other words, Theorem 5.6.1 asserts that the syntax of those regular languages that enjoy the noncounting property (5.29) can be handled by relying on binary features alone, a result that goes some way toward explaining the ubiquity of these features in linguistics.

To see this in more detail, we will first consider a toy language T based on the example *Which books did your friends say your parents thought your neighbor complained were/*was too expensive?* The transformational description of sentences like this involves an underlying structure *Your friends say ... the books were too expensive* from which the NP *the books* gets moved to the front, so that the agreement in number between *books ... were expensive* and *book ... was expensive* requires no further stipulation beyond the known rule of subject-predicate agreement that obtains already in the simple sentences *The book was expensive* and *The books were expensive*. This is easily handled by extending rule (5.7) with an agreement feature:

$$\mathrm{S} \rightarrow \mathrm{NP}\langle\alpha\mathrm{PL}\rangle\ \mathrm{VP}\langle\alpha\mathrm{PL}\rangle \tag{5.30}$$

where α is a variable that can take the values '+' or '−'. In other words, we replace (5.7) with a rule schema containing two rules: the assumption (further discussed in Section 7.3) is that such schemas are no more expensive in terms of the simplicity measure of grammars than the original (5.7), unadorned with agreement features, would be. Here we assume similarly abbreviated regular expressions: our language T is defined as the union of T_{sg} given by *which* N⟨−PL⟩ *did* Z* VP⟨−PL⟩ and T_{pl} given by *which* N⟨+PL⟩ *did* Z* VP⟨+PL⟩. Here Z stands for sentences missing an object: *your friends say* _, *your parents thought* _, *your neighbor complained* _, and the Kleene * operator means any number of these can intervene between the extracted subject *book(s)* and the predicate *was/were*. The regular expression *which* N *did* Z* VP, or what is the same, the automaton

$$0 \xrightarrow{\textit{which}} 1 \xrightarrow{\mathrm{N}} 2 \xrightarrow{\textit{did}} 3 \xrightarrow{\mathrm{VP}} 4 \quad (\text{loop } \mathrm{Z} \text{ on } 2) \tag{5.31}$$

will accept the desired language *without* proper agreement. To get the agreement right, we take the cascade product with the depth 2 bounded counter whose transition is defined for all clones as the identity mapping for all states and all symbols, except N⟨+PL⟩ will move clone 1 from the initial (accepting) state to the other (warning) state, and for clone 3 the input VP⟨+PL⟩ will force a move back to the initial state.

What this simple treatment of T shows is that cascade multiplication with a small reset automaton is sufficient to keep one bit of information around even though an arbitrary amount of material may intervene before we make use of this bit. If we measure the complexity of the automaton by the number of states it has, the cascade product is very expensive (multiplying with U_2 will double the size of the state space), but by the same general principle that treats (5.7) and (5.30) as having equal or nearly equal cost, a more proper measure of automaton complexity is the number of states in its cascade decomposition, which increases only linearly in the number of bits kept around.

5.7 Further reading

Distributional criteria for lexical categorization were advocated by Bloomfield (1926, 1933) and elaborated further by Bloch and Trager (1942) and Harris (1951). Definition (1) is known as the *Myhill-Nerode equivalence* or the **syntactic congruence** associated with a language L as it was Myhill (1957) and Nerode (1958) who proved, independently of one another, the key theorem that this equivalence, extended to Σ^*, has only finitely many classes iff L is regular. For a modern survey, see Pin (1997). For the use of noncounting in practical grammar design, see Yli-Jyrä (2003,2005) and Yli-Jyrä and Koskenniemi (2006).

The classical works on categorial grammars are Ajdukiewicz (1935), where the group is taken to be Abelian, Bar-Hillel (1953), and Lambek (1958). Some of the interest in the area was lost when Bar-Hillel, Gaifman, and Shamir (1960) proved the equivalence of CFGs to one form of categorial grammar, but the field has largely revived owing to its strong relation to semantics. The equivalence between CFGs and Lambek grammars has been proven by Pentus (1997) – a more accessible proof is given in Chapter 3 of Kracht (2003), which presents categorial grammars in greater depth.

Although X-bar theory clearly originates with Harris (1951), the name itself is from Chomsky (1970) and subsequent work, especially Jackendoff (1977). For a more detailed discussion, see Kornai and Pullum (1990). A good survey of the linguistic motivation for going beyond the CF domain is Baltin and Kroch (1989). The equivalence, both weak and strong, between alternative formulations of extended phrase structure and categorial systems has been proven in a series of papers by Joshi and his students, of which we single out here Vijay-Shanker et al. (1987) and Weir (1992).

Moravcsik and Wirth (1980) presents the analysis of (5.8) and (5.9) in a variety of syntactic frameworks. Some of these frameworks are still in use in essentially unchanged format; in particular *tagmemics*, being the standard theory at the Summer Institute of Linguistics is in wide use, and covers an immense variety of languages, many of which have no analyses in any other framework. Other frameworks, such as *Montague grammar*, *relational grammar*, and *role and reference grammar*, have changed mildly, but the reader interested in a quick overview can still get the basic facts from the respective articles in the Moravcsik-Wirth volume. Yet others, in particular mainstream transformational theory, have changed so radically that the presentation of *trace theory* in that volume is of historical interest only. The reader interested in more current developments should consult Chomsky (1995) and subsequent work.

The first proposal within generative linguistics to use case as the driving mechanism for syntax was by Fillmore (1968, 1977). The standard introduction to ergativity isDixon (1994). Dependency grammars are due to Tesnière (1959). Other modern formulations include Sgall et al. (1986), Mel'cuk (1988), Hudson (1990), and McCord (1990). Again, an important early result that served to deflect interest from the area was by Gaifman (1965), who proved the equivalence of one formulation of dependency grammar to CFGs, and again the area revived largely because of its strong connections to key notions of grammar. A variety of dependency and valency models are surveyed in Somers (1987) and Tapanainen and Järvinen (1997). A modern survey of argument linking theories is Levin and Rappaport (2005).

For linearization of Hayes-Gaifman-style DGs, see Yli-Jyrä (2005) and for linking a different formulation of DGs to mildly context sensitive grammars and languages see Kuhlmann (2005). The MIL formalization of DG is from Kornai (1987), except there it was given as an equational system in the sense of Curry and Feys (1958 Ch. 1E) rather than stated in the more widely used language of universal algebra. The ID/LP analysis of the Dutch crossed dependency is due to Ojeda (1988). The valence (clause) reduction analysis of *hate to smoke* and similar constructions originates with Aissen and Perlmutter (1983); see also Dowty (1985) and Jacobson (1990).

The method of mapping linguistic expressions to algebraic structures in order to capture some of their significant properties is not at all restricted to the simple examples of checking agreement or slot/filler relations discussed here. The modern theory of using types as a means of checking correctness begins with Henkin (1950). In *type-logical grammar* we can use proofnets to check the syntactic correctness of strings based on their types (Carpenter and Morrill 2005). Another theory of note, *pregroup grammar*, uses left and right adjoints to avoid the issues of group elements commuting with their inverses. Lambek (2004) defines pregroups as partially ordered monoids where the partial order is compatible with the monoid multiplication, and multiplication with every element has a left and a right adjoint. His example is the set of $\mathbb{Z} \rightarrow \mathbb{Z}$ functions that are unbounded in both directions, with function composition as the monoid operation. The relationship of pregroups to categorial grammar and (bi)linear logic is discussed in Casadio et al. (2005).

In Minsky's (1975) theory, the slots are endowed with ranges (used for error checking) and often with explicit algorithms that compute the slot value or update other values on an as-needed basis, but there are no clear natural language phenomena that would serve to motivate them. Altogether, the relationship between AI/KR and syntax is far more tenuous today than it was in the 1970s: on the one hand, AI/KR has largely given up on syntax as too hard, and on the other, its center of gravity has moved to machine learning, a field more immediately concerned with linguistic pattern recognition in speech, handwriting, and machine print recognition (see Chapters 8 and 9) than in syntax proper. For default inheritance and defeasible reasoning, see Ginsberg (1986a). The best introduction to thematic roles remains Dowty (1991).

The complex and often acrimonious discussion that followed Chomsky's (1957) introduction of arbitrary depth center-embedded sentences led to important methodological advances, in particular the introduction of the distinction between *competence* and *performance* (Chomsky 1965) discussed in Section 3.2. For the debate surrounding generative semantics, see Newmeyer (1980), Harris (1995), and Huck and Goldsmith (1995).

For the linguistic use of arbitrary semirings for weighted languages, automata, and transducers see Mohri et al (1996), Eisner (2001). The interpretation of the weights as degrees of grammaticality is due to Chomsky (1967). The density of languages was first discussed in Berstel (1973), who considered pairs of languages L and mappings f from Σ^* to $\mathbb{R}^+$. He also considered the ratio of the summed weights (summed for strings of length $\leq n$ in L in the numerator and for all strings in Σ^n in the denominator), while our definition (see Kornai 1998) uses the differences, segregated by sign. We believe that our choice better reflects the practice of computational language modeling, since in these models both underestimation and overestimation errors are present, and their overall effects are seldom determined in the limit (incorrect estimates for high-frequency strings are far more important than incorrect estimates for low-frequency strings). For natural numbers represented in binary, Minsky and Papert (1966) use density arguments to prove that e.g. the primes in base two are not regular; see also Cobham (1969) and Chapter 5 of Eilenberg (1974).

The early work on probabilistic FS and CF grammars is summarized in Levelt (1974 Ch. 3). Outside sociolinguistics, probabilistic theories of grammar had little influence because Chomsky (1957) put forward the influential argument that *colorless green ideas sleep furiously* is grammatical and **furiously sleep ideas green colorless* is ungrammatical, yet both have frequency zero, so probability has no traction in the domain of grammar (for a modern assessment of this argumentation, see Pereira 2000). Within sociolinguistics, however, the variability of the data is so overwhelming that there was never a question of abandoning the probabilistic method. The standard introduction to variable rules is Cedergren and Sankoff (1974). The free choice between the additive and the multiplicative models has been cogently criticized by Kay and McDaniel (1979). Logistic models were introduced to sociolinguistics by Rousseau and Sankoff (1978). HMMs as formal models were introduced by Baum and Petrie (1966) and Baum et al. (1970), and first put to significant use in speech recognition by Baker, Jelinek, and their colleagues at IBM. The standard tutorial introduction is Rabiner (1989); for a more extensive and deeper treatment, see Jelinek (1997). For a detailed discussion of probabilistic CFGs, see Charniak (1993).

The use of formal power series in noncommuting variables to describe ambiguity in CFLs and the concomitant use of semirings originates with Schützenberger (1960). Early developments are summarized in Chomsky and Schützenberger (1963). For the rational case, see Eilenberg (1974), and for general development of the semiring-oriented approach, see Kuich and Salomaa (1986). For further discussion of Parikh mappings in the weighted case, see Petre (1999). For initial estimates of state space size, see Kornai (1985) and Kornai and Tuza (1992) – the estimate we shall derive in Section 6.3.2 also takes ambiguity into account.

For a simple proof of Theorem 5.6.1 see Section 7.12 of Ginzburg (1968), written by Albert Meyer – the same ideas are presented in Meyer (1969). The classical study of this area is McNaughton and Papert (1971), which discusses both the well-known equivalence of noncounting and star height zero and its relation to temporal logic, also discussed in Maler and Pnueli (1994). For a modern treatment, see Thomas (2003). The idea of a prime minister and a cabinet is based on Bergeron and Hamel (2004), where the important relation between bit-vector operations and noncounting languages is explored. For closure of noncounting languages under a relaxed alternative of Karttunen's (1995) replacement operator, see Yli-Jyrä and Koskenniemi (2006).

A significant source of external evidence, not discussed in the main text as there is no mathematical treatment in existence, is language pathology. For the pioneering effort in this direction, see Jakobson (1941). Chomsky (1965) has distinguished the internal combinatorical evidence from the broader external considerations discussed here under the heading of *descriptive* vs. *explanatory adequacy.* The pivotal study linking the extensive literature of child language acquisition to formal theories of inductive learning is Wexler and Culicover (1980); see also Cussens and Pulman (2001). We shall return to inductive inference in Chapter 7.

As discussed in Chapter 2, grammarians before de Saussure and structuralism did not hesitate to consider evidence from another source: dialectal and historical variation of the language. Under the influence of de Saussure, such evidence has largely fallen into disfavor, since it is clear that people can, and do, learn language without access to such data. The same can be said of evidence from translation: though many language learners benefit from learning two or more languages in early childhood, monolingual language acquisition is obviously possible. Evidence from translation is generally accepted in the form of *paraphrases*, a method we will discuss further in Chapter 6.

6

Semantics

To the mathematician encountering linguistic semantics for the first time, the whole area appears as a random collection of loosely connected philosophical puzzles, held together somewhat superficially by terminology and tools borrowed from logic. In Section 6.1 we will discuss some of the puzzles that played a significant role in the development of linguistic semantics from a narrow utilitarian perspective: suppose an appropriate technique of mathematical logic can be found to deal with the philosophical puzzle – how much does it help us in dealing with the relationship between grammatical expressions and their meaning? Since the task is to characterize this relationship, we must, at the very least, provide a theory capable of *(A) characterizing the set of expressions* and *(B) characterizing the set of meanings*. By inspecting the Liar (Section 6.1.1), opacity (Section 6.1.2), and the Berry paradox (Section 6.1.3), we will gradually arrive at a more refined set of desiderata, distinguishing those that we see as truly essential for semantics from those that are merely nice to have. These will be summarized in Section 6.1.4.

In Section 6.2 we describe the standard formal theory that meets the essential criteria. Our point of departure will be Montague grammar (MG) in Section 6.2.1, but instead of formalizing the semantics of a largely artificial and only superficially English-like fragment, we set ourselves the more ambitious goal of exploring the semantics of everyday language use, which tolerates contradictions to a surprising degree. In Section 6.2.2, we introduce a version of paraconsistent logic, Ginsberg's (1986) system D, and survey the main construction types of English from a semantic perspective.

Finally, in Section 6.3, we begin the development of a formal theory that departs from MG in many respects. Our main concern will not be with the use of paraconsistent/default logic (although we see this as inevitable) but rather with replacing Tarski-style induction on subformulas by a strict left-to-right method of building the semantic analysis. We take this step because induction on subformulas presumes tree structure, while in natural language syntax there are clear technical reasons to prefer FSTs, HMMs, and other finite state methods of syntax analysis that do not naturally endow strings with tree structure.

6.1 The explanatory burden of semantics

While contemporary mathematical logic is indistinguishable from other branches of mathematics as far as its methods or driving esthetics are concerned, historically it has grown out of philosophical logic and owes, to this day, a great deal to its original philosophical motivations, in particular to the desire to eliminate ambiguities and paradoxes from the system. From a linguistic perspective, these are questionable goals since natural language is often ambiguous and clearly capable of expressing paradoxes. What linguistics needs is not a perfect language free of all ambiguity and contradiction, but rather a meta-theory that is capable of capturing these characteristic properties (we hesitate to call them imperfections, as there is no evolutionary pressure to remove them) of the object of inquiry.

6.1.1 The Liar

Pride of place among the philosophical puzzles readily expressible in natural language must go to the *Liar*:

$$\text{This sentence is false} \tag{6.1}$$

Perhaps the simplest resolution of the paradox would be to claim that the sentence (6.1) is simply not grammatical. But given its structural similarity to sentences such as *This man is asleep*, this is not a very attractive claim. A more sophisticated version of the same argument would use some version of the generative semantics view introduced in Section 5.2 and claim that sentence generation starts with some state of affairs to be put into words, and since there is no conceivable state of affairs corresponding to (6.1), the sentence just never gets generated, let alone interpreted. Again, this is not very attractive in the light of the fact that in rather simple contexts (6.1) is perfectly meaningful:

> The very first sentence in Fisher's biography of Lincoln asserts "Abe Lincoln was born in a log cabin in Idaho". This sentence is false.

Perhaps one could claim that *This* in (6.1) must refer to something external to the sentence, but again there is little to support this: examples such as *This sentence is in German* or *This sentence takes 56 characters to print* are perfectly ordinary. Since we cannot really blame the syntax of (6.1), we need to turn to its semantics. What is *this* referring to? Clearly, it refers to the entire sentence S in (6.1). Therefore, we may conclude that (6.1) asserts

$$\text{S is false} \wedge S = [\text{This sentence is false}] \tag{6.2}$$

The classical solution, Russell's theory of types, is based on the observation that (6.1) and (6.2) are not the same sentence and not entirely the same assertion. It is possible to build up a large and useful logical calculus that includes something like a T-scheme (the axiom scheme connecting assertions about the truth of sentences to the actual truth of sentences – more on this later) while carefully banning only a few self-referential statements. Here we are interested in a more linguistic solution, one that does not do away with self-referentiality, since there is no evidence that language treats linguistic objects such as words or sentences in any way different from other abstract nouns. We state this as our requirement *(C): expressions that have similar form should receive analyses by the same apparatus.*

The central observation is that predicates such as *true* and *false* are context-dependent, and in analyzing (6.1) there is no reason to use an eternal absolute notion of truth and falsehood. Since even the most eternal-looking mathematical expressions such as $6 \cdot 6 = 36$ depend on implicit assumptions (e.g. that of using base 10), there is every reason to suppose that natural language expressions such as (6.1) do also. Let us call a set of assumptions (closed formulas in a logical language) a *context* and denote it by C. If a statement s is true in the context (true in all models that the set of formulas C holds true), we write $s \in T(C)$, if false, we write $s \in F(C)$ – it is possible for s to be true but be outside the deductive closure of C. Now (6.1) and (6.2) can be reformulated as

$$S = [S \in F(C)] \tag{6.3}$$

$$S \in C \wedge S = [S \in F(C)] \tag{6.4}$$

respectively. Of these, (6.4) is manifestly self-contradictory, which, in an ordinary system of two-valued logic, is sufficient cause to declare it false. Kripke (1975) develops a three-valued logic with a *middle* truth value that offers an escape, but we do not follow this development because the 'strengthened' version of the Liar, *This sentence is not true*, would cause the same problem to a three-valued treatment that (6.1) causes in a two-valued system. Rather, we follow the Russellian insight and for each context C define $\widehat{C}$ as the set of those statements that refer to $T(C)$ or $F(C)$. Formally, we adjoin a predicate T meaning 'is true' to the language and make it subject to axioms (called the *T-scheme* in Tarski 1956) such as $x \Leftrightarrow T(x)$. Now we can capture the essence of the 'naive' analysis (formalized by Prior 1958, 1961) that (6.1) is not just contradictory, but false:

Theorem 6.1 For every context C, (6.1) is false in $\widehat{C}$.

Proof Suppose indirectly $S \in T(\widehat{C})$. By the T-scheme, we have $S \in F(C)$, and therefore C is a context for S. This, by S, leads to $S \in T(C)$, a contradiction that proves that our indirect assumption was false.

Discussion This is not to say that the Liar is not a valuable paradox: to the contrary, in the history of logic, the Liar and closely related paradoxes such as the barber paradox have led to many notable advances in logic, including Russell's theory of types and Tarski's undefinability theorem. However, from a linguistic perspective many of the issues brought up by the Liar are already evident in much simpler nonparadoxical sentences that have no self-referential or quotational aspect whatsoever. First, and most important, is the treatment of *nonintersective modifiers*. At the very least, we need a theory that covers *enormous flea* (Parsons 1970) and *lazy evaluation* before we can move on to *false sentence.*[1]

Once such a theory is at hand (and as we shall see, this is by no means a trivial task), there is the further question of whether we really need a linguistic semantic account of *true* and *false.* English, as a formal language, contains many expressions with these words, but formal semantics generally refrains from the explication of ordinary nouns and adjectives like *flea* and *enormous*, preferring to leave such matters either to specialists in biology and psychology or to lexicographers and knowledge representation experts. There is no compelling reason why *sentence* and *true* should occupy a more central position in our development of semantics: we conclude that for the purposes of linguistic semantics an analysis of these notions is nice but inessential. On the other hand, since self referential statements are formed by ordinary grammatical means, it follows from *(C)* that we should cover them.

Quotational sentences are also problematic from our perspective. Our intonational means to segregate quoted from direct material are limited: a sentence such as *'Snow is white' is true if and only if snow is white* is practically impossible to render differently from *'Snow is white' is true if and only if 'snow is white'*. Since we are attempting a formal reconstruction, a general appeal to the *universality* of natural language, (i.e. to the notion that everything can be discussed in natural language) is insufficient – what we need is a precise mechanism whereby matters such as quotation can be discussed. The use of quotation marks in written language provides the beginnings of such a mechanism, but as far as grammar is concerned these are not very well regulated, especially when it comes to quotations inside quotations beyond depth two. And even without embedding, the matter is far from trivial: compare *A 'quine' is a program that, when compiled and executed, prints its own source code* to *A quine is a program that, when compiled and executed, prints its own source code*. We conclude that a theory of quotation is nice but inessential, especially as the phenomenon is more characteristic of artificially regulated written communication than of natural language.

6.1.2 Opacity

Another important puzzle is that of *opacity* (used here in a very different sense from that in phonology or syntax). It has long been noted that certain predicates P about statements s and t do not allow for substitution. Even if $s = t$ it does not follow that $P(s) = P(t)$. Frege (1879) used *know* as an example of such a predicate, but the observation remains true of many other predicates concerning knowledge, (rational) belief, hope, preference, (dis)approval, etc., generally collected under the heading of *propositional attitudes*. Frege used as his example the knowledge state of ancient Greeks, who had not realized prior to Pythagoras that Hesperus (the evening star),

[1] What is common to the examples is that enormous fleas are not enormous things that are also fleas (hence the name nonintersective), lazy evaluation is not some lazy thing that happens to be an evaluation, and true sentences are not true things that happen to be sentences.

and Phosphorus (the morning star) were one and the same object, what we call Venus today. Thus, for an ancient Greek it was perfectly possible to know/believe that a bright object in the evening sky is Hesperus without knowing/believing that it is Phosphorus. In other words, *Believe(s)* does not follow from *Believe(t)* even though $s = t$.

In the tradition of philosophical logic, opacity is intimately linked to *a priori*, *analytic*, and *necessary* statements, and for those whose goal is to develop rational theories of beliefs, propositional attitudes, and judgments in general, such considerations are valuable. But from our perspective, the issue is nearly trivial: there is no denying that (i) sentences about beliefs, etc., are about the mental states of people and that (ii) mental models need not correspond to facts. The ancient Greeks had a model of the world in which there were two different entities where in reality there was only one. In other cases people, or entire civilizations, assume the existence of entities that do not correspond to anything in reality (the Western philosophical tradition generally exemplifies this by unicorns) or conversely use models that have literally no words for important entities. But imperfect knowledge and mistaken beliefs are part of the human condition: there is no logical error, and what is more important here, there is no *grammatical* error in thinking or saying things that are not so. Again, from a linguistic perspective the issues brought up by opacity are evident in much simpler cases that do not probe the edges of logical, philosophical, or physical necessity: ordinary fiction will suffice. What do we mean when we say *Anna Karenina is afraid of getting older*? How does it differ from *Jean Valjean is afraid of getting older*? Most people will agree that *Anna Karenina commits suicide* is true and *Jean Valjean commits suicide* is false. It is a highly nontrivial task to design a theory that provides the expected results in such cases, and although the desideratum follows from *(C)* it is worth stating it separately: *(D) expressions occurring in fiction should receive analysis by the same apparatus as expressions used in nonfiction contexts.*

6.1.3 The Berry paradox

In Section 1.5, we already alluded to the *Berry paradox* of finding

$$\text{the smallest integer not definable in } k \text{ words} \qquad (6.5)$$

For $k \leq 7$ the paradox is not apparent: we enumerate definitions of integers e.g. in lexicographic order, rearrange the list in increasing order of the numbers defined, and pick the first integer not on the list, say s_k. But for $k = 8$ there is a problem: (6.5) itself is on the list, so s_8 is defined by it, and we must pick $s_8 + 1$ (or, if it was defined elsewhere on the list, the next-smallest gap). But then, s_8 was not on the list, or was it? What makes this question particularly attractive is that desiderata *(A)* and *(B)* appear to be easily met: it looks trivial to define the syntax of number names and, for once, we have a well-agreed-upon semantic model of the domain. There are some core elements *one, two, ..., nine* that combine with other base ten core elements such as *ten, twenty, ..., ninety* and *hundred, thousand, million, billion, trillion, quadrillion, ...* all contributing to the comma-separated reading style

(CSRS) whereby $450, 789, 123, 450, 123$ is read off as *four hundred and fifty trillion seven hundred and eighty nine billion one hundred and twenty three million four hundred and fifty thousand one hundred and twenty three*. There is something of a performance issue in where to terminate the sequence: clearly, numbers such as *septendecillion* or *nonillion* only serve as party amusements, as the people who actually would need them will in real life switch over to scientific reading style (SRS) somewhere around 10^9. Once we have made a choice where to terminate the sequence, it is trivial to define a small finite automaton generating all and only the CSRS numbers and another automaton generating SRS numbers such as *three point two seven five times ten to the minus nineteenth*.

Exercise 6.1 Write a detailed grammar of the CSRS and SRS systems. Make sure that articles are correctly generated both in the conventional listing of numbers *one, two, . . ., ninety-nine, one hundred, one/a hundred and one, one/a hundred and two, . . ., two hundred, two hundred and one, . . .* and before round numbers. Define the appropriate notion of roundness. Consider grammatical responses to questions such as *How many students are in this school?*, explain why **hundred* is ungrammatical and *(exactly) one/a hundred* are acceptable.

Yet no sooner than we start specifying the syntax beyond these trivial fragments we run into difficulties. First, there is no one-to-one relationship between numbers and expressions composed entirely of number names: *seven thirty* generally refers to a time of the day, *nine eleven* to emergency services, and *forty fifty* or *two three thousand* to approximate quantities given by a confidence interval. This last class is so similar to CSRS that on occasion the lines between the two are blurred: *ninety nine hundred* can refer both to 9,900 and to 99–100. To be sure, in the written language hyphens are used with approximate quantities, but spoken English (which for the linguist has methodological primacy) shows no sign of this. When viewed from the perspective of spoken language, the whole enterprise of assigning semantics to CSRS and SRS is of dubious character: such numbers form highly stylized and regulated formal languages whose regulatory principles appear to be arithmetical, rather than linguistic, in nature. Arithmetic is not acquired with the same effortless ease as a first language. For most people the process starts later, and for many it remains incomplete.

To get the Berry paradox really working we also need our syntax to cover arithmetic expressions (*eight cubed* is a number expressible in two words instead of the four required by CSRS), other bases (*hex a a b nine* only requires five words), the ability to solve equations (*the smallest Mersenne prime with an eight digit exponent*) and in general to run algorithms (*the result of running md5sum on the empty file*). Unfortunately, as a technical device for the description of algorithms yielding a numerical output, natural language is spectacularly inadequate, requiring a variety of support devices such as end markers, quotation marks, and parentheses, which have little or no use in ordinary language, a fact that is painfully evident to anyone who has ever tried to debug source code over the phone. This makes writing a grammar that meets *(A)* a complex matter, as there is no easy test of whether an expression evaluates to a numerical value or not.

At the same time, when it comes to expressions with a numerical value, natural language overgenerates in two significant ways. First, there are a class of statements that become true just by means of saying so, and such statements can be used to define numbers. To illustrate the more general phenomenon, consider whether a statement such as *I promise to come tomorrow* can be false. I can break the promise (by not coming), but merely by saying *I promise* I have actually promised that I will come. There is a whole set of verbs such as *declare*, *order*, *request*, *warn*, and *apologize* that are similarly self-fulfilling – following Austin (1962), these are known as *performatives*. The only near-performative in mathematical usage is *let*: when we say *Let T be a right triangle*, *T* is indeed a right triangle. Compared with the performatives in everyday language, the power of *let* is rather limited. In particular, by saying things such as *let x be a prime between 8 and 10*, we have not succeeded in creating such a prime, while by saying *I name this ship Aoxamoxoa* one can indeed create a ship by that name. But in the subdomain of naming numbers, the full power of *let* is at our disposal: there are monomorphemic elements such as *eleven*, *twelve*, *dozen* in broad use, and others such as *gross*, *score*, *chiliad*, and *crore* known only to chiliads or perhaps crores of people. As examples such as *googolplex* show, all that is required for a successful act of naming is that the language community agrees that a word means a certain thing and not something else. The word *factoriplex* does not exist in current English but is easily defined as the product of all integers from 1 to googolplex, minus 42. Performatives trivialize the Berry paradox (there is no number meeting (6.5) since every number is nameable in one word) but yield another strong desideratum: *(E) the system must be flexible about naming.*

Second, natural language offers a wealth of algorithmic descriptions of numbers that are solvable only by encyclopedic knowledge. Clearly, *the year Columbus discovered America* evaluates to 1492, and equally clearly, knowing this is beyond the power of linguistic semantics. As far as the expression is concerned, it could refer to any year within Columbus' adult life, and no amount of understanding English will suffice to better pin down the exact date. But if we are reluctant to impute knowledge of medieval history to speakers of English, we should be equally reluctant to impute knowledge of Mersenne primes, or even knowledge of eight cubed. Rather than stating this as a negative desideratum, "the system should not rely on external knowledge", we state it as a positive requirement: *(F) the system must remain invariant under a change of facts.*

The remarks above are not meant to demonstrate the uselessness or irrelevance of the Berry paradox (which remains a valuable technical tool e.g. in Kolmogorov complexity; see Chapter 7) but rather to make clear that from the standpoint of linguistics the difficulties arise long before the diagonalization issues that are in the focus of the logical treatment would come to force. If we exclude the arbitrariness of number naming, references to encyclopedic knowledge, and the description of algorithms, what remains is some extension of the CSRS and SRS grammars. For these, (6.5) has a solution: we run the grammar, evaluate each expression, and pick the smallest one missing from the list of numbers so generated. The paradox disappears since this amounts to describing an algorithm rather than naming a number.

The hangman paradox

It is possible to interpret the Berry paradox in another manner: consider *Your obligation will be fulfilled only when you give Berry half of the money in your pocket.* A reasonable person may conclude that if he gives Berry half of the money in his pocket, his obligation is fulfilled. However, Berry may object, noting the amount of money still in the person's pocket, that the obligation is unfulfilled and demand another halving, another, and another, ad infinitum. Or consider *To complete this task, you need to sleep on it and you must report back to Berry that you are absolutely certain it's done.* The next day, you may entertain doubts. Since you have not so far reported back to Berry, you have not completed the task, and you need to sleep on it. This interpretation, traditionally presented under the name *hangman paradox*, has less to do with finding the smallest solution to (6.5) than with the more wide-ranging problem of imperatives that demand the impossible. There is nothing on the surface to distinguish these from imperatives that can in fact be met. A perfectly analogous problem is presented by questions that cannot be answered. For any given question, we do not know in advance whether an answer exists, and if we permit a question to be formulated in a sublanguage that is strong enough to refer to algorithmically undecidable issues, we know that some questions will not have answers. Again, there is no need to probe the edges of undecidability; ordinary adjectival modification will already furnish plenty of examples such as *prime between eight and ten* that receive no interpretation in any model, and embedding such references to impossible objects in questions or imperatives is a trivial matter. Carefully crafted realist fiction creates a world that appears possible, but to the extent that human language use, and indeed the whole human condition, is greatly impacted by narratives that defy rational belief, requirement D needs to be further strengthened: *(G) all expressions should receive analysis by the same apparatus as supposed statements of fact.*

6.1.4 Desiderata

Let us now collect the requirements on linguistic semantics that have emerged from the discussion. *(A)* asks for a characterization of the set of natural language expressions. In mathematical linguistics, this is done by generating the set (see Section 2.1), and in the previous chapters we have discussed in detail how phonology, morphology, and syntax, taken together, can provide such a characterization. *(B)* asks the same for the set of meanings, but our lack of detailed knowledge about the internal composition of the mental form that meanings take gives us considerable freedom in this regard: any set that can be generated (recursively enumerated) is a viable candidate as long as it can be used in a theory of semantics that meets the other desiderata. One particularly attractive approach is to represent meanings by well-formed formulas in some logical calculus, but other approaches, such as network representations, also have a significant following.

Requirement *(C)*, that expressions similar in form should receive analyses by the same apparatus, has far-reaching consequences. First, it implies a principle of homogeneity: to the extent meanings are typed, and this will be necessary at least to distinguish objects from statements, questions, and imperatives, *(C)* calls for expressions of the same linguistic type to be mapped on meanings of the same logical type. Second, to the extent the meanings of expressions are built recursively from the meanings of their constituents, *(C)* calls for direct compositionality in the form of some *rule to rule hypothesis*: each step in generating the expression must be paired with a corresponding step in generating the meaning. Third, to the extent that the truth, consistency, feasibility, adequacy, or even plausibility of an expression cannot be fully known at the time it is uttered, *(C)* implies our *(D)* and *(G)*: the meaning of expressions is independent of their status in the real world.

In model-theoretic semantics, the meaning representations (formulas) are just technical devices used for disambiguating expressions with multiple meanings. Since the formulas themselves are mapped onto model structures by an interpretation function, the step of mapping expressions to meanings can be composed with the interpretation function to obtain a *direct interpretation* that makes no reference to the intermediary stage of formulas. In this setup, requirement *(F)*, that the system must remain invariant under a change of facts, is generally taken to mean that the interpretation function is symmetrical (invariant under any permissible permutation of models). But *(F)* is asking for a considerably more fluid view of facts than what is generally taken for granted in model-theoretic semantics: from our perspective, the fact that *eight cubed* and *hex two hundred* are the same number, decimal 512, is an entirely contingent statement, perhaps true in some models where correct arithmetic is practiced but quite possibly false in others.

Exercise 6.2 From an early age, Katalin was taught by her older brother that three plus five is eight, except for chipmunks, where three chipmunks plus five chipmunks are nine chipmunks because chipmunks are counted differently. Two plus two is four, except for chipmunks, where it is five. One plus one plus one is three, except for chipmunks, where it is five. By age five, Katalin mastered both regular addition and multiplication and chipmunk arithmetic up to about a hundred. Describe her semantics of arithmetic expressions.

Finally, let us call attention to an important consequence of the mundane treatment of arithmetic expressions assumed here. In his discussion of Montague (1963), Thomason (1977) argues that any direct theory of propositional attitudes is bound to be caught up in Tarski's (1935) theorem of undefinability (Tarski 1956), rendering the resulting analysis trivial. However, as Thomason is careful to note, the conclusion rests on our ability to pass from natural language to the kinds of formal systems that Tarski and Montague consider: first order theories with identity that are strong enough to model arithmetic. Tarski himself was not sanguine about this: he held that in natural language "it seems to be impossible to define the notion of truth or even to use this notion in a consistent manner and in agreement with the laws of logic".

To replicate Tarski's proof, we first need to supplement natural language with variables. The basic idea, to formalize the semantics of a predicate such as *subject owns object* by a two-place relation $\zeta(s,o)$ is fairly standard (although, as we shall see in Section 6.3, there are alternatives that do not rely on variables at all). But the proposed paraphrases for first-order formulas, such as replacing

$\forall x[\exists y\zeta(x, y) \rightarrow \exists z\zeta(z, x)]$ by *for everything x, either there is not something y such that x owns y or there is something z such that z belongs to x* clearly belong in an artificial extension of English rather than English itself. Second, we must assume that the language can sustain a form of arithmetic; e.g. Robinson's Q. We have already expressed doubts as to the universality of natural language when it comes to logical or arithmetic calculi, but using Q we can narrow this down further: several of the axioms in Q appear untenable for natural language. Again, the key issues arise long before we consider exponentiation (a key feature for Gödel numbering) or ordering. Q comes with a signature that includes a successor s, addition $+$, and multiplication $\cdot$. By Q2 we can infer $x = y$ from $sx = sy$, Q4 provides $x + sy = s(x + y)$, and Q6 gives $x \cdot sy = xy + x$. All of these axioms are gravely suspect in light of the noncounting principle discussed in Section 5.5. There are many ways we can *start* counting in natural language: we can look at quotations of quotations *(Joe said that Bill said...)* and emphasis of emphasized material *(very very...)*, but there isn't a single way that takes us very far – whichever way we go, we reach the top in no more than four steps, and there Q2 fails.

Since this is one of the key points where the semantics of natural language expressions parts with the semantics of mathematical expressions, it is worth highlighting a consequence of the position that worlds, or models (from here on we use the two terms interchangeably), need not be consistent: our desideratum *(E)*, that the system must be flexible about naming, now comes for free. In the following, we employ *paraconsistent logic* to make room for entities such as unicorns, triangular circles, and numbers that remain unchanged after adding one. As readers familiar with modern philosophical logic will know, the use of paraconsistent logic opens the way to new solutions to the Liar and several other puzzles, but we will not pursue these developments here – rather, we will concentrate on problems that we have earlier identified as central, most notably the distinctions between *essential* and *accidental* properties introduced in Section 5.3.

6.2 The standard theory

In a series of seminal papers, Montague (1970a, 1970b, 1973, all reprinted in Thomason 1974) began the development of a formal theory that largely meets, and in some respects goes beyond, the desiderata listed above. In Section 6.2.1 we introduce this family of theories, known today as *Montague grammar* (MG). As there are several excellent introductions to MG (see in particular Dowty et al. 1981 and Gamut 1991), we survey here only the key techniques and ideas and use the occasion to highlight some of the inadequacies of MG, of which the most important is the use of a stilted, semiformalized (sometimes fully formalized) and regimented English-like sublanguage more reminiscent of the language use of logic textbooks than that of ordinary English. Linguists brought up in a more descriptive tradition are inevitably struck by the stock examples of MG like *Every man loves a woman such that she loves him* or *John seeks a unicorn and Mary seeks it.* While still avoiding the rough and tumble of actual spoken English, we take the object of inquiry to be a less regimented language variety, that of copyedited journalistic prose. Most of our examples in Section 6.2 will be taken from an American newspaper, the *San Jose Mercury News.* For reasons of expository convenience, we will often considerably simplify the raw examples, indicating inessential parts by [] wherever necessary, but in doing so, we attempt to make sure the simplified example is still one that could be produced by a reasonable writer of English and would be left standing by a reasonable copy editor.

6.2.1 Montague grammar

In Chapter 3, we defined signs as conventional pairings of sound and meaning and elaborated a formal theory of phonology capable of describing the sound aspect of signs. Developing a formal theory of semantics that is capable of describing the meaning aspect of signs will not proceed quite analogously since in phonology we could ignore the syntax, while in semantics it is no longer possible to do so.

The central idea of Montague (1970a) was to introduce two algebras, one syntactic and the other semantic, and treat the whole issue of linguistic semantics as a homomorphism from one to the other. As there are several minor technical differences in the way this idea was implemented in Montague's main papers on the subject (and subsequent research has not always succeeded in identifying which of the alternatives is really the optimal one), we will not remain meticulously faithful to any of the founding papers, but we will endeavor to point out at least the main strands of development in MG, which, construed broadly, is clearly the largest and most influential school of contemporary linguistic semantics.

As usual, an **algebra** is a set T endowed with finitely many operations $F_1, \dots, F_k$ of fixed arity $a_1, \dots, a_k$. By an **operation** of arity a, we mean a function $T^a \rightarrow T$, so by convention distinguished elements of the algebra are viewed as nullary operations. For Montague, the syntactic algebra is strongly typed but otherwise unrestricted: if p and q are expressions of types (categories) P and Q, an operation f will always (and only) produce an expression $f(p,q)$ of type R. In practice, Montague always used binary operations (a tradition not entirely upheld in subsequent MG work) and was very liberal as to the nature of these, permitting operations that introduce or drop grammatical formatives and reorder the constituents. Some later work, such as Cooper (1975) or McCloskey (1979), took advantage of this liberal view and permitted generative transformations as syntactic operations, while others took a much stricter view, permitting only concatenation (and perhaps wrapping; see Bach 1981). Either way, the syntactic and semantic algebras of MG provide the characterization of natural language expressions and meanings required by our desiderata *(A)* and *(B)*.

The mapping ϕ from the syntactic algebra to the semantic algebra must be a homomorphism: for each operation f of arity a in the syntactic algebra, there must be a corresponding operation g of the same arity in the semantic algebra, and if $r = f(p,q)$, we must have $\phi(r) = g(\phi(p), \phi(q))$. While in practice Montague used formulas of a higher-order *intensional logic* called IL as elements of the semantic algebra, with function composition (including applying an argument to a function) as the chief operation, it is clear from his work that he took a far more abstract view of what can constitute a proper semantic representation. In particular, the IL formulas are interpreted in model structures by another function ψ, which is also a homomorphism, so that the λ-calculus formulas, viewed by Montague as a mere pedagogical device, can be dispensed with entirely, interpreting natural language expressions directly in model structures. Later work explored several other choices for semantic algebra: Heim (1982) used *file change potentials*, Cresswell (1985) used structures that preserve a record of the way the expression was built up, and many other options are available, as long as the ultimate model-theoretic nature of the enterprise is preserved (i.e. the structures of the semantic algebra are interpreted in models).

Montague used IL first as a means of resolving issues of opacity. Instead of taking the meaning of terms to be their *extension*, the set of objects to which the interpretation function maps them in a single model structure, he chose to explicate meanings as *intensions*, the set of extensions in all modally accessible worlds. Since the extension of expressions may coincide (for example, in realis worlds both *the king of France* and *the king of Austria* denote the empty set), it is not evident how to express the clear meaning difference between *John aspires to be the king of France* and *John aspires to be the king of Austria* or even *John aspires to be a unicorn*.

With intensions, the problem is solved: since it is easy to imagine an alternative world where the French Revolution still took place but Austria, just like Belgium, retained the institution of monarchy, the intension of the two terms is different and we can ascribe the different meanings of the whole expressions to the different meanings of the NPs *king of France* vs. *king of Austria*. In MG, verbs denoting propositional or other attitudes are called *intensional* since they operate on intensions: their use solves many subtle interpretation problems already known to the Schoolmen, such as the *in sensu composito/in sensu diviso* readings of *I want a new car*, which may express that the speaker has her eye on a particular car or that she wants to replace the old one with a new one and it does not really matter which one. A similar treatment is available for intensional nouns such as *temperature*: by taking these to be functions that take different values in different possible words, we avoid concluding *thirty is rising* from *the temperature is thirty* and *the temperature is rising*.

One problem that this otherwise very satisfactory theory leaves open is known as the issue of *hyperintensionals*: there are expressions like *prime between 8 and 10* and *triangular circle* that denote empty sets in every modally accessible world; indeed, under most theories of the matter, in any world whatsoever. Clearly, *Pappus searched for a method to trisect any angle* presumes very different truth conditions from *Pappus searched for a method to square the circle* even if, in hindsight, it is clear that the two activities are equally futile.

Another technical device of MG worthy of special mention is the use of *disambiguated language*. Since ϕ (or $\phi \circ \psi$) is a function, only a single translation can attach to any expression, so those expressions that are ambiguous need to be assigned as many disambiguated versions as there are separate meanings. In syntax, the preferred method of disambiguation is by constituent structure, as in *[The man on the hill] with the telescope* as opposed to *The man on the [hill with the telescope]*. In semantics, however, we often find cases like (5.2), where the constituent structure is unambiguous, *Every man [loves a woman]*, yet the sentence has two distinct readings, (5.3) and (5.4), corresponding to the order in which the universal *every* and the existential *a* get to bind the variables in *loves(x,y)*. We can use derivation history to distinguish such cases: the same constituent structure is reached by applying (possibly different) operations in different orders.

To see how the ambiguity is handled in MG, we need to consider another characteristic feature of MG, the use of *generalized quantifiers*. Intuitively, individuals (proper nouns such as *John* or NPs such as *the dog*) should be interpreted as elements of model structures (type e), and properties (be they expressed adjectivally: *(is) red*, *(is) sleepy*; or by intransitive verbs: *sleeps*) are interpreted as functions from individuals to truth values (type $e \rightarrow t$). Yet quantified NPs, such as *some men*, *every dog*, *no tree*, etc., are syntactically congruent to simple NPs such as *John* or *the dog*, and our desideratum *(C)* demands that they should receive analysis by the same apparatus as these. Since quantified NPs can be easily conceptualized as sets of properties (those properties, be they essential or accidental, that are shared by all members of the set), we lift the type of simple NPs from e to $(e \rightarrow t) \rightarrow t$ and conceptualize them as the set of all properties enjoyed by the individual. Once this step is taken, the intuitive assignment of *sleeps* as function and *John* as argument can (indeed, must) be reversed for the function application to come out right. The translation of *every man* is $\lambda P(\forall x(\text{man}(x) \rightarrow P(x)))$, so that translation of *Every man sleeps* is obtained by applying this function to the translation of *sleeps*, yielding (by beta conversion) the desired $\forall x(\text{man}(x) \rightarrow \text{sleep}(x))$.

Similarly, *a woman* is translated $\lambda P(\exists x(\text{woman}(x) \rightarrow P(x)))$, and if *loves* is translated as a two-place relation $l(x, y)$, *loves a woman* will be $\exists y(\text{woman}(y) \rightarrow l(x, y))$. Some care must be taken to make sure that it is the second (object of love) variable that is captured by the existential quantifier, but once this is done, the result is again an $e \rightarrow t$ function ready to be substituted into $\lambda P(\forall x(\text{man}(x) \rightarrow P(x)))$, yielding $\forall x(\text{man}(x) \rightarrow \exists y(\text{woman}(y) \rightarrow l(x, y)))$, the reading given in (5.3). In Montague's original system, the other reading (5.4), $\exists y(\text{woman}(y)\forall x(\text{man}(x)\ l(x, y)))$ is obtained by radically different means: by introducing, and later deleting, a variable that is viewed as being analogous to a pronoun.

From here on, we do not follow Montague closely, since our primary concern is with natural language rather than with the semiformalized (sometimes fully formalized) and regimented English-like sublanguages used in most works of philosophical logic. In everyday usage, as evidenced e.g. by newspaper texts, quantified NPs such as *every Californian with a car phone*, *every case*, *every famous star*, etc., do not lend themselves to a strict interpretation of *every* x as $\forall x$ inasmuch as they admit exceptions: *every case, except that of Sen. Kennedy; every Californian with a car phone, except drivers of emergency vehicles; every famous star, including Benji*, etc. The problem, known as the *defeasability* of natural language statements, has given rise to a wide variety of *nonmonotonic logic* approaches (for an overview see Ginsberg 1986a). Of particular interest here are *generic* constructions such as *Sea turtles live to be over a 100 years old*, which can be true even if the majority of specific instances fail. At the extreme end, some generic statements such as *P6 processors are outdated* may be considered true without *any* individual instances holding true.

We define a **construction** as a string composed of nonterminals (variables ranging over some syntactic category) and terminals (fixed grammatical formatives and lexical entries) with a uniform compositional meaning, obtained by a fixed process whose inputs are the meanings of the nonterminals and whose output is the meaning of the construction as a whole. An example we saw in Section 5.3 was *X is to Y as Z is to W*, but we use the term here in the general linguistic sense, which covers the entire range from completely productive and highly abstract grammatical patterns such as

$$\mathrm{NP}\langle\alpha\mathrm{PERS}\ \beta\mathrm{NUM}\rangle\ \mathrm{VP}\langle\alpha\mathrm{PERS}\ \beta\mathrm{NUM}\ \gamma\mathrm{TENSE}\rangle \tag{6.6}$$

to highly specified and almost entirely frozen idioms such as

$$\mathrm{NP}\langle\alpha\mathrm{PERS}\ \beta\mathrm{NUM}\rangle\ \mathrm{kick}\langle\alpha\mathrm{PERS}\ \beta\mathrm{NUM}\ \gamma\mathrm{TENSE}\rangle\ \text{the bucket} \tag{6.7}$$

On occasion, when we are interested in the substitution of one construction into another, it will be necessary to assign a grammatical category (defined as including morphosyntactic features specified in angled brackets) to the construction as a whole, so a context free or mildly context-sensitive theory of constituent structure (see Section 5.1.3) roughly along the lines of GPSG (Gazdar et al. 1985) or early HPSG (Pollard 1984) is presupposed. For the purposes of this chapter, we can safely ignore transference across patterns, such as the phenomenon that the agreement portion of (6.7) is obviously inherited from that of (6.6), and concentrate more on the semantics of constructions.

As a limiting case, entirely frozen expressions (i.e. those constructions that no longer contain open slots, such as *go tell it to the Marines*) are simply taken as lexical entries; in this case, with meaning 'nobody cares if you complain'. (The indexicals implicit in the imperative *go* and explicit in the paraphrased *you* do not constitute open slots in the sense that interests us here.) Since compositionality cannot be maintained as a principle of grammar without relegating the noncompositional aspects of constructions to the lexicon, we introduce the following *Principle of Responsibility*: The semantics of any expression must be fully accounted for by the lexicon and the grammar taken together. To make this principle clear, consider constructions such as

$$\text{for all NP}\langle+\mathrm{DEF}\rangle,\ \mathrm{S} \tag{6.8}$$

as in the following examples: *For all the glamour of aerial fish planting, it was a mass production money-maker; The Clarence Thomas hearings, for all their import*... or *For all their efforts at parity and fairness, NFL officials*.... Informally, the construction means something like '*S*, in spite of the usual implications of *NP*⟨+*DEF*⟩'. In the case of *the glamour of aerial fish planting*, the implication that needs to be defeased is that glamorous things are restricted to the few, a notion incompatible with *mass production*. Thus, to make sense of (6.8), we need to rely on lexical information. Without doing so, the clear difference between the acceptability of the preceding examples and *???For all their protein content, eggs are shaped so as to ease passage through the duct* would remain completely mysterious.

Since universally quantified natural language NPs can actually have exceptions, we need to capture the notion of exceptionality some way; e.g. by saying that English *every man* means 'almost all men' in the sense that exceptions have measure zero. Unfortunately, there is no obvious way to define measure spaces over semantic objects such as *legal cases* or *California drivers with car phones* naturally, so to translate *Geraldo Rivera [reveals that he is an extremely attractive virile hunk of man who] has had sex with [] every famous star in the entertainment industry* we say that for all x such that x has no extra properties beyond being a famous star in the entertainment industry, Geraldo Rivera has had sex with x.

A less clumsy translation, in keeping with the standard treatment of generalized quantification, is to say that the property of having had sex with Geraldo Rivera is implied by the property of being a famous star in the entertainment industry. We say that *every N* is the set of *typical* properties that N has, where typicality is defined in the lexical entry of N. Since having four legs is typical of donkeys, *every donkey has four legs* will be true by definition and cannot be falsified by the odd lame donkey with three or fewer legs.

But if having four legs is an analytic truth for donkeys, how can we account for counterfactuals where five-legged donkeys can appear easily, or for the clear intuition that being four-legged is a contingent fact about donkeys, one that can be changed e.g. by genetic manipulation? The answer offered here is that to reach these we need to change the lexicon. Thus, to go from the historical meaning of Hungarian *kocsi* 'coach, horse-driven carriage' to its current meaning '(motor) car', what is needed is the prevalence of the motor variety among 'wheeled contrivances capable of carrying several people on roads'. A 17th century Hungarian would no doubt find the notion of a horseless coach just as puzzling as the notion of flying machines or same-sex marriages. The key issue in readjusting the lexicon, it appears, is not counterfactuality as much as rarity: as long as cloning remains a rare medical technique, we will not have to say 'a womb-borne human'.

To summarize our main departure from standard MG: under the treatment assumed here, *every man loves a woman* means neither (5.3), $\forall x \text{man}(x)\ \exists y \text{woman}(y)\ \text{loves}(x,y)$ nor (5.4), $\exists y \text{woman}(y)\ \forall x \text{man}(x)\ \text{loves}(x,y)$; it means that woman-loving is a typical property of men, just as donkey-beating is a typical property of farmers. Some of the typical properties of common nouns are analytic *relative to a given lexicon* while others are not. In fact, for every noun there are only a handful of defining properties (see Section 5.3), and these can change with time in spite of the inherent conservatism of the lexicon.

Ordinary adjectival modification means conjoining another property to the bundle (conjunction) of essential properties, so *brown dog* refers to the conjunction of all essential dog properties and brownness. *Enormous fleas* have the property of enormity conjoined to the essential properties of fleas, which include being rather small, so the notion is applied, without any special effort, on the flea scale. The same simple treatment is available for impossible objects such as triangular circles.

Figure 6.1 shows on the left a slightly triangular circle and on the right a slightly circular triangle. Whether the object in the middle, known as the *Reuleaux triangle*, is considered a triangle, a point of view justified by its having three distinct vertices, or a circle, a point of view justified by its having constant diameter, is a matter of perception.

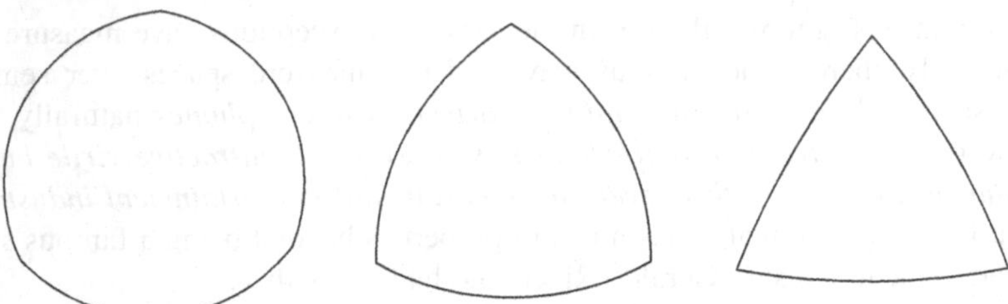

Fig. 6.1. The Reuleaux triangle and its cousins

What is clear from the linguistic standpoint is that adadjectives like *slightly*, *seemingly*, and *very* attach to adjectives like *circular*, *triangular*, and *equal* that have a strict mathematical definition just as easily as they attach to adjectives like *red*, *large*, and *awful* that lack such a definition. Clearly, what these adadjectives modify is the 'everyday' sense of these terms – the mathematical sense is fixed once and for all and not subject to modification. Just as we were interested in the everyday sense of *all* and *every* and found that these are distinct from the standard mathematical sense taken for granted in MG, here we are interested in the ordinary sense of *circular*. Working backward from typical expressions like *circular letter* and *circular argument*, we find that the central aspect of the meaning is not 'a fixed distance away from a center' or even 'fixed diameter' but rather 'returning to its starting point', 'being cyclic'.

In these examples, the morphologically primitive forms are nominal: the adjectival forms *circular* and *triangular* are clearly derived from *circle* and *triangle* and not the other way around. Since derivation of this sort changes only the syntactic category of the expression but preserves its meaning, we can safely conclude that *circle* in the everyday sense is defined by some finite conjunction of essential properties that includes 'being cyclic' and that the mathematical definition extends this conjunction by 'staying in an (ideal) plane, keeping some (exact) fixed distance from a point'. Similarly, *triangle* simply means 'having three angular corners' rather than the exact configuration of points and lines assumed in geometry.

This point is worth remembering, as there is often a somewhat naive tendency to treat the mathematical definition as the norm and assume that everyday language is 'sloppy' or 'fuzzy'. In reality, expressions like *a triangular patch of snow on the mountainside* are perfectly reasonable and convey exactly the amount of information that needs to be communicated: nobody would assume that the patch lies in a single plane, let alone that its edges are perfectly straight lines. Linguistics, as a scientific endeavor, centers on modeling the data provided by actual language use, as opposed to providing some norm that speakers should follow. This is not to say that the study of highly regulated technical language, as found in legal or scientific discourse, is of no use, but as these language varieties are acquired in a formal schooling process, over many years of adult study (as opposed to the acquisition of natural language, which takes place at an earlier age and generally requires no schooling whatsoever), there is nothing to guarantee that their properties carry over to natural language.

As everyday terms, *circular triangles*, *triangular circles*, *she-males*, or *wide awake sleepers* (spies who eagerly wait for the chance to be activated) are all illustrations of the same phenomenon, namely that adjectival modification is not simply a conjunction of some new property to the set of essential properties but a destructive overwrite operation. Thus, *brown dog* is simply an object that has the property (color) brown in addition to the properties essential to dogs, but *ownerless dog* is an object with the same properties *except* for lacking an owner. In other words, essential properties are merely defaults that can be defeased.

From this vantage point, English has far fewer hyperintensional constructions than assumed in the tradition of philosophical logic: certainly triangular circles and immaculate conceptions give rise to no logical contradictions. This renders the problem with hyperintensionals discussed above far less urgent, as we now only have to deal with cases in which the essential meaning of the adjective is in strict contradiction to the essential meaning of the noun it modifies, and the latter is given by a single conjunct. Thus, we need to consider examples such as

Mondays that fall on Tuesdays (6.9)

Mondays that fall on Wednesdays (6.10)

Does it follow that a (rational) agent who believes in (6.9) must also believe in (6.10)? While the matter is obviously somewhat speculative, we believe the answer to be negative: if we learn that *John believes Mondays can be really weird – he actually woke up to one that fell on Tuesday*, it does not follow that he also believes himself to have woken up on one that fell on Wednesday. Since he has some sort of weird experience that justifies for him a belief in (6.9), he is entitled to this belief without having to commit himself to (6.10), as the latter is not supported by any experience he has.

What makes this example particularly hard is that Mondays, Tuesdays, and Wednesdays are purely cultural constructs: there is nothing in objective reality to distinguish a Monday from a Tuesday, and it is well within the power of society to change by decree the designation a day gets. When the lexical knowledge associated to a term is more flexible, as in the case of triangles and circles, *accommodation*, finding an interpretation that defeases as little of the lexical meaning as needed to make sense of the expression, is much easier. Most lexical entries clearly contain more information than a purely logical definition. For example *bachelor*, does not just mean 'unmarried man' – the lexical entry must contain a great deal of default knowledge about preferring to live alone, eating TV dinners, etc., otherwise a sentence such as *In spite of having married recently, John remained a true bachelor* could make no sense.

So far, we have assumed that lexical entries contain some mixture of defeasible and strict (nondefeasible) information without committing ourselves as to their proportion. But whether lexical entries without defeasible content exist at all remains to be seen – if not, the problem of hyperintensionals is not pertinent to natural language and the standard MG treatment of opacity remains viable. An even more radical question, one that we shall pursue in Section 6.2.2, is whether strict content exists in the lexicon to begin with. If not, natural language is structurally incapable of carrying arguments with strict conclusions. To carry out this investigation, we need a system of logic that can sustain some distinction between essential and inessential properties and one that can sustain some notion of default. We find a suitable candidate for both in the seven-valued system D of Ginsberg (1986), to which we turn now.

6.2.2 Truth values and variable binding term operators

Recall that a **lattice** is an algebra with two commutative and associative binary operations $\vee$ and $\wedge$ satisfying the *absorption identities* $a \vee (a \wedge b) = a, a \wedge (a \vee b) = a$. Lattices are intimately related to partial orders: we say $a \leq b$ iff $a \vee b = b$ or, equivalently, if $a \wedge b = a$. If neither $a \leq b$ nor $b \leq a$ holds, we say a and b are not comparable. The lattice operations induce a partial order, and conversely, any partial order for which incomparable elements have a greatest lower bound and a least upper bound give rise to a lattice whose induced partial order is the same as the one with which we started. The least element of a lattice, if it exists, is called the *zero* or *false* element, and the greatest, if it exists, is called the *one* or *true* element.

A **bilattice** over a set B has binary operations $\vee, \wedge, +, \cdot$ such that $B, \vee, \wedge$ and $B, +, \cdot$ are both lattices, and the operations of one respect the partial order on the other (and conversely). Using traditional Hasse diagrams with bottom to top representing one ordering (say, the one induced by $+, \cdot$) and left to right representing the other (induced by $\vee, \wedge$), the smallest nontrivial bilattice comes out as a diamond figure. This is the four-valued system of Belnap (1977) shown in Fig. 6.2(i). As usual, t and f are the classical true and false, $\perp$ means 'both true and false', and u means 'unknown'. For system D, Ginsberg (1986) adds the values dt 'true by default', df 'false by default', and $\star$ 'both true and false by default', as shown in Fig. 6.2(ii):

Let us first see how to use these in lexical entries. We will briefly touch on most lexical categories admitted as primitives by the NSM school – pronouns, determiners,

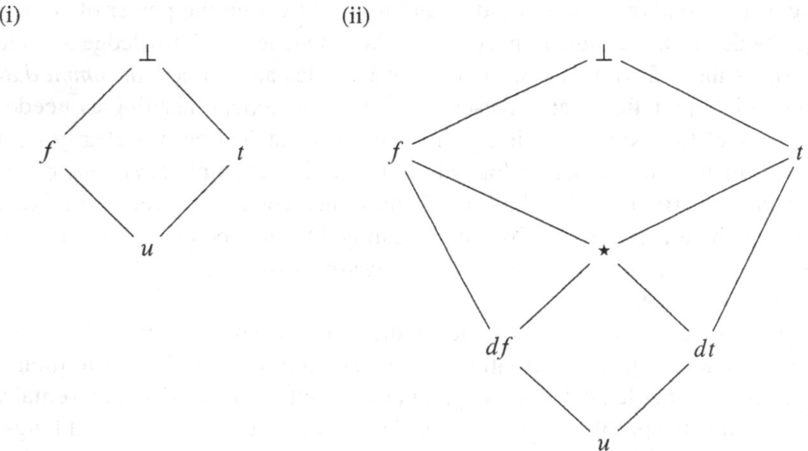

Fig. 6.2. (i) Paraconsistent logic (ii) with defaults

quantifiers, adjectives, verbs, adverbials, relations, and nouns – but defer adadjectives and connectives to Section 6.3. As we go along, we are forced to make a number of design choices, and the reader should not assume that the ones made here are the only reasonable ones. Our goal here is not to solve all outstanding problems of semantics but rather to present a single coherent system that supports a reasonable notion of everyday *paraphrase* and common sense *inference*. The key issue is our desideratum *(C)*, which calls for some uniform, mechanistic treatment of constructions. As long as we consider (5.9), *The duckling was killed by the farmer*, to be a reasonable paraphrase of (5.8), *The farmer killed the duckling*, we should be able to parse both these constructions, turn the crank of the semantic machinery, and obtain the same result either directly or perhaps by invoking inference steps along the way.

Relations, adpositions

While not considered a primitive category in the NSM system, many grammarians take the view that adpositions[2] form a major lexical category on a par with noun, verb, adjective, and adverb. Very often, they express a clear spatial relation: e.g. English *under* is a relation between two entities X and Y such that the body of X is, according to the vector set by gravity, below that of Y. To provide an adequate semantics of constructions involving *under*, we need to address a number of technical issues, but most of the machinery will be required elsewhere as well.

First, we need **model structures** that contain *entities* and *relations*. Entities have type e, and (binary) relations have type $e \times e$. The set of entities in a model structure is called the **universe** of the model. To fix ideas, we assume a distinguished model structure M_0 corresponding intuitively to reality, 'our world as of a fixed date', and some set of indexes I. There are model structures M_i with universe U_i for every $i \in I$, and at least some of these are accessible from M_0 by the evolution of time). Unlike MG, which fully integrates time into the structure of models, we leave matters of time outside a single model structure in this discussion, at least for significant durations (over a few seconds). We leave open the possibility that a local (possibly infinitesimally small) temporal environment is part of the model structure or, equivalently, a single model may be able to sustain statements about speed, acceleration, precedence, etc. We shall also ignore all complications stemming from different reference frames and relativity (see Belnap and Szabó 1996). We need to sustain two kinds of inferences:

$$\text{if X is under Y, and Y is under Z, then X is under Z} \tag{6.11}$$

$$\text{if X is under Y, Y can fall on X} \tag{6.12}$$

Inference (6.11) is clearly defeasible: if the rug is under the bed, and the coin is under the rug, it may still be that the coin is not under the bed (but rather under a portion of the rug that peeks out from under the bed). More importantly, even if the coin lies geometrically under the bed, once it is covered by the rug it requires a significant

---

[2] In English, prepositions, but in many languages such as Japanese or Uzbek [UZN], often postpositions: consider *soxil tomon* shore. NOM toward 'toward the shore'.

amount of reconceptualization to consider it being under the bed: this is similar to the smart aleck of Winograd and Flores (1986) who answers *yes* to the question *is there water in the fridge?* based on the fact that there are vegetables in the fridge and organic matter is composed of 50% water.

Somewhat surprisingly, it is (6.12) that appears to be nondefeasible: even in science-fictional contexts, the meaning of *under* is preserved. *The ship's gravity generators reversed. The gun under the bed was now over it and began to fall on it.* Although the phenomenon is already clear, with a little effort, even less arguable cases can be constructed:

$$\text{if X is under Y, Y is over X} \tag{6.13}$$

$$\text{if X is to the left of Y, Y is to the right of X.} \tag{6.14}$$

For a fuller theory of lexical semantics, several questions need to be answered. First, is there a need to distinguish a core (spatial) meaning of *under* from a more peripheral, metaphoric meaning? Clearly, from *France was under Vichy rule* it does not follow that **Vichy rule could fall on France*. Here we take the methodological stance that different meanings, as denoted by subscripts in lexicographic practice, are to be avoided maximally: the lack of a **Vichy rule could fall on France* construction is to be attributed to blocking (see Section 4.2) by the form *befell*. To be sure, there will always be cases such as $bank_1$ 'riverbank' vs. $bank_2$ 'financial institution' where the semantic relation between the two, if there ever was one, is beyond synchronic recovery, but for the most part we side with Jakobson, who in response to $bachelor_1$ 'unmarried man', $bachelor_2$ 'seal without a mate', and $bachelor_3$ 'young knight carrying the banner of an established one' summarized the meaning as 'unfulfilled in typical male role'.

Second, where is information such as (6.12) to be stored: in the lexical entry of *under*, in the entry of *fall*, both, or neither? Here we assume a neutral store of background knowledge we will call the *encyclopedia*, which is distinct from the lexicon, being devoted primarily to extragrammatical or *real-world* knowledge, but leave open the possibility that lexical entries may have pointers to encyclopedic knowledge. Our desideratum *(F), invariance under change of facts*, will be fulfilled by permitting modification (in the limiting case, full deletion) of the encyclopedia but keeping the lexicon, at least synchronically, unchanged. Note that arithmetic, including the chipmunk arithmetic of Exercise 6.2, is a matter of real-world knowledge. The main relational constructions that we need to cover include the copulative

$$S \to NP\langle\alpha\text{PERS }\beta\text{NUM}\rangle\ COP\langle\alpha\text{PERS }\beta\text{NUM }\gamma\text{TENSE}\rangle\ P\ NP \tag{6.15}$$

which is a full sentence such as *The cat was under the bed*, and the relative clause

$$NP \to NP\langle\alpha\text{PERS }\beta\text{NUM}\rangle\ P\ NP \tag{6.16}$$

which is an NP such as *the cat under the bed.*

Entities, constants

Mathematical logic makes routine use of a class of expressions, *constants*, that have no clear counterpart in natural language. Mathematical constants are *rigid designators* of the best kind: not only do they correspond to the exact same element in each model, but this cross-world identity comes at no cost: the name itself guarantees that the same cluster of properties is enjoyed by each instance. They are not subject to adjectival modification, **quick 5, *ancient π*, or relativization: *the e that Mary learned about yesterday* is the exact same $e, 2.718\ldots$, that we all learned about. Kripke (1972) treats all proper names as rigid designators, but this is not well-supported by the linguistic evidence. Proper names can shift designations as a result of adjectival modification (*the Polish Shakespeare* is not Shakespeare but 'the most distinguished Polish playwright') and similarly for possessive or relative clauses (*the Venice of the North* is some 'spectacular city built on islands in the North'), and so on. Here we take the stance that the relative constancy of proper names is primarily due to the lexical information associated to them, so that e.g. geographic names have a great deal more temporal constancy than names of organizations.

Similarly, Kripke (1972) argues that *natural kinds* such as *dog* or *water* cannot be defined in terms of paraphrases as we have suggested in Section 5.3 – a proper definition must ultimately rely on the scientific identification, in terms of DNA characteristics for dogs or the H_2O chemical formula for water. From a philosophical standpoint, this may be more satisfactory than the method of definition used here, which relies on conjoining essential properties. Indeed, it would be hard to argue that the ultimate reality of water is somehow distinct from H_2O. Yet from the natural language standpoint we are less concerned with ultimate reality than with actual usage, and H_2O is clearly *not* water but rather 'distilled water' or even 'chemically pure water' – adopting the scientific definition would lead to the rather undesirable consequence that *water quenches thirst best* would come out as highly questionable, as there are a range of experiments suggesting that chemically pure water is inferior to ordinary water in this regard. In general, those contexts that distinguish ordinary water from chemically pure water would all lead to paradoxical judgments: we would want to say *water generally has trace minerals* is true, but if we insist on defining water as H_2O the sentence is by definition false.

The MG treatment of generalized quantification takes not just quantified NPs to be sets of properties but all NPs. Proper names, in particular, denote not individuals but the set of all properties these individuals have. We have already seen that a more modest selection of properties, taking only those properties that are in some sense defining (essential, typical) of the quantified noun yields better results for the *every N* construction, and it is worth noting that proper names can be quantified the same way: *every Dr. Johnson finds his Boswell* 'great men will find admiring biographers'. By our requirement *(C)*, which calls for homogeneous semantic treatment of syntactically homogeneous constructions, NPs composed of a determiner and a noun should also be sets of properties. We will assume here that *the* adds a single property, that of definiteness, to the set of properties that make up common nouns. This situation provides a clear example of the autonomy of syntax thesis discussed in Section 5.3: if 'definiteness' is just like 'redness', there is no semantic explanation why *the red boy* is grammatical but **red the boy* is not. To summarize the major constructions we have to account for:

$$\text{NP} \rightarrow \text{every/the N} \tag{6.17}$$

which accounts for singular NPs such as *the point*, *every point*, and

$$\text{NP}\langle\text{PL}\rangle \rightarrow \text{(every/the) Num N}\langle\text{PL}\rangle \tag{6.18}$$

which accounts for plural NPs such as *every three points* [determine a plane] or *the three points* [determining a plane].

Adjectives

If all proper names, common nouns, and NPs are bundles of properties, is there still a need for a simple *e* type (*e*ntities, *e*lements of model structures)? To be sure, it generally requires a significant amount of definitional work to get to a specific entity: consider *the third nail from the top in the fence segment starting at the northern end of my backyard at Hiawatha Lane, Arlington, MA*. In definitions like these, the whole expression is anchored to proper names like *MA*, *Arlington*, *Hiawatha Lane*, to indexicals like *I* or *my*, to demonstratives like *this*, and to essential properties such as the cardinal directions *north*, *top*. Once we have a specific entity, we can name it, but the two model structures before and after the act of naming are different: in one, the nail in question is like any other nail, and in the other it enjoys the property of having a name.

Remarkably, even though a whole set of culturally regulated naming conventions are deployed for the purpose (there are not likely to be two Hiawatha Lanes in one town, or two Arlingtons in one state), the act of naming the nail still requires the presence of the definite article *the*. This is not true for proper names, which are inherently definite: we say *London, you know, the one in Ontario* rather than **the Ontario London*. The main insight of Kripke (1972), that proper names behave like rigid designators, can therefore be salvaged by means of assigning the definiteness property to proper names in the lexicon and by allowing for some relationship between entities (type *e*) and the bundle of properties that define them, traditionally taken as type $(e \rightarrow t) \rightarrow t$.

In the Leibniz/Mostowski/Montague tradition, there is a direct, in a philosophical sense definitional, one-to-one relation between entities and the sets of their properties: if two entities are not the same, there must be some property that holds for one but not for the other. If no such property is found, the two entities cannot be individuated and must therefore be counted as one and the same.[3] This opens the way toward

[3] The philosophical problem of individuation is particularly acute for elementary particles of the same kind, which have no distinguishing properties once their quantum state is determined. We make no pretense to be able to contribute to the philosophy of quantum mechanics and simply assume that things are large enough so that there will be a whole range of location properties to distinguish between any two tokens of the same type.

characterizing adjectives by their intension. Following Belnap (1977), we define the domain of a property P in the universe U_i by two sets, called the *positive* and the *negative* support P_+ and P_-. In classical logic, P_- would just be the complement of P_+, but here we permit the two to overlap and their union to leave some of U_i uncovered. The four truth values, denoted u, t, f, and $\bot$, are assigned to entities x as expected: $P(x) = \bot \Leftrightarrow x \in P_+ \wedge x \in P_-$; $P(x) = t \Leftrightarrow x \in P_+ \wedge x \notin P_-$; $P(x) = f \Leftrightarrow x \notin P_+ \wedge x \in P_-$; $P(x) = u \Leftrightarrow x \notin P_+ \wedge x \notin P_-$. We will need two entailment relations, $\models_+$ and $\models_-$ (read *true-entails* and *false-entails*), with the expected properties such as $I \models_- U \wedge V \Leftrightarrow I \models_- U \vee I \models_- V$.

We have two candidate definitions for common nouns: on the one hand, we could identify the meaning of a noun such as *candle* with the property of 'being a candle', i.e. as a subset of U_i (a single $e \rightarrow t$ function at any index). This is the traditional logical approach. Under the traditional linguistic approach, to find the meaning of *candle*, one consults a dictionary, where it is defined as

a light source, now used primarily for decoration and on festive occasions, made of wax, tallow, or other similar slowly burning material, generally having a cylindrical shape, but also made in different shapes. (6.19)

Our goal here is to formalize this second approach (i.e. to treat common nouns as bundles of their essential properties). Since properties are also subsets of U_i, this amounts to defining the set of candles by the intersection of the essential candle properties: serving as a light source, being used on festive occasions, being made of wax or a similar substance, etc. This is not to say that dictionary definitions are directly applicable as they are: for example, the definition above does not mention the wick, while we take wicks to be essential for candles. The main construction that we have to account for is

$$\mathrm{N} \rightarrow \mathrm{A\ N} \tag{6.20}$$

which describes adjectival modification as in *cylindrical shape*.

Pronouns, indexicals, quantifiers

The standard MG method is to treat pronouns as variables. While the metaphor of substituting specific instances is attractive, giving pronouns variable status would leave a gaping hole in the edifice: why cannot we quantify over them? There is no **every me* or **every you*. Alternatively, in languages with morphological person/number marking, it should be easy to use these marks on quantifiers as well (this is especially clear for languages such as Hungarian that reuse the same set of markers across verbs and nouns), yet cross-linguistically we do not find person/number marking on quantifiers at all.

On the whole, pronouns have a distribution very close to that of proper nouns: perhaps the only situation where a pronoun cannot be replaced by a proper noun is in *resumptive* positions as in *I wonder who they think that if Mary marries him/*John then everybody will be happy* or in lexical entries such as *make it clear* that contain pronouns as in *Studying Chapter 2 makes it/*the book very clear that the author* *rejects the global warming hypothesis.* Therefore, assigning pronouns and proper nouns radically different types would be hard to justify. If proper names are of type e, pronouns will have type $e \rightarrow e$ and will correspond to a distinguished function among all $e \rightarrow e$ functions, namely the identity. As there is only one identity function, this method goes a long way toward explaining why we find only a very restricted set of pronouns (only as many as there are gender classes), while there is an infinite supply of variables. If proper nouns are treated as bundles of their essential features, pronouns will be the identity function over such bundles, except possibly adding a gender feature to the conjunction.

Another aspect of pronouns worth noting is that *I* is uniquely identified by the speaker who utters it, but *you* generally requires some gesture or equivalent cue to pick up its referent. This phenomenon is even more marked with other pro-forms such as *here*, *there*, *now*, etc., whose content clearly depends on the context they are uttered. Following Kaplan (1978, 1989), these are generally collected under the heading of *indexicals*. The standard MG treatment is to use n-tuples of speaker, hearer, time, place etc., to structure the set I of indexes, and make the content of indexicals dependent on these.

In translating natural language expressions to formulas of some logical calculus, the standard method is to automatically supply variables to quantifiers. (As discussed earlier, the standard logical interpretation of $\forall$ is not the appropriate one: *generic* sentences such as *Hares outrun foxes* express states of affairs believed to be true of whole classes, even though individual exceptions may exist, and the same is true of the typical use of *every*. For the moment, we ignore the issue and proceed the way MG does.) Clearly, the sentence *Everyone thinks he will win* is reasonably paraphrased as

$$\forall x \text{man}(x)\text{think}(x, \text{will\_win}(x)) \tag{6.21}$$

and if *he* is viewed as a variable x, use some mechanism to make sure that *everyone* binds x. Needless to say, if *he* is viewed as a variable y, the translation must become

$$\forall y \text{man}(y)\text{think}(y, \text{will\_win}(y)) \tag{6.22}$$

As long as we treat the formulas as mere abbreviatory devices and take the interpretation in model structures as our eventual goal, alphabetic variants such as (6.21) vs. (6.22) make no difference, as they will hold in the exact same models. Correct bookkeeping is more of an issue in theories such as discourse representation theory (DRT) (Kamp 1981, Heim 1982), where the intermediate formulas play a more substantive role, but even in classic MG, some care must be taken in the choice of alphabetic variants to avoid the accidental capture of one variable by an unintended quantifier; otherwise we could end up with $\forall x[\text{woman}(x)\forall x[\text{man}(x)\text{kissed}(x, x)]]$ as the translation of *every woman kissed every man.*

To present the variable binding mechanism in a standard format, we define a **variable binding term operator** (VBTO) as an operator that binds one or more free variables in a term or formula. An example immediately familiar to the reader will be the definite integral $\int_a^b \_ \, dx$, which binds the variable x of the function to whicj it is applied, or the λ abstractor λx. As usual, we require the equivalence of alphabetic variants: if $\vee$ is a VBTO, $F(x)$ and $F(y)$ are terms with x free in $F(x)$ exactly where y is free in $F(y)$, and we require $\vee x F(x) = \vee y F(y)$. We also require extensionality: $\forall x F(x) = G(x) \Rightarrow \vee x F(x) = \vee y F(y)$ (see Corcoran et al. 1972).

Once VBTOs are at hand, they can be applied to other phenomena as well, including questions such as *Who did John see?*, where it is the interrogative particle *who* that is treated as a VBTO wh so that the translation becomes wh$x$ see(John,x), and also to 'moved' and 'gapped' constituents as in *John saw everyone* or *John read the first chapter of everything that Mary did [read the first chapter of]*. While it is possible to translate long distance dependencies such as *Who did Mary see that John said that Bill resented?* as

$$\text{wh } x \text{ saw(Mary,x) \& said(John,resented(Bill,x))} \tag{6.23}$$

the first occurrence of the variable is now very far from the last, and the bookkeeping required to keep track of what should bind what where gets very complicated.

While the pure syntax of first-order formulas is context free (this is what makes Tarski-style induction over subformulas possible), getting the semantics right requires full context-sensitive (or at least indexed, see Exercise 5.5) power, so the complexity of the logic apparatus overshadows the syntactic complexity of natural language. This gives us a vested interest in finding mechanisms of semantic description that are combinatorically less demanding – these will be discussed in Section 6.3.

In reality, there are significant performance limitations to nesting these constructions, and in practice speakers tend to break them up e.g. as *What did John say, who did Mary see that Bill resented?* Even though the meaning of these sentences is somewhat different from (6.23) in that Mary's seeing someone now appears as part of what John said, the added simplicity is well worth the lost expressiveness to most speakers. Altogether, the semantics should account at least for the cases where pronouns (including interrogative, reflexive, and quantified expressions) appear in the positions normally filled by NPs, as in *He saw John, John saw him, Everyone saw John, John saw everyone, Who saw John? John saw himself*, etc.

The syntax should account for English-specific facts, e.g. that in situ object interrogatives, without an incredulous intonation, *John saw WHO?*, are far more rare than their *do*-support counterparts, *Who did John see?*, or that constructions such as **Himself saw John* are ungrammatical. The interaction between singular and plural forms, definite and indefinite articles, and quantifiers is very complex, with seemingly specific constructions such as *The hare will outrun the fox* still offering generic readings. Another complicating factor, in English and many other languages, is the use of the copula. So far, linguistic semantics has not advanced to the stage of providing a detailed grammar fragment covering these interactions in English, let alone to a parametric theory that contains English and other languages as special cases.

We emphasize here that the issue of eliminating variables from the statement of semantic rules is independent of the issue of generics. Even if no variables are used, the translation 'hares are fox-outrunners' (formulated as $H \subset FO$ in Peirce

Grammars; see Böttner 2001) implies there are no exceptions, while in fact the critical observation is that there can be exceptions (and in extreme cases such as *the P6 processor* discussed above, all cases can be exceptional).

Verbs

As the reader will have no doubt noticed, definition (6.19) of *candle* already goes beyond the use of adjectives, with binary or more complex relations such as *use for decoration*, *use on festive occasions*, *made of wax*, *having a cylindrical shape*, etc. With relationships such as

$$\text{use(people, candle, for decoration)} \tag{6.24}$$

$$\text{use(people, candle, on festive occasions)} \tag{6.25}$$

several questions present themselves. Is the *use* of (6.24) the same as the *use* of (6.25)? In general, the question of how lexical entries are to be individuated is particularly acute for verbs: consider *John ran*$_1$ *from the house to the tree*, *the fence ran*$_2$ *from the house to the tree*, *Harold ran*$_3$ *for mayor*, and *the engine ran*$_4$ *for a full week.* Using paraphrases such as *the fence/*John extends from the house to the tree*, it is easy to determine that run_1 and run_2 mean different things. For multilingual speakers, translations also give a good method for separating out dictionary senses: when the same word must be translated using two different words into another language, we can be virtually certain that there are two distinct meanings, but the converse does not hold. For example English run_1 and run_2 receive the same Hungarian translation, *fut*, while run_4 is a distinct motion verb *jár* 'walk', and run_3 has no single verb to translate it (more complex phrases such as 'participates in the election for' must be used).

Having separated out at least four senses of *run*, we want to make sure that we do not go overboard and treat *ran* differently in *John ran the Boston Marathon* and *John ran the Boston Marathon yesterday.* To do this, we will invoke the traditional distinction between *arguments*, which correspond to bindable slots on the relation, and *adjuncts*, which do not. We say that *Boston Marathon* fills an argument slot while *yesterday* does not, and it is only arguments that appear in lexical entries.

Formally, we will distinguish six VBTOs or *kāraka* (deep cases) called Agent, Goal, Recipient, Instrument, Locative, and Source. The following table summarizes their English and Sanskrit names, the basic description of the argument they bind, and the place in the Ashṭādhyāyī where they are defined.

Agent	*kartṛ*	the independent one	(1.4.54)
Goal	*karman*	what is primarily desired by the agent	(1.4.49)
Recipient	*saṃpradāna*	the one in view when giving	(1.4.32)
Instrument	*karaṇa*	the most effective means	(1.4.42)
Locative	*adhikaraṇa*	the locus	(1.4.45)
Source	*apādāna*	the fixed point that movement is away from	(1.4.24)

In constructing lexical entries, we thus need to specify both the arguments and the way they are linked. Since *run*$_4$ is intransitive (cf. *The engine runs, *The engine* *runs to Boston*) we can construct a formula Ax run(x), where A (Agent) is a VBTO. When the semantic relation has two arguments, as in $AxGy$ run(x, y), it is the type of the binder that decides the role in which the substituted element will appear. In the case of *run*$_3$, the Agent is *Harold* and the Goal is *mayor*.

In the process of generating the sentence, we begin by setting the (generally interlinked) tense, voice, and mood features and deciding on what to use for Agent and Goal. These get *expressed* by the appropriate morphological and syntactic devices: tense/aspect by the choice between *runs, ran, is running, will run*, etc., the Goal by the preposition *for* (as is typical in many languages), and the Agent by appearing in the preverbal position (a parochial rule of English). In the case of *run*$_2$, there is no fixed point that movement is away from. There is no movement to begin with, but even if we allow for some generalized, symbolic notion of movement, it is clear that *the fence runs from the house to the tree* describes the exact same situation as *the fence runs from the tree to the house*, while *John runs from the house to the tree* is truth-conditionally distinct from *John runs from the tree to the house*. Thus, we treat *run*$_1$ as $AxSyGz$ run(x, y) and $AxLyGz$ run(x, y, z).

Here we cannot even begin to survey the variety of constructions that the system should account for (Pāṇini's dhātupāṭha has about two thousand verbal roots, and Levin (1993) has nearly a thousand verb classes) and will restrict ourselves to the prototypical action sentence (5.8) as provided by the lexical entry $AxGy$ kill(x, y). Unlike many modern theories of the passive, Pāṇini does not require a separate lexical entry for the passive verbal complex *be killed*, not even one that is generated by a lexical rule from the active form.

6.3 Grammatical semantics

In Section 5.2 we surveyed a range of syntactic theories we called *grammatical* because they rely on notions such as (deep) case, valence, dependency, and linking, which are expressed only indirectly in the manner in which words are combined. Our goal here is to develop a formal theory of semantics that fits these theories as well as standard MG fits combinatorial theories of syntax. In Section 6.3.1, we summarize the system of semantic types and contrast it with the far richer system of syntactic types. In Section 6.3.2, we introduce *signs* and describe the mechanisms of their combination. We consider the use of system D instead of classic two-valued logic only a minor departure from MG and one that will not play an important role in what follows, beyond helping us define the range of semantic phenomena one should consider critical. A more significant departure from the tradition is that no MG-style fragment covering all the major constructions surveyed will be presented. To draw the limits of such a fragment, one would need to survey the frequency of the construction types in question, an undertaking beyond the scope of this book.

The central innovation in Section 6.3.2 is the introduction of a strict left-to-right calculus of combining signs (presented as a parser but equally valid as a generation method). To the extent that MG relies on induction over subformulas, and thus on a notion of a parse tree for a formula, we significantly depart from the MG tradition here. But to the extent that our goal is still to provide a truth-conditional formulation, as opposed to a 'language of thought' model, the reader may still safely consider the development here as part of MG.

6.3.1 The system of types

From the semantic viewpoint, we have found a need for only two basic types: e (entities) and t (truth values). Of these, e is used directly only for certain proper names, if at all. Most entities are treated as bundles of properties $(e \to t) \to t$. Derived types were also kept simple: two-place relations are of type $e \times e$ rather than $(e \to t) \to t \times (e \to t) \to t$, three-place relations are of type $e \times e \times e$, and so on. Given the well-known problems with hyperintensionals, we remained neutral on the use of intensional types for solving the problems of opacity, even though some scattered arguments in their favor (most notably, the existence of 'intensional' nouns such as *temperature*, and the ability to treat indexicals by manipulating indexes) have been noted.

This is in sharp contrast to the syntactic viewpoint, where we see a profusion of types, both basic and derived. In addition to the basic NP and S types, there are adverbs S/S (for the time being, we ignore issues of directionality – we return to this matter in Section 6.3.2), adjectives NP/NP, intransitive verbs S/NP, transitive verbs S/NPxNP, ditransitive verbs S/NPxNPxNP, and so on. Quantifiers and determiners require either some type Det so that bare nouns can be treated as NP/Det or some bare noun type N so that quantifiers and determiners can be assigned type NP/N. Without loss of generality, we can take the second option and redefine adjectives as N/N, clearly more appropriate than NP/NP in light of the observation that, at least in the plural, bare nouns readily take adjectival modifiers (without the need to add a determiner or quantifier later). Also, nonintersective adjectives are clearly functions from nouns to nouns.

Let us see how the remaining syntactic categories fit into this basic scheme. Since the distribution of pronouns is near-identical to that of NPs, we can perhaps ignore the rare resumptive cases and assign the type NP to pronouns as well. Since adverbials modify verbs, their type must be S/S, or perhaps V/V, where V is some abbreviatory device for the main verb types S/NP^k. Obviously, adverbs are a class distinct from adjectives: compare *brown fox*, *soon forgets* to **soon fox*, **brown forgets*. In English (and many other languages), there is an overt morphological suffix *-ly* that converts adjectives to adverbials: *happy fox*, *happily forgets*. By the logic of category assignment, adadjectives (modifiers of adjectives such as *slightly*, *seemingly*, *very*) must have category A/A, or, if A is analyzed as N/N, adadjectives must be (N/N)/(N/N). Remarkably, adadjectives can also function as adadverbs (that is, as modifiers of adverbs): consider *slightly happy*, *very soon*, etc. By the same logic, adadverbs are (V/V)/(V/V), so we have a defining relation

$$(N/N)/(N/N) = (V/V)/(V/V) \tag{6.26}$$

This makes perfect sense if both common nouns and verbs are treated as unary relations, but such a treatment makes no sense for transitive and higher-arity verbs.

A similar problem is observed for conjunctions such as *and* or *or*, which operate the same way across the whole system of categories: if X_1 and X_2 are both of category C, the phrase "X_1 and X_2" will also be of category C. Conjunctions therefore must have type C/CxC where C is some abbreviatory device for all categories S, NP, N, V, etc., possibly including conjunctions themselves, as in the expression *and/or.* It is worth noting that conjunctions work so well that reducing X_1 *and* X_2 makes sense even in cases where the X_1 would not get types assigned: consider *Mary wanted, and Bill obtained, every album of John Coltrane*. This phenomenon, known as *nonconstituent coordination* (NCC), argues either for a broad view of category combination or for the adjunction of zero elements to the structure. The former approach is taken in combinatory categorial grammar (Steedman 2001), but the latter is also defensible since the zero elements (called *traces* in transformational grammar) can also be used to derive the peculiar intonation pattern of NCC sentences.

One way of investigating the apparatus of category combination is to consider near-synonymous verbs that satisfy defining relations similar to (6.26). Take

$$AxGyRz \text{ give(x,y,z)} = AzGySx \text{ get(x,y,z)} \tag{6.27}$$

which expresses the idea that the truth conditions of *John gives a book to Bill* and *Bill gets a book from John* are identical. Similar relations hold for *sell/buy*, *fear/frighten*, *kill/cause to die*, etc. However, adverbs do not affect the arguments uniformly. Compare *This book will sell well* to **This book will be well-bought*, and similarly *to kill the duckling easily* does not mean *to cause the duckling to die easily*, and it is not clear what, if anything, the expression *?to easily cause the duckling to die* should mean.

In Chapter 3 we described signs as conventional pairings of sound and meaning. Here we refine this notion, defining a **sign** as a triple (sound, structure, meaning), where *sound* is a a phonological representation of the kind discussed in Chapters 3 and 4 (for ease of reading, we will use orthographical representations instead), *structure* is a categorial signature of the kind discussed in Chapter 5, and *meaning* is a formula in a logical calculus similar to that used in extensional versions of MG but taking truth values in system D. There are two kinds of elementary signs we need to consider, lexical entries and constructions. (Most elementary signs can be further analyzed morphologically, but from the perspective of syntax they are elementary in the sense that they have to be learned as units since their overall properties cannot be deduced from the properties of their parts.)

By a **directional category system** we mean an algebra with three binary operations $\backslash$, $/$, and $.$, which have the following properties:

$$y.(y\backslash x) = (x/y).y = x \tag{6.28}$$

$$z\backslash(y\backslash x) = (y.z)\backslash x \tag{6.29}$$

$$(x/y)/z = x/(z.y) \tag{6.30}$$

The concatenation operation . is marked only for clarity in (6.28)–(6.30), generally it is not written at all. By a *nondirectional categorial system* we mean an algebra with two binary operations, – (typeset as a fraction) and ., that have the following properties:

$$\frac{x}{y}.y = y.\frac{x}{y} = x \tag{6.31}$$

$$\frac{\frac{x}{y}}{z} = \frac{x}{y.z} = \frac{x}{z.y} \tag{6.32}$$

In nondirectional (also called unidirectional) systems, the typography is generally simplified by using slashes instead of fractional notation – this causes no confusion as long as we know whether a directional or a nondirectional system is meant.

Given a set of basic categories B, from $x, y \in B$ we freely form derived directional categories $x\backslash y, y/x$ (in the nondirectional case, $\frac{x}{y}$) and the concatenated category $x.y$. The full set of categories C is then obtained as the smallest set closed under these formation rules and containing B as a subset. **c-categorial grammars** map the lexicon one-to-many on category systems: we say that a sequence $l_1, \ldots, l_k$ of lexical elements is grammatical if there exist values of these mapping $c_1, \ldots, c_k$ and some order (not necessarily left to right) of performing the simplifications given in (6.28)–(6.32) that yield a designated element of the category system.

Discussion The definition given here is slightly different from the one given in Example 2.2.2 and developed in Section 5.1.2. Categorial grammar, as the term is generally understood, does not use concatenated categories at all, only those categories obtained from the basic categories by (back)slashes. However, the linguistic use of categories is better approximated by the variant presented here – we call the resulting system *c-categorial* 'concatenation categorial' to preserve the distinction, to the extent it is relevant, between these and the more standard categorial systems.

As (6.26) makes clear, in the cases of linguistic interest, we are not mapping onto the free algebra but rather on relatively small and well-behaved factors. In Section 5.2, we noted that the entire system of (sub)categories is finite, on the order of a few thousand categories, modified by morphosyntactic features that can themselves take only a few thousand values. One important way of making sure that free generation of derived categories does not yield an infinite set of categories is to enforce the noncounting property (5.29) as defining relations: for each category $x \in C$ we add the equation

$$x.x.x.x = x.x.x.x.x \tag{6.33}$$

Another important means of controlling the size of the category system is to take the system of basic categories B to be small. As our survey in Section 6.3.1 indicates, we can generally make do with only three basic categories: *S*, *NP*, and N. This set is hard to reduce further, as there are clear monomorphemic lexical exponents of each: *sentential* expressions such as *yes*, *ouch*, etc., are category S; *proper names* like *John* are category *NP*; and common nouns like *boy* are category *N*. As for the remaining major categories, adjectives like *red* are $N\backslash N$, intransitive verbs like *sleep* are $NP\backslash S$, transitive verbs are $NP\backslash S/NP$, ditransitives are $NP\backslash S/NP.NP$, and so on. It is often convenient to abbreviate the whole class $\{NP\backslash S/NP^i \mid i = 0, 1, 2, \ldots\}$

as V, but the category V need not be treated as primitive, just as the category A, which is simply an abbreviation for $N\backslash N$, does not need to be listed as part of the nonderived categories comprising B. Adverbs, adadjectives, and adadverbs are also derived categories, which leaves only one candidate, the set of adpositions P, as a potential addition to B. Here we take the view that adpositional phrases differ from adjectival phrases only in that verbs can subcategorize for them: semantically, they express two-place relations, so with the adpositional NP filled in they correspond to adjectives. In other words, we will treat being *on the hill* as a property, just like being *red*.

To complete the argument for $B = \{S, NP, N\}$, we need to discuss a variety of other category-changing (monomorphemic) elements, ranging from inflectional affixes such as the plural *-s* to 'particles' such as the infinitival *to* and the relativizer *that*. For inflectional material, we can use direct products: if $\langle \alpha F \rangle$ is a morphosyntactic feature and x some category, we freely form the category $x\langle \alpha F \rangle$. Ideally, this would imply that whenever a morphosyntactic distinction such as number or person is available in one category, such as that of verbs, the same distinction should also be available for all other categories. This is far from true: while nouns are indeed available in different numbers, only pronouns have different persons, and cross-linguistically adverbials do not show person or number variants.

In other words, the full system of categories is obtained not as a direct product of the basic and the inflectional categories but rather as a homomorphic image of this direct product. There are a variety of elements, ranging from the marker of definiteness *the* to the marker of infiniteness *to*, that languages with more complex morphology express by inflectional, rather than syntactic, means; cf. Romanian *frate/fratele* 'brother/the brother', Hungarian *eszik/enni* 'he eats/to eat' – different languages make the cut between the two parts of the direct product differently.

To make sure that we cover at least the same range of basic facts that MG covers, we need to consider relative clauses such as *[the boy] that Mary saw, [the boy] that saw Mary*. As these are clearly nominal postmodifiers, their category must be N/N, but note that their semantics differs from that of adjectives (category $N\backslash N$) significantly in that relative clauses are always intersective: a *lazy evaluation* need not be lazy, but an *evaluation that is lazy* must be. This observation is given force by examples such as *P: red paint that is blue* vs. *Q: blue paint that is red* – whatever substances P and Q may be (perhaps improperly manufactured paints that left the factory without quality control?), it is evident that P is in fact blue and Q is red, and not the other way around. To get a relative clause, we need a verb with one argument missing; e.g. Gx saw(Mary,x) or Ax saw(x,Mary). In categorial terms, such expressions are $\frac{S}{N}$ (with directionality, at least in English, tied to which argument is missing), so for subjectless relatives we have to assign *that* the category $(N\backslash N)/(NP\backslash S)$ and for objectless relatives $(N\backslash N)/(S/NP)$. In languages with overt case marking, the relativizer will actually carry the case of the *NP* that is missing from the S. In English, this is marginally observable in the requirement to use *whom* in *The boy whom Mary saw* as opposed to **The boy whom saw Mary*.

To simplify the notation, in what follows we will give category information in a nondirectional fashion (using forward slashes / instead of fractions) and encode the directionality restrictions in the *phonology* portion of the ordered triples. Also, we generalize from the set of six VBTOs (deep cases) listed in Section 6.2 to a broader concept of typed VBTOs that can bind only arguments of a given syntactic type T: for these we use the uniform notation λ_T. Thus, λ_{Vi} is a VBTO that can only bind intransitive verbs, $\lambda_{NP\langle PL.ACC\rangle}$ is one that can only bind forms that have category $NP\langle PL,ACC\rangle$, and so on. In the semantics, whether coupled with a nondirectional or a directional system of categories, we take this typing, rather than the linear order of the VBTOs, to be the determining factor in substitution. What we require is the analog of Fubini's theorem that order of execution does not matter. Formally, if x is of type U, y is of type V, and P is some two-place relation of type $U \times V$, we have

$$\lambda_U x \lambda_V y P(x,y) = \lambda_V y \lambda_U x P(x,y) \tag{6.34}$$

and if a and b are of types U and V, respectively, we take

$$\begin{aligned} &\lambda_U x \lambda_V y P(x,y)ab = \lambda_V y \lambda_U x P(x,y)ab \\ &= \lambda_U x \lambda_V y P(x,y)ba = \lambda_V y \lambda_U x P(x,y)ba = P(a,b) \end{aligned} \tag{6.35}$$

Before reformulating the main constructions surveyed in Section 6.2 in terms of signs, let us see some example of triples, first without VBTOs. The lexical entry of the sign for *after* in constructions such as *After the rain, Mary went home* will be

$$(\text{after}, s_1, s_2; S/(S.S); `\tau(s_1) < \tau(s_2)') \tag{6.36}$$

Here s_1, s_2 are strings (of type S). We make no apologies for the ad hoc notation 'the temporal value of s_1 is less than that of s_2' that we provided for the third (semantics) part of the triple. To make this less ad hoc, one would need to develop a full theory of temporal semantics, a very complex undertaking that would only detract from our current purpose. We also gloss over the important problem of how an NP such as *the rain* is to be construed as an S to which a temporal value can be reasonably assigned. An attractive solution is to assume a deleted verb *fell*, *stopped*, etc. Turning to phrasal constructions such as the 'arithmetic proportion' discussed in Section 5.3, we have

$$(s_1 \text{ is to } s_2 \text{ as } s_3 \text{ is to } s_4; S/(NP.NP.NP.NP); `\sigma(s_1)/\sigma(s_2) = \sigma(s_3)/\sigma(s_4)') \tag{6.37}$$

To see how lexical and phrasal signs combine, consider *London is to Paris as John Bull is to Marianne*. As discussed in Section 6.2.2, *London* and *Paris* are not simply some rigid designators but contentful lexical entries 'capital of England, large city in England', and similarly *John Bull* and *Marianne* are 'person typifying English character, personification of England' and 'person typifying French character, personification of France', which at once makes the proportionality of (6.37) evident: both sides express some relation of England to France. Were we to ignore the lexical information, it would be a mystery why *London is to Berlin as John Bull is to Marianne* is, under the ordinary interpretation of this sentence, false.

6.3.2 Combining signs

While the theory of combining signs is in principle neutral between parsing (analysis) and generation (synthesis), in practice there is a dearth of truly neutral terminology and the discussion needs to be cast either in generation- or in parsing-oriented terms. Here we take the latter option, but this implies no commitment to interpretative semantics – generative terminology would work just as well.

In the analysis of computer programs, two major strategies are available: either we begin by a pure syntax pass over the code and translate only a parsed version of the program (one that has already been endowed by a tree structure), or we build the entire semantics as we go along. Computer science terminology varies somewhat, but these strategies are often called *compilation* and *interpretation*. Here our focus is with compilation, assigning as many meaning representations to an entire sentence (or larger structure) as its degree of ambiguity requires.

As a simple example, consider *Time flies*, which means 'tempus fugit' or, if interpreted as an imperative, as a call to measure the speed of a common insect. The ambiguity rests on the ambiguity of the lexical signs (time; N; 'tempus') vs. (time; V; 'mensuro') and (flies; V; 'fugito') vs. (flies; N⟨PL⟩; 'musca'), and on the existence of multiple paths through the grammar (state diagram) as shown schematically below:

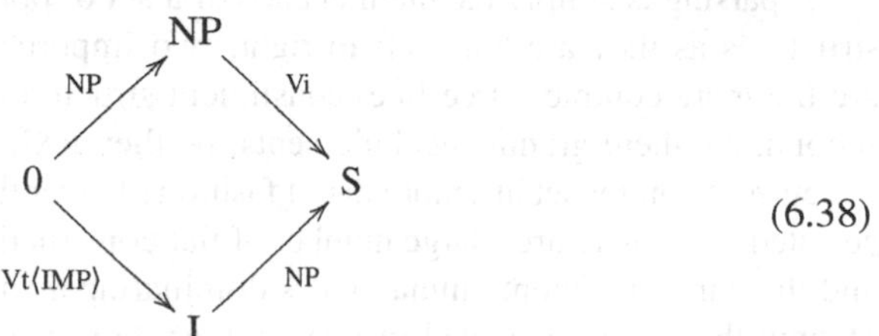

(6.38)

Ignoring matters of agreement for the moment, the top path corresponds to the rule (5.7) and the bottom path to the imperative construction as e.g. in *Eat flaming death!* For clarity, we added node labels corresponding to the category of the construction reached by traversing the arcs from the start state 0 and accepting the sequence of symbols on them.

Parsing with some FSA consists in tracking paths through the state diagram in accordance with the lexical entries scanned. We begin by placing a pointer at the start state and look up the lexical entry (or entries) that have phonological (in our case orthographical) components that match the first word of the string to be parsed. In this example there are two such entries, so we nondeterministically advance to both states at the end of these arcs, collecting the semantics of each as we go along. In general, the same nondeterministic update step is to be performed on all members of the active pointer set: for each we consider what outgoing arcs match the incoming lexical entry or entries and advance the pointer, possibly nondeterministically, to the ends of the respective arcs – if no such arc can be found, the pointer is removed from the active set. One important possibility to consider at each step is an update that matches a silent (nonemitting) arc. For example, in English, object relativizers are typically optional; e.g. *the man that I saw* is equivalent to *the man I saw*, so that the entire network responsible for these contains (under a homomorphism that deletes agreement features) a subnetwork such as (6.39):

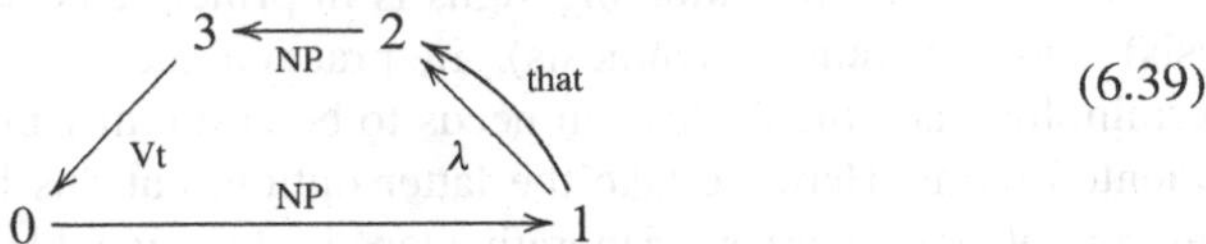

(6.39)

In parsing a sentence such as (5.28), one path takes us through a relative clause with a silent relativizer *fat (that) people eat*, yielding an NP subject for the predicate *accumulates.* The formal model itself shows no garden path effect: having matched *fat people* to the NP path of (6.38) and subsequently *eat* to the predicate, the pointer that embodies this analysis is at node S, corresponding to a complete sentence. The subsequent word *accumulates* has no outgoing match, so this pointer is eliminated from the pointer set, leaving only the correct path, which is as it should be, given the fact that (5.28) is unambiguous. However, we may speculate that the human parser does not maintain a full nondeterministic set of pointers and partial parses the same way a machine could but will prune after awhile all nonpreferred paths. The garden path phenomenon is suggestive of a pruning strategy corresponding to a very narrow (perhaps only a couple of words) lookahead.

If parsing is simply the maintenance of a set of pointers and the partial semantic structures as they are built left to right, two important questions arise. First, how are the more complex (tree-like) constituent structures created and maintained, and second, are there grammatical elements, be they lexical or phrasal, that manipulate the current pointer set in a more direct fashion? Under the construction grammar view adopted here, there are a large number of flat constructions such as (6.36) and (6.37), and the number of nonterminal nodes dominating a construction is a direct function of how these are embedded in one another. For example, in *The king of England opened Parliament* there are two constructions to consider: the familiar S $\rightarrow$ NP VP (5.7), and the NP$\langle$POS$\rangle$ $\rightarrow$ NP of NP construction. In rewriting terms, we apply (5.7), rewrite the NP using the possessive rule, and rewrite the VP using

$$\text{VP} \rightarrow \text{V NP} \tag{6.40}$$

(or rather, with the VP on the left carrying the same plural agreement feature as the V on the right, and the NP carrying the ACC case feature; see Section 5.7.)

This means that to understand the contribution of *The king* we need to first understand how it contributes to *The king of England* (as discussed in Section 5.2.3, this is not necessarily a straightforward possessive relation) and next understand how the whole NP contributes to the sentence (by binding the Agent valency of *open*). Altogether, the depth of the phrase structure tree corresponds exactly to the degree that constructions are recursively substituted in one another, and there seems to be no principled limitation on this. To the extent we see a limitation, it is on the number of pointers k that can be maintained simultaneously in human memory. What the garden path phenomenon suggests is that this limit is at two. (Even $k = 2$ is something of a stretch: in reality, when the first parse dies out, what we need to do is to recall the whole sentence from short-term auditory memory and restart the process – the second pointer is there to keep the mechanism from taking the path it took the first time by marking that particular analysis as dead).

In the strict left to right parsing model assumed here, each partial parse leads to a deterministic state, with different parsing alternatives typically corresponding to different states, though we leave open the possibility of pointer paths merging. Consider the initial sequence *Who did* which can be completed in many ways: by an in situ question particle *what?*; by an objectless clause *Mary see?*; by far more complex constructions such as *her parents think would be the best match for Mary?*; or, if we are willing to treat question intonation as a boundary melody, simply by this melody (orthographically represented as *?* in these examples). The last case corresponds to *sluicing* questions, typically uttered by a person different from the one who uttered the lead sentence: *A: John decided to buy an airplane. B: Who did?* As there are at least two different senses of *who* to consider (one of which can be replaced by *whom* in a stylistically marked variant of English, but the other cannot) and there are at least two different senses of *did*, we can entertain four different paths through the state machine. At this state of the parse, the string *Who did* has a nondeterministic analysis (pointer set) composed of these four possibilities: depending on the continuation, only one will be an initial segment of the complete parse.

One peculiarity of the state machines arising in natural language syntax is that at any given stage of the left to right parse, at least one of the pointers in the currently active set is close to a final state. Consider the initial sequence *After a good.* A complete continuation could be *boy does his homework, he may play* – this requires seven words. However, a much shorter continuation, with *meal, leave*, is also available: we take advantage of the fact that in English an intransitive verb alone is sufficient to form an imperative sentence (a form of the lexicality constraint discussed in Section 5.1.3) and also of the observation (see Erdélyi-Szabó et al. 2007) that nouns such as *storm*, *meal*, etc., supply temporal coordinates (typecasting *meal* to S is a silent move). Since the construction (6.36) involves two sentences, in the *homework* continuation a pointer different from that used in the *meal* case will survive, but the situation is analogous to chess, where we may consider deeply embedded structures of gambits and counter-gambits, but a fast way out, by one player resigning, is always available.

How big is the state machine? In Corollary 4.4.1, we demonstrated that vocabulary size is infinite, but in Section 5.1.1 we argued that the set of strict (sub)categories C is finite – in Section 5.2.2 we estimated, rather generously, $|C|$ to be below $7.7 \cdot 10^{23}$. A more realistic estimate is 10^6, subdirectly composed of some 10^3 lexical classes with some 10^3 inflectional possibilities. By the preceding observation, any active pointer set has at least one member that is within a few (say, five) steps of becoming final. As there are at most $|C|^5 \leq 10^{30}$ five-step continuations, and there is a pointer in any active set (with at most k members) that leads to a final state by such a continuation, there can be at most $k \cdot 10^{30}$ states, given that those states that can never be reached by any member of an active set can obviously be pruned. Again, the estimate is rather generous: in reality, much shorter continuations suffice and no more than 10^{12} states are expected, even if we permit $k > 2$. While this number may still look too large, what really matters is not the raw number of states and transitions but rather the manner in which these are organized – the overall complexity of the system (see Chapter 7) is considerably smaller than these numbers may suggest.

Lexicality, taken in a broad sense, amounts to the statement 'if it can be done by a phrase, it can be done by a single word'. So when we ask the general question whether phrases can manipulate the active pointer set in a manner more direct than the continuation tracking discussed so far, what we would really like to see are lexical entries capable of such manipulation. A striking example is provided by coordinating conjunctions, in particular *and*. Consider the initial segment *Mary wanted and*. This has two major continuations of interest here: *obtained records* and *Bill obtained records*. Absent the rather characteristic intonation pattern (orthographically represented as ,) of NCC, the preferred path is with constituent coordination: *wanted and obtained* become parallel verbs whose Agent valency is filled by *Mary* and whose Goal valency is filled by *records*. However, when we see the next element, *Bill*, this path is no longer available, and it is *Mary wanted* that needs to be made parallel with *Bill* _. This can only be done if *and* has the power to introduce new active pointers nondeterministically at each previously traversed state. The continuation *Bill* is then free to take the first of these. One argument in favor of treating *and* in this fashion is that it can occur in the initial position: *And now, ladies and gentlemen,...*

How are the semantics portions of signs to be combined? Since constructions such as (6.37) cannot be recognized until the formatives *is to*, *as*, *is to* that define them are supplied, the relevant semantic function $\sigma(s_1)/\sigma(s_2) = \sigma(s_3)/\sigma(s_4)$ cannot be fully invoked at the stage *London is*. To invoke it partially, $\sigma(\text{London})/\sigma(s_2) = \sigma(s_3)/\sigma(s_4)$ must already at this stage be maintained as one of the nondeterministic analyses, only to discard it when the next word, *foggy*, arrives. Psychologically it may make more sense to assume that processing is deferred at least until a complete constituent is collected, just as in the processing of predicate calculus formulas, where interpretation proceeds through well-formed subformulas, but we have no intention here of presenting a psychologically realistic parsing model (see Section 5.6); our goal is simply to build a formal account couched in parsing terms.

The real issue, then, is whether disjoint pieces must be kept in partial parses or whether a single combined sign must be available at each turn. The problem is well illustrated by the case of English transitives, which, according to most theories, are built using the rules (5.7) and (6.40). In other words, the construction Vt NP itself is equivalent to an intransitive verb, and since its semantic representation can only be built after the object is available, the whole construction cannot be interpreted earlier. To implement this view as a parsing strategy would require keeping track of *The king of England* and *opened*, and once *Parliament* comes under scan, combine it first with *opened* and combine only the result with *The king of England*. In the proposal made here, (6.35) makes the formalism strong enough to carry out the substitutions in any order, so that the Agent VBTO of $AxGy$ open(x, y) can bind *The king of England* without assuming the existence of a constituent *The king of England opened*. To the extent scope-taking does not follow the linear order, we may still want to invoke a delay mechanism such as *Cooper storage* (see Cooper 1975), but, given our observations about the defeasability of quantifiers, much of the standard evidence in favor of such a mechanism needs significant reexamination.

6.4 Further reading

The Liar paradox goes back to the 4th century BCE. The solution adopted here is essentially that of Barwise and Etchemendy (1987), using the elementary formulation of Walker (2003) rather than the more technical ZFC^{-}+ Anti-Foundation Axiom-based version (Aczel 1988) that is now standard. Other notable efforts at a solution include three-valued logic (for a thorough discussion, see Visser 1989) and paraconsistent logic (Priest 1979, Priest et al. 1989). Read (2002) and Restall (2007) reconstruct the work of the medieval logician Thomas Bradwardine, who reaches, by rather different means, the same conclusion, that (6.1) is not contradictory, it is false. Our desideratum *(C)*, which amounts to some strong form of direct compositionality, is somewhat controversial, see Janssen (1997) and Barker and Jacobson (2007).

For the interaction of language and number systems, see Hurford (1975) and Wiese (2003). For propositional attitudes, see in particular Frege (1879), Quine (1956), Barwise and Perry (1983), or the short tutorial summary by Bäuerle and Cresswell (1989). Although our conclusions are diametrically opposed, our discussion is greatly indebted to Thomason's (1980) summary of the prerequisites for Tarski's undefinability theorem. For an alternative view on the viability of quotational theories see des Rivières and Levesque (1986). Another alternative treatment of opacity, based on *structured meanings*, was developed in Cresswell (1985). See Ojeda (2006a) for detailed argumentation why a nonintensional treatment is to be preferred both on grounds of simplicity and grounds of adequacy.

With the Principle of Responsibility, our discussion departs somewhat from the MG tradition in that we make lexical semantics carry a great deal of the explanatory burden. MG in general is silent on the meaning of lexical entries: aside from some meaning postulates that tie the meaning of nonintensional verbs and nouns to their intensionalized meanings, only a few function words are ever assigned translations, and attempts to push the basic techniques further (see in particular Dowty 1979) go only as far as purely logical meanings can be assigned. For nominals see Pustejovsky (1995). The method of defining lexical entries in terms of their essential properties goes back to Aristotle and is perhaps best articulated in contemporary terms by Wierzbicka and the NSM school. For a recent critique of this approach, fueled largely by the ideas of Kripke (1972) already discussed here, see Riemer (2006). Although the use of generalized quantifiers is often viewed as characteristic of MG, the idea goes back to Leibniz and was stated in modern terms as early as in Mostowski (1957).

A good starting point for generics is Carlson and Pelletier (1995). That generics admit exceptions and thus require a mechanism greatly different from that of standard quantification has long been noted (Jespersen 1924). For a summary evaluation and quick dismissal of nonmonotonic approaches to generics, see Pelletier and Asher (1997), and for a more sympathetic view, see Thomason (1997) in the same volume. The standard treatment of exceptions is Moltmann (1995), who takes the exception domain to be subtracted from each element in the set of properties that make up a generalized quantifier (see also Lappin 1996). In our case, only essential properties are used, and the exception domain could be subtracted from them, but our discussion here is also compatible with the view that those essential properties that are contradicted by the exception get dropped out entirely. The possibility of using paraconsistent logic has already been considered by Bäuerle and Cresswell (1989);

> Possibly we would need in addition a nonstandard propositional logic, perhaps e.g. the kind that Belnap (1977) thinks a computer should use when reasoning from inconsistent information.

but they dismissed it promptly on the grounds that "it is hard to see how any approach of this kind can guarantee that we have enough impossible worlds". We use Ginsberg's (1986) system D, which combines paraconsistency with default reasoning.

The idea of eliminating bound variables from logic goes back to Schönfinkel (1924), but fuller development begins with Curry and Feys (1958) and Quine (1961). Suppes (1973) and Suppes and Macken (1978) credit the idea to Peirce; see also Purdy (1992) and Böttner (2001). For a summary of the main ideas from both the linguistic and the mathematical perspectives, see Barker (2005), where both the standard analyses with variables and the variable-free approach pioneered by Szabolcsi (1987), Jacobson (1999), and others are discussed. Another approach, doing away with variables in favor of *arbitrary objects*, is introduced in Fine (1985).

Kārakas are discussed in Staal (1967) and Kiparsky (2002). The use of VBTOs (in particular, lambda-operators) to capture argument linking was suggested in lectures by Manfred Bierwisch (1988). For c-categorial grammars, we could not locate any specific reference, but the idea of using concatenated (direct product) categories is part of the folklore. Some of the ideas about pointer set maintenance go back to unpublished work of the author and László Kálmán, in particular Kornai and Kálmán (1985). The sluicing phenomenon was identified (and named) by Ross (1969); for a recent overview see Merchant (2001).

Historically, only the MG tradition of linguistic semantics has been presented in a fully formalized manner, but many of the ideas presented in this chapter have been stated quite clearly, if informally, in studies such as Wierzbicka (1985), Langacker (1982, 1987, 1991), Fauconnier (1985), and Jackendoff (1972, 1983, 1990).

7 Complexity

Grammars are imperfect models of linguistic behavior. To the extent that we are more interested in competence than in performance (see Section 3.1), this is actually desirable, but more typically discrepancies between the predictions of the model and the observables represent serious over- or undergeneration (see Section 2.2). There is, moreover, an important range of models and phenomena where it is not quite obvious which of the cases above obtain. Suppose the task is to predict the rest of the series $2, 3, 5, \ldots$. A number of attractive hypotheses present themselves: the prime numbers, the Fibonacci numbers, square-free numbers, the sequence $2, 3, 5, 2, 3, 5, 2, 3, 5, \ldots$, and so on. The empirically minded reader may object that the situation will be greatly simplified if we obtain a few more data points, but this is quite often impossible: the set of actual human languages cannot be extended at will.

Therefore, it would be desirable to have an external measure of simplicity, so that we can select the best (most simple) hypothesis compatible with a given range of facts. Starting with Pāṇini, linguists tend to equate simplicity with shortness, and they have devoted a great deal of energy to devising notational conventions that will make the linguistically attractive rules and generalizations compactly expressible. The central idea is that such notational conventions get amortized over many rules, so in fact we can introduce them without seriously impacting the overall simplicity of the system.

In this chapter, we begin to develop such a theory of simplicity. In Section 7.1 we introduce the basic notions of *information* and *entropy*, and in Section 7.2 we present the basic theory of Kolmogorov complexity originally put forth in Solomonoff (1964). Section 7.3 deals with the more general problem of *inductive learning* from the perspective of complexity.

7.1 Information

In the classical model of information theory (Shannon 1948), the *message* to be transmitted is completely devoid of internal structure: we are given a (finite or infinite) set $A = \{a_1, a_2, \ldots\}$ of elementary messages and a probability distribution P over A.

More complex messages are formed by concatenating elementary messages, and it is assumed that this is a Bernoulli experiment, so that the choice of the next elementary message is independent of what went on before. To transmit one of the a_i, we need to encode it as a bitstring $C(a_i)$. We need to make sure that C is invertible and that $C^+ : A^+ \to \{0, 1\}^+$, defined by lifting C to be a (concatenation-preserving) homomorphism, will also be invertible. This is trivially satisfied as long as no codeword is a prefix of another codeword (i.e. the code is **prefix-free**) since in that case we know exactly where each codeword ends (and the next one begins) in the stream of bits transmitted.[1]

Exercise 7.1 Construct a code C such that C is not prefix-free but C^+ is nevertheless invertible.

An important subclass of prefix-free codes are fixed-length codes, where the length n_i of $C(a_i)$ is constant (usually a multiple of eight). However, in the cases that are of the greatest interest for the theory, the number of elementary messages is infinite, so no fixed-length code will do. Prefix-free variable-length binary codes will satisfy the following theorem.

Theorem 7.1.1 Kraft inequality. In a prefix-free code, if c_i are codewords of length n_i,

$$\sum_i 2^{-n_i} \leq 1 \tag{7.1}$$

Proof A prefix-free set of codewords can be depicted as a binary tree, where each sequence of zeros and ones corresponds to a unique path from the root to a leaf node, zero (one) meaning turn left (right). For the tree with two nodes, (7.1) is trivially satisfied (as equality). Since all prefix codes can be obtained by extending a leaf one or both ways, the result follows by induction for finite codes and, by standard limit arguments, for infinite codes as well.

Given a probability distribution *P*, *Shannon-Fano codes* are computed by an algorithm that constructs this tree top-down by dividing the total probability mass in two parts recursively until each node has only one elementary message. Shannon-Fano codes are of historical/theoretical interest only: in practical applications, the number of elementary messages is finite, and the more efficient *Huffman codes*, where the tree is created from the bottom up starting with the least probable message, have replaced Shannon-Fano codes entirely.

Exercise 7.2 Specify the top-down algorithm in more detail using raw (cumulative) probabilities. Compare your algorithms with the actual Fano (Shannon) algorithms. Specify the bottom-up procedure, and compare it with the Huffman algorithm. Do any of these procedures fit the notion of a *greedy* algorithm?

Now we introduce the quantity $H(P) = -\sum_i p \log_2 p$, known as the **entropy** of the distribution P. While the definition of H may at first blush look rather arbitrary, it is, up to a constant multiplier (which can be absorbed in the base of the logarithm

[1] Somewhat confusingly, codes enjoying the prefix-free property are also called *prefix codes*.

chosen) uniquely defined as *the* function that enjoys some simple and natural properties we expect any reasonably numerical characterization of the intuitive notion of *information* to have.

Theorem 7.1.2 (Khinchin 1957) Uniqueness of entropy. We investigate nonnegative, continuous, symmetrical functions $I(p_1, p_2, \ldots, p_k)$ defined for discrete probability distributions $P = \{p_1, p_2, \ldots, p_k\}(\sum p_i = 1)$. (i) If P' is formed from P by adding another outcome a_{k+1} with probability $p_{k+1} = 0$, we require $I(P) = I(P')$. (ii) We require I to take the maximum value in the equiprobable case $p_i = 1/k$. (iii) For P, Q independent we require $I(PQ) = I(P) + I(Q)$ and in general we require $I(PQ) = I(P) + I(Q|P)$. The only functions I satisfying the conditions above are $-c \sum p_i \log(p_i) = cH$ for arbitrary nonnegative constant c.

Proof Let us denote the maximum value $I(P_k)$ taken in the equiprobable case $P_k = \{1/k, 1/k, \ldots, 1/k\}$ by $l(k)$. By (i) we have $l(k) \leq l(k+1)$. By property (iii), taking r independent distributions P_k, we have $l(k^r) = l(P_k^r) = l(P_{k^r}) = rl(k)$. Letting $k = e^z$ and $l(e^x) = B(x)$, this functional equation becomes $B(zr) = rB(z)$, which, by a well-known theorem of Cauchy, can only be satisfied by $B(z) = cz$ for some constant c, so $l(e^z) = cz$, and thus $l(k) = c\log(k)$ (and by the monotone growth of l established earlier, $c > 0$). Turning to the nonequiprobable case given by rational probabilities $p_i = n_i/n$, we define Q to contain n different events, divided into k groups of size n_i, such that the conditional probability of an event in the ith group is $1/n_i$ if a_i was observed in P and zero otherwise. This way, we can refer to the equiprobable case for which I is already known and compute $I(Q|P) = c\sum_{i=1}^{k} p_i \log(n_i) = c\sum_{i=1}^{k} p_i \log(p_i) + c\log(n)$. In the joint distribution PQ, each event has the same probability $1/n$, so that $I(PQ) = c\log(n)$. Given our condition (iii), we established $I(P) = -c\sum_{i=1}^{k} p_i \log(p_i)$ for p_i rational and thus by continuity for any p_i. That H indeed enjoys properties (i)–(iii) is easily seen: (i) is satisfied because we use the convention $0\log(0) = 0$ (justified by the limit properties of $x\log(x)$), (ii) follows by Jensen's inequality, and finally (iii), and the more general chain rule $H(P_1P_2\ldots P_k) = H(P_1) + H(P_2|P_1) + H(P_3|P_1P_2) + \ldots + H(P_k|P_1P_2\ldots P_{k-1})$ follows by simply rearranging the terms in the definition.

Discussion There are a range of other theorems that establish the uniqueness of H (see in particular Lieb and Yngvason 2003), but none of these go all the way toward establishing H as *the* appropriate mathematical reconstruction of the intuitive notion of information. One issue is that entropy is meaningful only over a statistical ensemble: the information content of individual messages is still a function (negative log) of their probability. Many authors find this counterintuitive, arguing e.g. that *There is a leopard in the garden* provides the same amount of information as *There is a dog in the garden*, namely the presence of an animal. Even if we grant the point that probabilities are important (e.g. a message about a surprising event such as winning the lottery is more informative than a message about a likely event such as not winning it), outside the games of chance domain it is not at all trivial to assign reasonable background estimates to probabilities. Here we take the view that this is inevitable inasmuch as information content depends on our expectations: the same message, *He* *had steak for dinner*, is more informative about a supposed vegetarian than about a meat-and-potatoes guy.

Exercise 7.3† Obtain a sample of English and count the frequency of each character, including whitespace. What is the *grapheme entropy* of the sample? How much is the result changed by using e.g. Perl or C program texts rather than ordinary (newspaper) text in the sample?

Exercise 7.4† Write a program that parses English into syllables, and count each syllable type in a larger sample. What is the *syllable entropy* of the text?

While it is easiest to consider phonemes, graphemes, syllables, and other finite sets of elementary messages, the definition of entropy is equally meaningful for infinite sets and in fact extends naturally to continuous distributions with density $f(x)$ by taking $H(f(x)) = -\int_{-\infty}^{\infty} f(x)\log_2(f(x))dx$. Here we will consider English as being composed of words as elementary messages and estimate its *word entropy*. The task is made somewhat harder by the fact that words, being generated by productive morphological processes such as compounding (see Chapter 4), form an infinite set. Although the probabilities of *the*, *of*, *to*, *a*, *and*, *in*, *for*, *that*, and other frequent words can be estimated quite reliably from counting them in samples (corpora) of medium size, say a few million words, it is clear that no finite sample will provide a reliable estimate for all the (infinitely many) words that we would need to cover.

Therefore we start with Zipf's law (see Section 4.4), which states that the rth word in the corpus will have relative frequency proportional to $1/r^B$, where B, the *Zipf constant*, is a fixed number slightly above 1. To establish the constant of proportionality C_k, recall Herdan's law that a corpus of size N will have about $N^{1/B}$ different words. Let us denote the cumulative probability of the most frequent k words by P_k and assume Zipf's law holds in the tail, so that we have

$$1 - P_k = C_k \sum_{r=k+1}^{N^{1/B}} r^{-B} \approx C_k \int_k^{N^{1/B}} x^{-B}dx = \frac{C_k}{(1-B)}[N^{\frac{1-B}{B}} - k^{1-B}] \quad (7.2)$$

For large N, the first bracketed term can be neglected, and therefore we obtain $C_k \approx (1 - P_k)(B - 1)k^{B-1}$. The first k words, for relatively small fixed k, already cover a significant part of the corpus: for example, the standard list in Volume 3 of (Knuth 1971) contains 31 words said to cover 36% of English text, the 130–150 most frequent collected in Unix `eign` cover approximately 40% of newspaper text, and to reach 50% coverage we need less than 256 words. To estimate the entropy, we take

$$H = -\sum_{r=1}^{k} p_r \log_2(p_r) - \sum_{r=k+1}^{N^{1/B}} p_r \log_2(p_r) \quad (7.3)$$

The first sum, denoted H_k, can be reliably estimated from frequency counts and of course can never exceed the maximum (equiprobable) value of $\log_2(k)$. The second sum can be approximated by integrals:

$$\frac{C_k B}{\log(2)} \int_k^{N^{1/B}} \log(x) x^{-B} dx - \frac{C_k \log(C_k)}{\log(2)} \int_k^{N^{1/B}} x^{-B} dx \tag{7.4}$$

The value at the upper limit can be neglected for large N, so we get the following theorem.

Theorem 7.1.3 The word entropy H of a language with Zipf constant B is given by

$$H \approx H_k + \frac{1 - P_k}{\log(2)}(B/(B-1) - \log(B-1) + \log(k) - \log(1 - P_k)) \tag{7.5}$$

H_{256} can be estimated from medium or larger corpora to be about 3.9 bits. P_{256} is about 0.52, and B for English is about 1.25, so the estimate yields 12.67 bits, quite close to the $\sim$ 12 bits that can be directly estimated based on large corpora (over a billion words). In other languages, the critical parameters may take different values. For example, in Hungarian, it requires the first 4096 words to cover about 50% of the data, and the entropy contributed by H_{4096} is closer to 4.3 bits, so we obtain $H \leq 15.41$ bits. Equation (7.5) is not very sensitive to the choice of k but is very sensitive to B. Fortunately, on larger corpora, B is better separated from 1 than Zipf originally thought. To quote Mandelbrot (1961b:196):

> Zipf's values for B are grossly underestimated, as compared with values obtained when the first few most frequent words are disregarded. As a result, Zipf finds that the observed values of B are close to 1 or even less than 1, while we find that the values of B are not less than 1.

As we shall see in the following theorem, for prefix codes the entropy appears as a sharp lower bound on the expected code length: no code can provide better "on the wire" compression (smaller average number of bits). Our estimate therefore means that English words require about 12 bits on average to transmit or to store – this compares very favorably to using 7-bit ascii, which would require about 35 bits (the frequency-weighted average word length in English is about five characters).

Theorem 7.1.4 (Shannon 1948) Let a_i be arbitrary messages with probability p_i and encoding $C(a_i) = c_i$ of length n_i such that $\sum_i p_i = 1$ and the set of codewords is prefix-free,

$$L(P) = \sum_i p_i n_i \geq H(P) \tag{7.6}$$

with equality iff the codewords are all exactly of length $\log_2(1/p_i)$.

Proof

$$H(P) - L(p) = \sum_i p_i \log_2 \frac{2^{-n_i}}{p_i} = \log_2 e \sum_i p_i \ln \frac{2^{-n_i}}{p_i} \tag{7.7}$$

The right-hand side can be bound from above using $\ln x \leq x - 1$, and by the Kraft inequality we get

$$H(P) - L(p) \leq \log_2 e \sum_i p_i (\frac{2^{-n_i}}{p_i} - 1) \leq 0 \tag{7.8}$$

When probabilities are very far from binary fractions, direct encoding may entail considerable loss compared with the entropy ideal. For example, if $p_1 = .9, p_2 = .1$, the entropy is 0.469 bits, while encoding the two cases as 0 vs. 1 would require a full bit. In such cases, it may make sense to consider blocks of messages. For example, three Bernoulli trials would lead tp 111 with probability .729; 110, 101, or 011 with probability .081; 001, 010, or 100 with probability .009; and 000 with probability .001. By using 0 for the most frequent case, 110, 100, and 101 for the next three, and finally 11100, 11101, 11110, and 11111 for the remaining four, we can encode the average block of three messages in 1.598 bits, so the average elementary message will only require $1.598/3 = 0.533$ bits.

Exercise 7.5 Prove that no block coding scheme can go below the entropy limit but that with sufficiently large block size the average code length can approximate the entropy within any $\varepsilon > 0$.

Exercise 7.6 Prove that a regular language is prefix-free iff it is accepted by a DFSA with no transitions out of accepting states. Is a prefix-free language context-free iff it is accepted by a DPDA with the same restriction on its control?

In real-life communication, prefix-free codes are less important than the foregoing theorems would suggest, not because real channels are inherently noisy (the standard error-correcting techniques would be just as applicable) but because of the peculiar notion of *synchrony* that they assume. On the one hand, prefix-freeness eliminates the need for transmitting an explicit concatenation symbol, but on the other, it makes no provision for `BEGIN` or `END` symbols: the only way the channel can operate is by keeping the sender and the receiver in perfect synchrony. In Section 7.2 we will discuss a method, *self-delimiting*, that makes any set of codewords prefix-free.

Exercise 7.7 Research the role of the ascii codes 0x02 (STX), 0x03 (ETX), and 0x16 (SYN).

Variable-length codes (typically, Huffman encoding) therefore tend to be utilized only as a subsidiary encoding, internal to some larger coding scheme that has the resources for synchronization. We will discuss an example, the G3 standard of fax transmission, in Section 9.3.

Exercise 7.8 Take the elementary messages to be integers i drawn from the geometrical distribution ($p_i = 1/2^i$). We define a complex message as a sequence of n such integers, and assume that n itself is geometrically distributed. How many bits will the average complex message require if you restrict yourself to prefix-free codes (no blocking)? With blocking, what is the optimal block size? What is the optimum average message length if the restriction on prefix-freeness is removed?

Human language, when viewed as a sequence of phonemes, shows very strong evidence of *phonotactic regularities* (i.e. dependence between the elementary messages). If we choose syllables as elementary, the dependence is weaker but still considerable. If we use morphemes as our elementary concatenative units, the dependence is very strong, and if we use words, it is again weaker, but far from negligible. Besides its uses in the study of coding and compression, the importance of entropy comes from the fact that it enables us to quantify such dependencies. For independent variables, we have $H(PQ) = H(P) + H(Q)$ (see condition (iii) in

Theorem 7.1.2 above), so we define the **mutual information** between P and Q as $H(P) + H(Q) - H(PQ)$. Mutual information will always be a nonnegative quantity, equal to zero iff the variables P and Q are independent. We also introduce here **information gain**, also known as **relative entropy** and **Kullback-Leibler (KL) divergence**, as

$$\sum_i p_i \log_2(p_i/q_i) = -H(P) - \sum_i p_i \log_2(q_i) \tag{7.9}$$

where the last term, denoted $H(P, Q)$, is known as the **cross entropy** of P and Q. The importance of K-L divergence and cross entropy lies in the fact that these quantities are minimal iff $P = Q$, and thus methods that minimize them can be used to fit distributions.

All forms of communication that are parasitic on spoken language, such as writing, or exercise the same fundamental cognitive capabilities, such as sign language, are strongly non-Bernoullian. Other significant sources of messages, such as music or pictures, also tend to exhibit a high degree of temporal/spatial redundancy, as we shall discuss in Chapters 8 and 9.

7.2 Kolmogorov complexity

The model described above is well suited only for the transmission of elementary messages that are truly independent of one another. If this assumption fails, redundancy between successive symbols can be squeezed out to obtain further compression. Consider, for example, the sequence of bits 010100101011... that is obtained by taking the fractional part of $n\sqrt{2}$ and emitting 1 if greater than .5 and 0 otherwise. It follows from Weyl's theorem of equidistribution that P(0) = P(1) = .5. The entropy will be exactly 1 bit, suggesting that the best we could do was to transmit the sequence bit by bit: transmitting the first n elementary messages would require n bits. But this is clearly *not* the best that we can do: to generate the message at the other side of the channel requires only the transmission of the basic algorithm, which takes a constant number of bits, plus the fact that it needs to be run n times, which takes $\log_2 n$ bits.

We have not, of course, transcended the Shannon limit but simply put in sharp relief that entropy limits compressibility *relative* to a particular method of transmission, namely prefix-free codes. The faster method of transmitting 010100101011... requires a great deal of shared knowledge between sender and recipient: they both need to know what $\sqrt{}$ is and how you compute it, and they need to agree on `for`-loops, `if`-statements, a compare operator, and so on. Kolmogorov complexity is based on the idea that all this shared background boils down to the knowledge required to program Turing machines in general or just one particular (universal) Turing machine.

Turing's original machines were largely hardwired, with what we would nowadays call the program burned into the finite state control (firmware) of the machine.

For our purposes, it will be more convenient to think of Turing machines as universal machines that can be programmed by finite binary strings. For convenience, we will retain the requirement of prefix-freeness as it applies to such programs. We say that a partial recursive function $F(p, x)$ is **self-delimiting** if for any prefix q of the program p, $F(q, x)$ is undefined. This way, a sequence of programs can be transmitted as a concatenation of program strings. The second variable of F, which we think of as the input to the machine programmed by p, is a string of natural numbers or rather a single (e.g. Gödel) number that is used to encode strings of natural numbers.

Definition 7.2.1 The **conditional complexity** $C_F(x|y)$ of x given y is the length of the smallest program p such that $x = F(p, y)$, or ∞ if no such program exists.

To remove the conditional aspects of the definition, we will need two steps, one entirely trivial, substituting $y = 0$, and one very much in need of justification, replacing all Fs by a single universal Turing machine U.

Definition 7.2.2 The **complexity** $C_F(x)$ of x relative to F is the length of the smallest program p such that $x = F(p, \lambda)$, or ∞ if no such program exists.

Theorem 7.2.1 (Solomonoff 1960, Kolmogorov 1965) There is a partially recursive function $U(p, y)$ such that for any partially recursive $F(p, y)$ there exists a constant c_F satisfying

$$C_U(x|y) \le C_F(x|y) + c_F \tag{7.10}$$

Proof We construct U by means of a universal Turing machine $V(a, p, x)$ that can be programmed by the appropriate choice of a to emulate any $F(p, x)$. To force the prefix-free property, for any string $d = d_1 d_2 \dots d_n$, we form $d^0 = d_1 0 d_2 0 \dots 0 d_n 1$ by inserting 0s as concatenation markers and 1 as an end marker. Since any binary string p can be uniquely decomposed as an initial segment a^0 and a trailer b, we can define $U(p, x)$ by $V(a, b, x)$. In particular, for $F(p, x)$ there is some f such that for all p, x $F(p, x) = V(f, p, x)$. In case x can be computed from y by some shortest program p running on F, we have $U(f^0 p, y) = V(f, p, y) = F(p, y) = x$, so that $C_U(x, y) \le 2|f| + C_F(x|y)$.

Discussion There are many ways to enumerate the partial recursive functions, and many choices of V (and therefore U) could be made. What Theorem 7.2.2 means is that the choice between any two will only affect $C_U(x|y)$ up to an additive constant, and thus we can suppress U and write simply $C(x|y)$, keeping in mind that it is defined only up to a $O(1)$ term. In particular, relative to the Turing machine T that prints its input on its output and halts, we see $C_T(x) \le l(x) + c_T$ for some constant c_T that is independent of the bitstring x (or its length $l(x)$).

The claim is often made (see e.g. Chaitin 1982) that $C(x)$ measures the complexity of an individual object x, as opposed to entropy, which very much presumed that the objects of study are drawn from a probability distribution. However, this claim is somewhat misleading since the focus of the theory is really the asymptotic complexity of a sequence of objects, such as initial substrings of some infinite string, where the $O(1)$ term can be really and truly neglected. For a single object, one could always find a U that will make $C_U(x)$ zero, just as if our only interest was in compressing a single file, we could compress it down to 1 bit, with an uncompress function that prints out the object in question if the bit was set and does nothing otherwise.

To take advantage of asymptotic methods, one typically needs to endow familiar unordered objects with some kind of order. For example, formal languages are inherently unordered, but it is no great stretch to order Σ^* (or any $L \subset \Sigma^*$) lexicographically. Once this is done, we can talk about the nth string and replace sets by their characteristic function written as a (possibly infinite) bitstring whose nth bit is 1 or 0, depending on whether the nth string y_n enjoyed some property or not. In order to discuss regular languages, we need to capture the structure of the state machine in bitstrings. Given some language $L \subset \Sigma^*$ and any string x, we define $\chi = \chi_1\chi_2\ldots$ to be 1 on the nth bit χ_n iff the string xy_n is in L – the first n bits of χ will be denoted by $\chi{:}n$. Clearly, two strings x and x' have the same χ iff they are right congruent, so that different χ correspond to different states in the DFSM. For L regular, the first n bits of any χ can be transmitted by transmitting the DFSA in O(1) bits, transmitting the state (out of finitely many) to which the χ in question corresponds (again O(1) bits), and transmitting the value of n, which requires no more than $\log_2(n)$ bits.

Theorem 7.2.2 (Li and Vitányi 1995) $L \subset \Sigma^*$ is regular iff there exists a constant c_L depending on L but not on n such that $\forall x \in \Sigma^* \;\; C(\chi{:}n) \leq \log_2(n) + c_L$.

Discussion We have already seen the only if part – the converse depends on a lemma (for a proof, see Li and Vitányi 1997, Claim 6.8.1) that for any constant c_L there will be only finitely many *infinitely long* bitstrings that have no more than $\log_2(n) + c_L$ asymptotic complexity – once this is demonstrated, the rest follows by the usual construction of FSAs from right congruence classes.

Suppose our goal is to transmit bitstrings over a channel that transmits 0s and 1s in an error-free (noiseless) manner. Unless we have some out-of-band method of indicating where the transmission of a given bitstring begins and ends, we need to devote extra bits to the boundary information. An inefficient but simple method is to prefix each string x by its own length $l(x)$. For this to work, we need to encode $l(x)$ in a manner that makes it recoverable from the code. This can be accomplished e.g. by giving it as a string of 1s (base one) and using 0 as the end marker – this is called the **self-delimiting code** S of x. For example, $S(001) = 1110001$. In this encoding, $l(S(x)) = 2l(x) + 1$. Not only is S prefix-free, but the codeword and a following string y can be unambiguously reconstructed from any $S(x)y$ by counting the number of 1s with which it begins and slicing off as many bits following the first 0 as there were 1s preceding it. We can use this property of S to create a more efficient code $D(x)$, where the payload x is suffixed not to the base one encoding of the length but rather to $S(l(x))$, where $l(x)$ is given in base two rather than base one; for example, $D(001) = S(11)001 = 11011001$. While this particular example comes out longer, in general

$$l(D(x)) = l(S(\lceil \log_2 l(x) \rceil)) + l(x) \leq 2\log_2 l(x) + l(x) + c \tag{7.11}$$

i.e. the overhead of the transmission is now only $2\log_2 l$ plus some small constant.

People (mathematicians) have relatively clear intuitions about the randomness of everyday (resp. mathematical) objects – certainly we feel that **furiously sleep* *ideas green colorless* is much more random than *colorless green ideas sleep furiously*, which at least has largely predictable ordering at the part of speech level. If the sender and the recipient share this knowledge, e.g. because they are both speakers of English, it may be possible to transmit the second sentence (but not necessarily the first) in fewer bits than it would take to transmit a random string of words. Kolmogorov complexity offers a way to replace intuitions about randomness by rigorous definitions, but this comes at a price. As we have emphasized repeatedly, $C(x)$ is defined only up to an additive constant. One may think that by fixing a small 'reference' universal Turing machine U this fudge factor could be removed, but this is not quite so: for any fixed U, the function $C_U(x)$ is uncomputable.

Theorem 7.2.3 (Kolmogorov 1965) $C(x)$ is uncomputable.

Proof Suppose indirectly that $C(x) : \{0,1\}^* \to \mathbb{N}$ is computed by some TM: it is then possible to program another TM that outputs, for each $n \in \mathbb{N}$, some string x_n that has $C(x_n) > n$. Let the length of the program emulating this TM on the universal machine we chose be p: this means that $C(x_n) \leq \log_2(n) + p$ since we found a program of total length $\log_2(n) + p$ that outputs x_n. Since p is fixed but the program by definition outputs an x_n with $C(x_n) > n$, we have a contradiction for n sufficiently large, e.g. for $n = 2^p (p > 2)$.

While Kolmogorov complexity oscillates widely and uncomputably, on the whole $C(x)$ is well-approximated by $l(x)$. We have already seen that $C(x) < l(x) + O(1)$, and for any constant k, only a small fraction of the bitstrings of length l can have $C(x) \leq l - k$. More precisely, as there are at most $2^{l-k+1} - 1$ programs of length $\leq l - k$, these can encode at most $2^{l-k+1} - 1$ of the 2^l bitstrings of length l, so there must be at least one string x_l for every l that is truly incompressible (has $C(x) = l$), at least half of the strings of length l have $C(x) \geq l - 1$, at least three-quarters have $C(x) \geq l - 2$, and in general at least $1 - 2^{-k}$ have $C(x) \geq l - k$.

To put this in perspective, in Section 7.1 we estimated the word entropy of English to be around 12–13 bits. The readers who worked on Exercise 5.8 will know that in journalistic prose the median sentence length is above 15 words, so for more than half of the sentences the simple encoding scheme would require over 180 bits. Were these typical among the bitstrings of length 180, only 2^{-30} (about one in a billion) could be compressed down to 150 bits or less. As we shall see in Chapter 8, sentences can be compressed considerably better.

Another way to show that only a few strings can have low Kolmogorov complexity is by reference to the Kraft nequality (Theorem 7.1.1) if we encode each string with its shortest generating program relative to some fixed string y. To make sure this is a prefix-free code, we use $K(x|y)$, defined as the shortest self-delimiting program on some universal TM and called the **prefix complexity** instead of $C(x|y)$, which was defined as the shortest program, not necessarily self-delimiting. With the prefix notion of Kolmogorov complexity, which differs from our previous notion at most by $2\log_2 C(x|y)$, we have

$$\sum_{x \in \{0,1\}^*} 2^{-K(x|y)} \leq 1 \tag{7.12}$$

The most important case is of course $y = \lambda$. Here the Kraft inequality means that the numbers $2^{-K(x)}$ sum to less than 1. Therefore, we can turn them into a probability measure either by adding an *unknown event* that has probability $1-\sum_{x\in\{0,1\}^*} 2^{-K(x)}$ or we can normalize by multiplying all values by a constant. Using this latter method, we define the **universal probability distribution** by taking the probability of a string x to be $2^{-K(x)}$ (times some constant). Solomonoff (1964) arrived at this notion by considering the idea of programming a TM by a random sequence of bits. If the shortest program that leads to x has n bits, the probability of arriving at x by a randomly selected program is roughly 2^{-n} since longer programs can contribute very little to this value.

The standard **geometrical probability distribution** over bitstrings with parameter $r > 1$ simply assigns probability $(1 - 1/r)(1/2r)^l$ to any string of length l – in particular, the probability of the empty string λ will be $1 - 1/r$. This corresponds to an experiment in which 0s and 1s are chosen by tossing a fair coin, and at each step the experiment is continued with probability $1/r$. Transmitting a random (incompressible) string x of length l actually requires *more* than l bits since we also need to transmit the information that the transmission has ended. We can use (7.11) as an upper bound on the length of the full transmission, which would yield $(1/2)^{l+2\log_2(l)+c}$ for a string of length l. For each fixed r, when l is large enough, the universal distribution generally assigns smaller probability to a string than the geometrical distribution would – the remaining probability mass is spent on the few strings that have low complexity.

7.3 Learning

Returning to our original example of predicting the next term of the series $2, 3, 5, \ldots$, we can see that to make this more precise we need to make a number of choices. First, what is the domain and the range of the function to be learned? Can the next term be π? In mathematical linguistics, our primary interest will be with strings, but this is not a significant limitation since strings can encode more complex data structures such as k-strings (see Section 3.3), parse trees (see Section 5.1), or even grammars (see below). A considerably harder issue is brought up by weighted (probabilistic) structures since it is not at all obvious that even a single real number, say the frequency of some feature of interest, can be learned.

Second, we need to specify the *hypothesis space* that delimits the choice of solutions. In the sequence learning and the closely related sequence prediction tasks, if we know that the only hypotheses worth considering are quadratic polynomials, the answer is already given by knowing the value of the function at three points. If, on the other hand, arbitrary degree polynomials or all computable functions are acceptable answers, no finite amount of data will uniquely identify one. In mathematical linguistics, the main issue is to identify a grammar that can generate a given data set, and this, as we shall see, is so hard that it is worth considering simplified versions of the problem.

Third, we must specify the method of providing examples to the learning algorithm. For example, if the target to be learned is some formal language, it makes a big difference whether it gets presented to the algorithm in a fixed (e.g. lexicographical) order, completely randomly, or perhaps following some prescribed probability distribution. In the first case, we can be certain after a while that an example not encountered so far will never be encountered later (because we are past it in the lexicographical ordering); the two other methods of providing data have no such closed-world assumption.

Fourth, we need to have some criteria for success. We will consider two main paradigms: identification in the limit (Gold 1967) and probable approximate correctness (Valiant 1984). By an algorithm capable of **identification in the limit (iitl)**, we mean an algorithm that will produce a hypothesis in each step such that after some number of steps it will always produce the same hypothesis, and it is the correct one. If the algorithm can signal that it converged, we are speaking of *finite identification* – this is obviously a stronger criterion than iitl. By a **probably approximately correct (pac)** learning algorithm we mean one that is capable of approximating a family of distributions to an arbitrary degree with the desired (high) probability. But before turning to these, we need to capture the idea that the more complex (in the limiting case, entirely random) the material, the harder it is to learn.

7.3.1 Minimum description length

The mathematical theory of learning, or *inductive inference*, is very rich, though by no means mature. Depending on how we specify a problem in the four dimensions above, a broad variety of results can be obtained. In applying the ideas of Kolmogorov complexity to natural language phenomena, we are faced with two technical problems. First, the *languages* of greatest interest in mathematical linguistics are noncounting, both in the informal sense of being free of arithmetic aspects and in the precise sense given by the noncounting property (5.29). In spite of the impressive size of the state space (estimated at 10^{12} in Section 6.3.2), counter-free automata are mathematically *simpler* than the full FSA class, while Theorem 7.2.2 characterizes the latter as having, for all $x \in \Sigma^*$, a constant c_L conditional prefix complexity $K(\chi{:}n|n)$ (or, what is the same, $\log_2(n) + c_L$ Kolmogorov complexity). Since the whole machinery of Kolmogorov complexity is defined only up to a constant (the $\log_2(n)$ in C comes from the need to transmit n and thus cannot be improved upon), there is no easy way to analyze languages simpler than regular. Second, the *grammars* used in mathematical linguistics are just axiom systems (though highly tuned to the subject matter and clearly impressive in size), and as such they retain the unpredictability of smaller axiom systems. Specifically, the great hopes of Chaitin (1982)

> I would like to measure the power of a set of axioms and rules of inference. I would like to be able to say that if one has ten pounds of axioms and a twenty-pound theorem, then that theorem cannot be derived from those axioms

remain unfulfilled since we can construct axiom systems of increasing proof-theoretic strength whose Kolmogorov complexity is the same (see van Lambalgen 1989,

Raatikainen 1998). What we need is some refinement of Kolmogorov complexity that incorporates a notion of **universal grammar**, which we define here simply as a set of permissible models A. First we assume, as in principles and parameters theory (see Section 5.3), that there are k binary parameters, so that $|A| = 2^k$. The complexity of describing some data set x relative to A is therefore the complexity of describing A, say $K(A)$, plus the k bits required to select the appropriate model parameters.

In general, we are interested in all sets A such that $K(A) + \log_2 |A|$ is, up to an additive constant, the same as $K(x)$. If among these $\hat{A}$ has the minimal prefix complexity $K(\hat{A})$, the shortest description of $\hat{A}$ is called the **minimal sufficient statistic** for x. Because $K(x|\hat{A}) = \log_2 |\hat{A}|$ up to a constant, $\hat{A}$ is the optimal model of x in the sense that x is maximally random with respect to $\hat{A}$.

In applying this framework, known as *minimum description length* (MDL), to grammatical model selection, we encounter a number of difficulties. First, the data x, strings encoding the grammar of some natural language, are not completely at hand: at best, we have *some* grammatical description of *some* languages. Second, even to the extent the data are at hand, they are not presented in a normalized format: different grammars use different notational conventions. Third, and perhaps most important, the devices used for eliminating redundancy are not universally shared across grammars. Here we consider three such devices, *anuvṛtti*, *metarules*, and *conventions*.

The Pāṇinian device *anuvṛtti* relies on a characteristic of the formal language Pāṇini employs in writing grammatical rules that is not shared by later work, namely that the rules (*sūtras*) are given in a technically interpreted version of Sanskrit without recourse to any special notation. Since the rules are given in words (and were for many centuries transmitted orally without the benefit of writing them down), it is possible to take a rule set

A B C D E (7.13)

P Q C D E (7.14)

R S C D (7.15)

and abbreviate it by deleting those words that were already mentioned in previous rules to yield

A B C D E (7.16)

P Q (7.17)

R S (7.18)

even though this device leaves some doubts whether (7.18) should really be interpreted as R S C D or rather as R S C D E. Such ambiguities are resolved by several principles. Some of these are highly mechanical, such as the principle that if a main element is discontinued, so are all its dependents. This can be used to reconstruct many rules in their unabbreviated format just by listing what is and what isn't considered a main element. Other principles (e.g. that a deleted element is to be thought of as being present as long as it is compatible with the rule statement) make an appeal to the meaning of rules. As such, these cannot be considered fully automatic by today's standards, or, at the very least, the challenge to program a Turing machine based on such a principle is wide open. Be that as it may, anuvṛtti shortens the statement of the grammar by over a third, a very significant compression ratio.

Contemporary formal grammars use a different abbreviatory device, *metarules*, which are rules to generate rules. In Section 5.7, we have already seen one instance, replacing two rules, the singular and the plural versions of (5.7), by a single rule schema (5.30). Such a schema comes with the interpretation that all variables in it must be uniformly (in all occurrences) replaced by all specific values the variable can take to yield as many rules as there are such replacement options. Of particular interest here are rules that depend on individual lexical entries: there are thousands of these, and there are generally very good reasons to group several of them together according to various criteria such as shared elements of lexical meaning, shared (sub)category, etc. Once we start using a hierarchically organized lexicon (see e.g. Flickinger 1987), it is natural to use network inheritance as an abbreviatory device, and in computerized systems this is common practice. Formally, it requires some system of metarules to unroll all the inheritance and present lexical entries (and the grammar rules that use them) in their unabbreviated form. The rate of compression depends greatly on the way rules are formulated, but again we expect very significant compression of the rule system, perhaps as much as 50%.

Another major device used by many grammarians is the distinction between *conventions* and rules. This can take many forms, but the key idea is always to designate some forms or rules as being intrinsically simpler than their representation would allow. In Section 3.2, we already mentioned the phonological theory of *markedness*, which takes e.g. the natural class of high vowels to be defined only by their [+high] and [+syll] features: the redundant values such as [−low] or [+voiced] are supplied by an automatic set of *markedness conventions*. Markedness conventions, much like anuvṛtti, supply missing values and can interact with the statement of rules as a whole. Rules simplified under anuvṛtti are simpler in the direct sense of being shorter (requiring fewer words to state), while rules simplified under markedness are just *regarded* as being shorter, inasmuch as we explicitly define a *simplicity measure* that only counts marked values. For example, context-sensitive voice assimilation in obstruent clusters is common, and a rule like

$$[+\text{obst}] \rightarrow [\alpha\text{voice}]/\_\begin{bmatrix}+\text{obst}\\ \alpha\text{voice}\end{bmatrix} \tag{7.19}$$

which requires cluster members to assimilate to their right neighbor in voicing, is regarded as intrinsically simple. However, since a rule like

$$[+\text{syll}] \rightarrow [\alpha\text{high}]/\_\begin{bmatrix}+\text{obst}\\ \alpha\text{voice}\end{bmatrix} \tag{7.20}$$

is unlikely to crop up in the phonology of any language, we make sure that the simplicity measure penalizes it, e.g. for assimilation of values across features. Generally, a complicated simplicity measure is seen as a sign of weakness of the underlying theory. For example, if assimilation is viewed as spreading (see Section 3.3) of an autosegment, rules like (7.20) are precluded, and this obviates the need to patch matters up by a post hoc simplicity measure. However, the working linguist is rarely in a position to anticipate future breakthroughs, so stipulation of simplicity remains a part of the descriptive apparatus.

To some extent, these problems can be circumvented by considering grammars piecemeal rather than in their entirety. For example, stating the stress system of a language would require (i) finding its place in a typological system such as StressTyp (see Section 4.1.4) and (ii) describing the stress patterns relative to some slot in the typology. Describing the ideally clean systems prescribed by the typology is an expense that gets amortized over many languages, so the term $K(A)$ enters the overall complexity with a multiplier considerably lower than the term $K(x|A)$, which is specific to the language. This is very much in line with traditional linguistic thinking, which values universal rules far more than parochial ones.

Unfortunately, decomposition of the grammar in this sense is not entirely unproblematic since one can often simplify the statement of one kind of rule, say the rules of stress placement, at the expense of complicating some other part of the grammar, such as the rules of compounding. To apply the MDL framework would therefore require a problem statement that is uniform across various aspects of the system; for example, by taking phonology, morphology, syntax, and semantics to be given by FSTs and assuming that universal grammar is a list of transducer templates that can be filled in and made part of an intersective definition at no cost (or very little cost, given amortization over many languages).

One final issue to consider, not just for stress systems but for any situation where there are only a finite number of patterns, is whether generating the patterns by a rule system provides a solution that is actually superior to simply listing them. The issue is particularly acute in the light of theorems such as

Theorem 7.3.1 (Tuza 1987) For every n there exists finite languages L_n containing $n^2 - n$ strings such that it requires at least $O(n^2/\log(n))$ rules in a regular, context-free, or length-increasing (context sensitive) grammar to generate L_n.

Discussion Since this is a worst case result, a possible objection is that the specific construction employed in Tuza (1987), namely languages based on n different terminals collected in a terminal alphabet Σ and $L_n = \{xy|x, y \in \Sigma, x \neq y\}$, is unlikely to appear in natural language grammars. In phonology, rules of *dissimilation* that forbid the occurrence of more than one token of some type are quite common (for a brief overview see Idsardi 2006), and there is no easy way to rule these out entirely. In syntactic rules, the phenomenon of *lexicality* (see Section 5.1.3) is quite pervasive, so much so that for each preterminal (strict lexical subcategory) p we expect at least one rule unique to p. With the size of the preterminal set estimated at 10^6 (see Section 6.3.2), Theorem 7.3.1 limits the compression achievable by generation to a factor of $\sqrt{(}\log(10^6))$ or to about 27% of the length of the raw pattern list (actually less, since a rule will, in general, take more symbols to encode than a pattern would). Generative grammar, if viewed as a compression device which has the potential to save three quarters of the space that would be required to list the patterns, is clearly a valuable tool in its basic (finite state, context free, or context sensitive) form, but the other space-saving devices discussed here, anuvṛtti, metarules, and conventions, can enhance its compression power considerably.

7.3.2 Identification in the limit

To specify a learning problem in iitl terms, we need to specify the domain and the range of the functions to be learned, the hypothesis space, and the order in which data will be presented to the learner. As an example, let us consider the issue of how infants acquire the phoneme inventory (see Section 3.1) of their language. The function they must identify is one that has at least acoustic input, utterances, and provides, for each utterance, a string (or k-string), over a yet to be determined phonemic alphabet (or tier alphabets). We say the input consists at least in acoustic data because it is clear that the infant also has, almost all of the time, access to visual cues such as lip rounding – we ignore this fact here but return to the matter in Section 8.3. We also simplify the problem statement by assuming the existence of automatic (no learning or training required) low-level acoustical *feature detectors* that digest the raw acoustic signal into parallel streams of discrete feature sequences (see Section 8.3).

While this may look like drastic oversimplification, defining away a core part of the problem, in fact there is massive evidence from psycholinguistics that speech perception operates in terms of discrete units (a phenomenon known as *categorical perception*, see Liberman 1957), that this mechanism is operative in infants even before they learn to speak (Eimas et al. 1971), and that it leverages deep perceptual abilities that were acquired evolutionarily long before primates (for chinchilla perception of voicing and syllable structure see Kuhl et al. 1975). High sensitivity to distinctive features in the acoustical signal has been demonstrated for all the thirty or so features that act distinctively in the phonology of some languages. We defer the perceptual problem, how to create detectors for voicing or other features, to Chapters 8 and 9, and consider only the learning problem of finding the right phonemic alphabet based on the output of this exquisitely sensitive perceptual apparatus, given in feature vectors. Instead of asking how acoustic waveforms (a class of $\mathbb{R} \rightarrow \mathbb{R}$ functions) get mapped onto (k-)strings, we ask how strings of feature vectors get so mapped.

As for the hypothesis space, clearly the most desirable outcome would be to find a static mapping g that maps feature vectors to phonemes and obtain the sequence-to-sequence mapping by lifting g (applying it pointwise) to strings. Such a restriction of the hypothesis space comes with a price: there are many string-to-string mappings that cannot be so obtained. For example, if the inputs are 0 and 1 (vectors of length 1) and the outputs are A and B, if f is defined as always A except when the previous three inputs were 000 or 111, there isn't a single g that can be lifted to f. A much broader, but still restrictive, hypothesis would be to assume that the solution is to be identified with a (multitape) finite state (k-)transducer that outputs a phoneme (or k-string) based on the feature vector under scan on its input tapes and on its current state.

Finally, let us consider the order in which the data are presented to the learner. Even if the task is to find a function g that operates on individual vectors, it is not realistic to assume (except, perhaps, for vowels that can be uttered in isolation) that the data will be presented pointwise – rather, the input is given in longer strings, syllables at the very least but more likely words or full sentences. In particular, stop consonants will never occur in isolation, but their phonemic value must be learned just the same. Since there are only finitely many phonemes, a random presentation of words will sooner or later contain every one of them, and we do not particularly need to put any constraint on the presentation of the data other than those excluding a certain kind of *child-directed speech* (CDS, also called *baby talk* and *motherese*) that purposely avoids the sounds considered hard or stigmatized. Simplifying matters somewhat, in modern-day Israel, Sefardic parents, whose phonemic inventory contained guttural consonants, may not have passed these on to their children since the Ashkenazi variety of Modern Hebrew, which has higher prestige, lacks them. An obvious restriction on data presentation, then, is to require that it be *semicooperative* in the sense that all relevant data (everything in the domain of the function to be learned) are eventually presented.

In reality, a great deal of language change can be attributed to imperfect learning, but the Modern Hebrew case cited above is more the exception than the rule: phonemic inventories can be very stable and remain virtually unchanged over many generations and centuries. This is because the phoneme inventory can be iitl learned upon any reasonable (semicooperative) presentation of the data. The chief learning strategy appears to be selective forgetting. For example, Ntlaka'pamux [THP] has glottalized voiceless stop consonants that phonologically differ in place of articulation (uvular vs. velar), while English lacks this contrast and these consonants altogether. Learners of English who are 6–8 months oldrecognize the distinction just as well as Ntlaka'pamux infants of the same age, but by 11–12 months their ability fades, while the Ntlaka'pamux infants of course retain it (Werker and Tees 1984). In spite of the selective forgetting of those features that play no distinctive role in the language, the full learning algorithm seems to involve a great deal of memorization.

The algorithms used by linguists to describe a new language, called *discovery procedures*, generally require more than semicooperation and assume full (two-sided) presentation of the data in the form of an informant who can provide negative information (grammaticality judgments) as well, asserting if needed that a certain form hypothesized by the learner is *not* in the language. Both negative and very low probability positive examples help to accelerate the learning process, and this is why in descriptive grammars we often find references to contrasts not easily exemplified. For example, in English, the difference between unvoiced *š* and voiced *ž* is seen only in pairs like *Aleutian/allusion*, *Confucian/confusion*, *mesher/measure*, *Asher/azure*, and *dilution/delusion*, which are very unlikely to be heard by infants at an early age. That said, by age four, when children begin to show signs of acquiring morphology, the algorithm that yields the phonemic alphabet has clearly converged on the basis of positive data alone. How a phoneme like *ž* gets created by the infant without a great deal of positive evidence is something of a mystery unless we presume a learning algorithm that can only take a few options and must live with the resulting overgeneration. If we find that voicing is distinctive in English, for which there is a great deal of positive evidence outside the *š/ž* pair, and we find that *š* is present in the system, for which again there is overwhelming positive evidence, we must live with the consequences and admit *ž* in the system.

As our next example, let us consider iitl of finite automata. Given an alphabet T, a regular language $L \subset T^*$, and a *text*, defined here as a series of examples $s_1, s_2, \ldots$ drawn from L semicooperatively (each $s \in L$ sooner or later appears in the series) but possibly with repetitions, we are looking for an algorithm that produces at each step i a DFSA A_i that generates all the s_j for $1 \leq j \leq i$ and converges to the DFSA A_0 that generates L in the sense that there exists some k such that $A_i = A_0$ for $i > k$.

Stated thus, the problem is not solvable. Let us first consider this for an important algorithm known as *identification by enumeration*. This is a lazy algorithm based on some notion of complexity or a priori probability that can be used to linearly order all possible hypotheses (DFSAs). At step i, it outputs the first (according to the pre-specified order) hypothesis that is still compatible with the data points $s_1, s_2, \ldots, s_i$ seen so far. If the text is presented in an adversarial fashion, the algorithm can at no stage be certain that a string t not presented so far will not be presented later. Whatever complexity ordering the algorithm embodies, the DFSA U with one (accepting) state and loops for all symbols will come early in the enumeration in the sense that we need to entertain other (more complex) hypotheses as well, and the adversary can pick any language L that is generated by one of these more complex DFSA V. Since the language T^* generated by U is compatible with any finite amount of data presented semicooperatively, a lazy algorithm will never have a reason to move away from the hypothesis U and will never reach the correct hypothesis V.

An analogous situation can be found in the game of twenty questions: if there are more than 2^{20} animals (as there are), an answerer who knows the whole taxonomy can always win. The key idea is that the answerer need not settle on a particular animal at the outset; it is enough that at any point there remain animals that fit the sequence of answers provided so far. Each question cuts the set of still available animals in two parts, and one of these sets will have more than 2^{20-i} members: the answerer answers so as to select this set. After 20 questions, the answerer still has more than one animal compatible with the questions asked so far and the questioner has lost. In other words, since each question can secure at best one bit of information, and it requires more than 20 bits to code the full set of animals, the questioner cannot always win (since a winning strategy would amount to a coding of the set in 20 bits) and an answerer can exploit the information deficit to make sure the questioner never wins. Gold (1967) presents a similar strategy with adversarial (but semicooperative) data presentation to prove the following theorem.

Theorem 7.3.2 (Gold 1967) No set of languages that contains all the finite languages and at least one infinite language is iitl.

Discussion Note that the theorem is sharp: the set of all finite languages is iitl, in fact the trivial algorithm that simply guesses the language to be the union of the strings presented so far will learn the correct language from any semicooperative text. A family of languages containing all finite languages and at least one infinite one is called **superfinite**, and since all the classical language families in the Chomsky hierarchy are superfinite, the theorem is often taken to mean that iitl with semicooperative data presentation is simply not the right model for grammar learning. Yet there is no reason to assume that the set of languages permissible by universal grammar will contain all finite languages; for example, the language $\{a, aba, abba, abbba, abbbba\}$ lacks the noncounting property (5.29) and is thus not a (potential) natural language.

An indexed family of nonempty languages $L_1, L_2, \ldots$ satisfies Angluin's **Condition 1** iff there is an effective procedure that will create, for each i, a finite subset T_i of L_i such that for all j, $L_j \not\subset L_i$ follows from $T_i \subseteq L_j$. The T_i is called a **telltale** for L_i because once we know from semicooperative data presentation that T_i is part of the target language, no subsets L_j of L_i need to be considered anymore. For such families, we have the following theorem.

Theorem 7.3.3 (Angluin 1980) An indexed family of nonempty languages is iitl iff it satisfies Condition 1.

For alphabets Σ with at least three letters, a classic construction of Thue (1906, reprinted in Nagell 1977) asserts the existence of an infinite word $\xi = x_1x_2x_3\ldots$, which is **square-free** in the sense that no substring of it has the form $\alpha\alpha$. For any threshold $k > 1$, any regular set of subwords of ξ will be counter-free. Let us now consider the language $P_0 = \Sigma^*$ and the languages P_i obtained from Σ^* by removing from Σ^* the prefix $x_1, \ldots x_i$ of ξ. This is an indexed family of nonempty languages that are all regular (complement of a finite language) and noncounting (complement of a noncounting language) with threshold k for any $k > 1$. If this family is iitl, by Theorem 7.3.3 there is a finite telltale T_0 such that for all j, $L_j \not\subset \Sigma^*$ must follow from $T_0 \subseteq L_j$. Since the conclusion is false but the premiss will be true for any L_j with j greater than the longest string in T_0, we have a contradiction that proves the following theorem.

Theorem 7.3.4 (Kracht 2007) For $k > 1$, the family of regular noncounting languages over an alphabet Σ that has at least three elements is not iitl.

Discussion To see that Theorem 7.3.3 is applicable, we need to assert that the languages P_i are constructible. Thue's original proof relies on the homomorphism h given by $0 \mapsto 01201; 1 \mapsto 020121; 2 \mapsto 0212021$ over the alphabet $\{0, 1, 2\}$ and constructs ξ as $0xh(x)h^2(x)h^3(x)\ldots$, where $x = 1201$ (i.e. as the result of the infinite iteration of h from starting point 0). Since h increases the length of any string at least by a factor of 5 (and at most by a factor of 7), it requires at most $\lceil \log(l)/\log(5) \rceil$ iterations to compute the prefix $x_1 \ldots x_l$ of ξ. Since the algorithm $A(i, w)$ that decides the membership of w in P_i needs only to test whether w is a prefix in ξ, it can run in time and space linear in $|w|$.

Obviously, if A is an iitl learnable family, so is every $B \subset A$, and if C is not iitl learnable, neither will any $D \supset C$ be iitl learnable. But this is not enough to characterize the iitl families since there are many incomparable iitl families, such as the set of all finite languages vs. the set of all languages that can be generated by CSGs with a bounded number of rules (Shinohara 1990).

If we are prepared to relax the requirement of semicooperation (positive evidence only) and admit discovery procedures that rely on more data, there are several important results concerning regular languages and DFSAs. With a fully cooperative text (given in lexicographic order), Trakhtenbrot and Barzdin (1973) provided a polynomial algorithm that produces a DFSA consistent with all data (positive and negative) up to length n. If full cooperation is relaxed (positive and negative examples are presented but not all up to a given length), the problem is NP-hard (Gold 1978). If cooperation is extended by the answerer (informant) providing not just yes/no answers but also a full set of strings that reach every state of the automaton, the exact DFSA can be learned in finite time (Angluin 1981). If *grammar comparison* is allowed in the form of questions 'Is this DFSA equivalent to the target?', polynomial learning of DFSAs is possible (Angluin 1987). It is equally possible to tighten the requirement of semicooperation by assuming a certain amount of downright misinformation or just noise.

7.3.3 Probable approximate correctness

Assuming that each language learner is exposed to semicooperatively presented text from the older generations and identifies a grammar that can account for the text, there is still no guarantee that the learner's grammar will be the *exact* equivalent of the grammars that were used in generating the text. In fact, languages can and do change in aspects that go far beyond a superficial updating of vocabulary: whole grammatical constructions fall into disuse and eventually disappear, paradigms simplify, new constructions and new paradigms enter the language at a surprising rate. Sometimes the process can be attributed to contaminated text, especially if there is large-scale migration or change in ruling class, but even languages reasonably isolated from these effects change over time. To capture this phenomenon, we need to consider some notion of *approximate* learning that involves some measure of similarity between what is learned and what should have been learned: the key idea of Valiant (1984) was to express approximation in probabilistic terms.

To specify a pac learning problem, we again need to specify the domain and the range of the functions to be learned. These are generally the characteristic functions of some sets called *concepts*. We also need to specify the hypothesis space, generally as a family H of concepts (sets) that are all subsets of the same universe called the *sample space* S. As the name suggests, S will be endowed with a fixed (but not necessarily known) probability measure P that dictates both how data will be presented to the learning algorithm and how the goodness of fit is to be measured between the target concept C and a hypothesis C' proposed by the algorithm. We say that C' **approximates** C within ε if $P(C \Delta C') < \varepsilon$ (here Δ is used to denote symmetric set difference). As for our criterion of success, we say that an algorithm δ, ε **pac-learns** C if, after being presented with a sufficient number n of randomly (according to P) chosen labeled examples, it produces, with probability $>1-\delta$, a concept C' that approximates C within ε. Our chief interest is with algorithms that are polynomial in n, $1/\delta$ and $1/\varepsilon$, and ideally we'd want algorithms that are robust under change of the distribution P or even *distribution-free* (i.e. independent of P).

As an example, let us again consider regular languages: the sample space is T^* endowed with a suitable (e.g. geometrical) probability distribution, a concept to be learned is some regular language $C \subset T^*$, and a hypothesis produced by the algorithm is some DFSA c', which generates the language C'. Stated in pac terms, the problem is still hard. In particular, Kearns and Valiant (1989) demonstrate that a polynomial pac learner for all regular languages would also solve some problems generally regarded as cryptographically hard. However, *interleaving languages* (the musical analog of k-strings; see Ross 1995) are pac learnable, and if membership queries are allowed, a polynomial pac discovery procedure for DFSAs exists (Angluin 1987).

The **Vapnik-Chervonenkis (VC) dimension** of a space H of concepts is the maximum number of samples that can be labeled any way by members of H. Each member of H creates a binary decision (labeling) on S, and if a set of n is such that all 2^n labelings can be obtained by a suitable choice of H, we say the set of points is *shattered* by H – the VC dimension of the hypothesis space is the cardinality of the largest set that can be so shattered. Obviously, for a finite class H, the VC dimension will be $\leq \log_2 |H|$. The VC dimension of the hypothesis space is closely related to pac learnability: the number of examples n needed for δ, ε pac learning satisfies

$$\Omega\left(\frac{1}{\varepsilon}\log\frac{1}{\delta} + \frac{VC(H)}{\varepsilon}\right) \leq n \leq O\left(\frac{1}{\varepsilon}\log\frac{1}{\delta} + \frac{VC(H)}{\varepsilon}\log\frac{1}{\varepsilon}\right) \tag{7.21}$$

While the two bounds are very close and the VC dimension of many classical learning models is known, the error bounds computed from (7.21) are rather pessimistic both because the results are distribution-free and because they are, as is common in theoretical computer science, worst-case results. In principle, many linguistic problems could be recast as pac learning, but in a practical sense the algorithms that are most important for the linguist owe little to the pac framework both because the VC bounds do not characterize the problem well and because linguistic pattern matching, to which we turn in Chapter 8, is generally concerned with n-way, rather than 2-way, classification problems.

7.4 Further reading

Although there is some interesting prehistory (Nyquist 1924, Hartley 1928), information theory really begins with Shannon (1948) – for a modern treatment, see MacKay (2003). The insufficiency of the classical *quantitative* theory of information has been argued in Bar-Hillel (1964), whose goal was to replace it by a *semantic theory of information* – the same goal is restated in Dretske (1981) and Devlin (1991). Devlin (2001) argues that the semantic theory that fits the bill is *situation semantics*; for a short introduction, see Seligman and Moss (1997), and for a detailed exposition see Barwise and Perry (1983).

Li and Vitányi (1997) provide an encyclopedic treatment of Kolmogorov complexity. For a succinct presentation of the central ideas, see Gács (2003) and for a discussion of the relation of Kolmogorov complexity and learnability see Clark (1994). The MDL framework originates with Rissanen (1978). For a thorough discussion of anuvṛtti, invoking a full metagrammar of over a hundred principles, see Joshi and Bhate (1984). The idea of using grammars to generate grammars goes back to Koster (1970). A significant fragment of English grammar, with heavy use of metarules, is presented in Gazdar et al. (1985). The first major discussion of simplicity measures in generative linguistics is Chomsky and Halle (1968 Ch. 9); for a comparison with the earlier structuralist ideas, see Battistella (1996).

On the status of *ž*, see McMillan (1977). The acquisition of phonology offers a particularly rich storehouse of phenomena that support a universal grammar-based view of language acquisition. For example, infants employ rules such as reduplication and final devoicing that the adult grammar of the language they learn may entirely lack. However, the idea that the acquisition of phonology relies heavily on hardwired UG has its detractors; see e.g. (Zamuner et al. 2005). The same can be said for syntax, where the *poverty of stimulus* argument offered in Chomsky (1980), that certain facts about natural language could not be learned from experience alone, though widely accepted, has significant detractors; see e.g. Pullum and Scholz (2002).

The classic sources on iitl and pac learning remain Gold (1967) and Valiant (1984). For inductive inference in general, see Angluin and Smith (1983), Angluin (1992), and Florencio (2003). For the lower bound in (7.21), see Ehrenfeucht et al. (1989), and for the upper bound see Blumer et al. (1989) and Anthony et al. (1990).

8

Linguistic pattern recognition

In general, the pattern recognition task is defined as one where an infinite, continuous set of inputs is associated with a finite variety of outputs. A typical example is *face recognition*, where the goal is to identify the face as belonging to the same person in spite of changes in viewing angle, distance, light, makeup and hairdo, facial expression, etc. We speak of *linguistic* pattern recognition when the set of outputs is structured linguistically. This means both that the output units of linguistic significance follow each other in discrete time (e.g. a temporal succession of letters, words, or sentences) and that these units themselves come from a finite (or finitely generated) set. We could stretch the definition to include data that lack temporal organization. For example, the recognition of isolated characters is considered by many to be a linguistic pattern recognition task, especially in the case of Han and Hangul characters, which can be decomposed spatially though not necessarily temporally (see Sproat 2000). However, no amount of stretching the definition will allow for face or fingerprint recognition, as the output in these domains can be made finite only by imposing some artificial cutoff or limitation on the system.

Both linguistic and nonlinguistic pattern recognition involve an early stage of signal processing, often referred to as the *front end*. In speech recognition, the front end is generally based on acoustic principles, while in optical character recognition image processing techniques are used. In Chapter 7, we looked at codes that were fully invertible. Here we will distinguish *low-* and *high-level* signal processing based on whether the input signal is largely recoverable from the output or not. In low-level signal processing, we encode the input in a largely invertible (lossless) manner, while in high-level signal processing, so much of the information that was present in the input gets discarded that the output can no longer serve as a basis for reconstructing the input. It should be clear from the foregoing that the distinction between low- and high-level signal processing is a matter of degree, especially as many signal processing algorithms have tunable parameters that control the degree of information loss. Nevertheless, there are some general guidelines that can, with proper care, be used to distinguish the two kinds of signal processing in nearly all cases of practical interest.

On one side, any transformation that limits losses to such a degree that they are below the threshold of human perception is definitely low-level. A typical example would be the algorithm used in ripping music CDs: here the input is twice 16 bits (stereo) sampled at 44.1 kHz, for a total of 1411 kbps, and the output is MP3, requiring only 128 kbps (stereo) or 64 kbps (mono). On the other side, any *feature extraction* algorithm that produces categorial (especially 0–1) output from continuous input is considered high-level. A typical example would be an algorithm that decides whether a stretch of speech is voiced or unvoiced (see Chapter 9). Here the key issue is not whether the output is categorial but rather the number of categories considered. Two categories (one bit) is definitely high-level, and 64k categories (16 bits) are typical of low-level signal processing. We discuss this issue, *quantization*, in Section 8.1. Because of their special importance in linguistic pattern recognition, in Section 8.2 we revisit the basic notions of Markov processes (chains) and hidden Markov models (HMMs) that were introduced in Section 5.5 both for the discrete (quantized) and the continuous cases.

Another criterion for distinguishing low- and high-level processing is whether the output of the front end is still suitable for human pattern recognition. For example, aggressive signal processing techniques can reduce the number of bits per second required for transmitting speech signals from 64 kbps (MP3 mono) to 2 kbps (MELP) with considerable degradation in subjective signal quality but little or no loss of intelligibility. Surprisingly, there exist pattern recognition methods and techniques that work best when the signal is degraded beyond the point of human recognizability. We discuss such cases, and the reasons for their existence, in Section 8.3. Topic detection, and the general problem of classifying documents, has all characteristics of linguistic pattern recognition when there is a single concept such as 'spam' to be learned. As the number of topics increases (complex topic hierarchies often have tens of thousands of categories) the problem gradually takes on characteristics of nonlinguistic pattern recognition. In Section 8.4 we discuss the tools and techniques used in topic classification in some detail because these have broad applicability to other linguistic problems as well.

8.1 Quantization

Historically, quantization techniques grew out of analog/digital signal conversion, and to some extent they still carry the historical baggage associated with considerations of keeping the A/D circuitry simple. We begin with air pressure $s(t)$ viewed as a continuous function of time, and we assume the existence of an amplitude bound A such that $-A \leq s(t) \leq A$ for all t considered.

Exercise 8.1 Research the empirical amplitude distribution of speech over long stretches of time. If this has variance σ, how do you need to set k in $A = k\sigma$ such that less than one in a hundred samples will have $|s| > A$? How do you need to set k so that on the average no more than one in a thousand samples gets clipped?

Our goal is to quantize both the time axis and the amplitude axis, so that $s(t)$ is replaced by a stepwise constant function $r(t_i)$, where r can take on only discrete values $r_0, r_1, \ldots, r_N$. In the simplest case, both the r_i and the t_j are uniformly spaced. The reciprocal of $t_{i+1} - t_i$ is called the **sampling frequency** – in the case of speech, it is typically in the 6–40 kilohertz range. For simplicity, N is usually chosen as a power of 2, so that instead of N-way sampling we speak of b-bit sampling, $N = 2^b$. In the electrical engineering literature, this is known as **pulse code modulation**, or PCM. In the case of speech, b is typically between 1 and 24. At high sampling rates (20 kHz or above), one bit is already sufficient for low-quality, but generally intelligible, speech coding.

Uniform spacing means that the quantization levels are the centers of the 2^b intervals of size $\Delta = 2A/2^b$. The squared error of the stepwise approximation, called the *quantization noise*, can be estimated for any elementary interval of time by taking b sufficiently large so that the error becomes a random variable with zero mean that is uniformly distributed over the range $-\Delta/2, \Delta/2$. Since the squared error of such a variable is obviously $\Delta^2/12$, the more commonly used **signal to quantization noise ratio**, or SQNR, is $3P_x 2^{2b}/A^2$, where P_x is the **signal energy**, defined as the area under the curve $s^2(t)$. Typically, speech engineers express such energy ratios using the **decibel scale**, 10 times their base 10 logarithm, so we obtain the conclusion that adding an extra bit improves SQNR by about 6.02 decibels for PCM.

There is, of course, no reason to assume that the uniform quantization scheme is in any way optimal. There are many methods to decrease the number of bits per second required to transmit the signal or, equivalently, to increase signal quality for the same number of bits, and here we provide only a thumbnail sketch. In one family of methods, the original signal gets warped by the application of a concave function W so that we uniformly quantize $W(s(t))$ instead of $s(t)$. If W is chosen linear for $|s| < A/87.56$ and logarithmic for larger s, we speak of **A-law** PCM, commonly used in digital telephony in Europe since the 1970s. If $W = \text{sign}(s)A\log(1 + 255|s|/A)/3$, we speak of **$\mu$-law** PCM, commonly used for digital telephony in the United States and Japan. These *log PCM* methods, predating MP3 by over three decades, provide high-quality speech at the same 64 kbps rate.

The quality of speech is a somewhat subjective matter: the standard way of evaluating it is by mean opinion scale (MOS), ranging from 1 (bad) to 5 (excellent). A MOS grade of 4.5 or higher is considered *broadcast* quality; 4.0–4.5 is *toll* or *network* quality (phone networks aim at this level for toll services); 3.5–4.0 is considered *communications* quality (also known as *cell grade* quality); and 2.5–3.5 is *synthetic* quality (a historical name that no longer reflects the actual quality of modern synthetic speech, which can be indistinguishable from broadcast-quality human speech). Log PCM has MOS 4.3 – below the broadcast level but well within the toll range.

Exercise 8.2 What must be the amplitude distribution of speech over long stretches of time for A-law (μ-law) to provide the optimal warping function? Compare these with the result of Exercise 8.1.

Beyond A-law or μ-law, in practical terms very little can be gained by matching the warping function more closely to the long-term characteristics of speech. *Adaptive* methods that exploit redundancies in the short-term characteristics of speech will be discussed in Section 9.1, but before turning to these, we need to gain a better understanding of the primary motivational example of quantization, phonemes.

In Section 3.1, we defined phonemes as mental units belonging in a phonemic alphabet that is specific to a given language. Since humans are capable of transcribing speech into phonemic units and the resulting string of symbols by definition preserves all linguistically significant (potentially meaning-altering) contrasts, using as many quantization levels (actually, quantization cells in n-dimensional space; see Section 9.1) as there are phonemes in the language is a particularly attractive proposition. Here we are referring not just to the small group of humans trained in phonemic transcription but to the much larger set of literate humans who can provide orthographic transcriptions (an arguably harder task, especially for complex orthographies with large historical baggage, but one that all cognitively unimpaired humans seem to be capable of) and even to illiterate people inasmuch as there is a wealth of psycholinguistic evidence that they, too, use phonemic units in all language comprehension and production tasks.

The phoneme recognition problem is made hard by three factors. First, only a small number of phonemes, the vowels, appear freely in isolation, and human recognition is optimized to deal with sequences, where contextual cues are available, rather than with isolated phonemes. Second, the linear sequence recognition problem is often complicated by tempo, pitch, volume, and other autosegmental effects. For example, the phonetic sequence */tš/* maps on a single phoneme *č* in Spanish but on a sequence of two phonemes *t* and *š* in German, so the problem is recovering not just the sequence but also the autosegmental linking pattern. Finally, the mapping from mental units to physical units cannot be inverted: cases of *contextual neutralization* such as Example 3.2.1 (Russian final devoicing) abound, and even more disturbingly, there seem to be many cases of *absolute neutralization* where a contrast never surfaces.

A familiar example is the 'silent *e*' of English orthography in words like *ellipse*: it is never pronounced, but a system such as Chomsky and Halle (1968) that assumes an underlying *e* that gets deleted from the surface seems better equipped to explain the regularities of English stress than a system that makes no recourse to such devices. Another example familiar to phonologists is 'velar' vs. 'palatal' *i* stems in Hungarian: until we add a suffix, there is absolutely no way to distinguish the *i* sound found in *híd* 'bridge' from that of *víz* 'water', but once a suffix governed by vowel harmony is added the distinction becomes evident: *hidat (*hidet)* 'bridge.ACC' but *vizet (*vizat)* 'water.ACC' and similarly with the other harmonizing suffixes (Vágó 1980). A system that assumes a rule that wipes out (neutralizes) the distinction between the two kinds of *i* is better suited to explaining the facts of the language than one that does not (see Vágó 1976) – so much so that nobody quite succeeded in constructing a rule system without any hidden contrast.

In a classic experiment, Peterson and Barney (1952) removed all three confounding factors by restricting attention to the ten steady state vowels of English. By instructing the speakers to produce clear examples and marking in the data set all instances where at least one of 26 listeners could not clearly identify the vowel, they removed the 'performance' factors that could lead to neutralization; by using steady

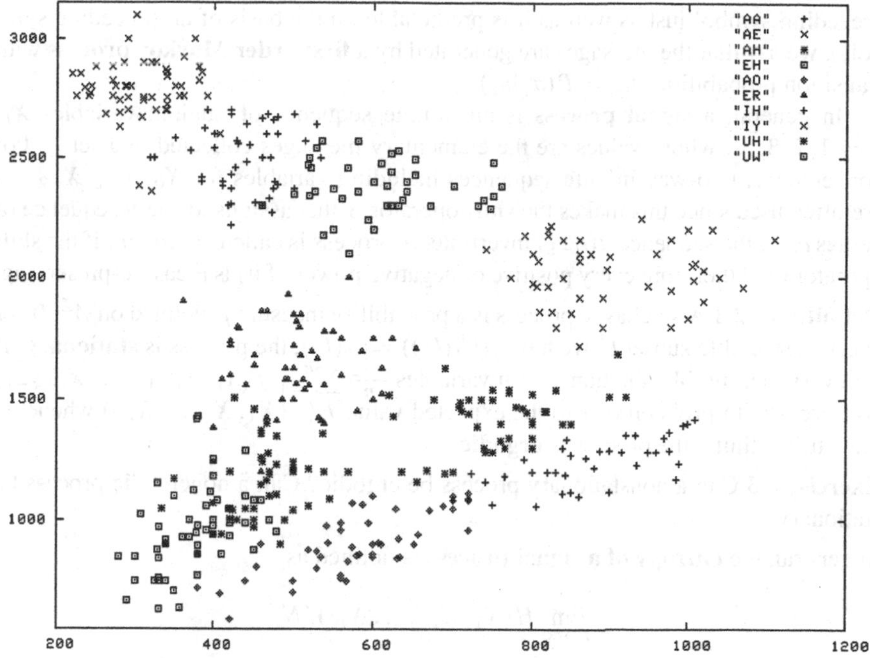

Fig. 8.1. Peterson-Barney (1952) 1st and 2nd formant data, female speakers

state vowels, they removed all tempo issues; and by using only clearly monophonemic vowels, they removed all issues of segmentation and autosegmental association. One would hope that on such a clean set of data it would be trivial to find ten canonical waveforms $p_1, \dots p_{10}$ and a distance measure d such that for each waveform q the p_i such that $d(q, p_i)$ is minimal (in standard notation, $\text{argmin}_i d(q, p_i)$) would serve to identify q. Unfortunately, the original acoustic data are lost, so we cannot test the performance of contemporary recognizers on this particular set, only the fundamental frequencies and the first three *formants* (resonant frequencies of the vocal tract) have been preserved (see Watrous 1991).

Figure 8.1 shows the first two formants for unambiguous vowels produced by female speakers. In formant space, the data show a number of overlapping clusters, which makes it clear that variability (both across speakers and across repeat utterances of the same speaker) is a major issue. This finding has deeply influenced the design of modern speech recognition systems, to which we turn now.

8.2 Markov processes, hidden Markov models

We begin with the simple case where the message to be coded exhibits first order Markovian dependence. Consider a finite set of elementary messages (symbols) $a_i (1 \leq i \leq k)$ with probabilities p_i. If in all complex messages $P(a_{i_n}|a_{i_{n-1}}) = P(a_{i_n}|a_{i_1}a_{i_2}\dots a_{i_{n-1}})$ holds (i.e. a_{i_n} is predictable on the basis of the immediately preceding symbol just as well as it is predictable on the basis of all preceding symbols), we say that the messages are generated by a **first order Markov process** with transition probabilities $t_{ij} = P(a_j|a_i)$.

In general, a **signal process** is an infinite sequence of random variables X_t, $t = 1, 2, 3, \ldots$ whose values are the elementary messages collected in a set A. For convenience, two-way infinite sequences including variables for $X_0, X_{-1}, X_{-2}, \ldots$ are often used since this makes the **shift** operator S that assigns to every sequence of values $a(t_i)$ the sequence $a(t_{i-1})$ invertible. A process is called *stationary* if the shift operator (and therefore every positive or negative power of it) is measure-preserving.

Definition 8.2.1 A stochastic process is a probability measure μ defined on $A^{\mathbb{Z}}$. If for every measurable subset U we have $\mu(S(U)) = \mu(U)$, the process is **stationary**. If for every measurable function f of n variables $\frac{i=1}{n} \sum_{i=1}^{n} f(X_i, X_{i+1}, \ldots, X_{i+k-1})$ converges with probability 1 to the expected value $E(f(X_1, X_2, \ldots X_k))$ whenever the latter is finite, the process is **ergodic**.

Exercise 8.3 Can a nonstationary process be ergodic? Can a nonergodic process be stationary?

In general, the **entropy of a signal process** is defined as

$$\lim_{n \to \infty} H(X_1, X_2, \ldots, X_N)/N$$

if this limit exists: in the case of word-unigram based models, it is often referred to as the *per-word entropy* of the process. For the Bernoulli case studied in Section 7.1, the random variables X_i are independently identically distributed, so this definition reduces to $H(X_1) = H(X)$. In the non-Bernoulli case, by the chain rule we have $\frac{1}{N}(H(X_1) + H(X_2|X_1) + H(X_3|X_1X_2) + \ldots + H(X_N|X_1X_2 \ldots X_{N-1}))$, and if the process is Markovian, this reduces to $\frac{1}{N}(H(X_1) + H(X_2|X_1) + H(X_3|X_2) + \ldots + H(X_N|X_{N-1}))$. If the process is stationary, all terms except the first one are $H(X_2|X_1)$, and these dominate the sum as $N \to \infty$. Therefore we obtain the following theorem.

Theorem 8.2.1 The word entropy of a stationary Markov process is $H(X_2|X_1)$.

To fully define one-sided first-order Markov chains we only need a set of *initial probabilities* I_i and the transition probabilities t_{ij}. In two-sided chains, initial probabilities are replaced by the probabilities T_i that the chain is in state i. These obviously satisfy $T_j = \sum_{i=1}^{n} T_i t_{ij}$ and thus can be found as the eigenvector corresponding to the eigenvalue 1 (which will be dominant if all t_{ij} are strictly positive). By a classical theorem of A. A. Markov, if the process is **transitive** in the sense that every state can be reached from every state in finitely many steps (i.e. if the transition matrix is irreducible), the state occupancy probabilities T_i satisfy the law of large numbers:

Theorem 8.2.2 (Markov 1912) For any ε, δ arbitrarily small positive numbers, there exists a length N such that if m_i denotes the absolute frequency of the process being in state a_i during a trial of length N, we have

$$P(|m_i/N - T_i| > \delta) < \varepsilon \qquad (8.1)$$

The current state of the chain can be identified with the last elementary message emitted since future behavior of the chain can be predicted just as well on the basis of this knowledge as on the basis of knowing all its past history. From Section 8.2.1, the word entropy of such a chain can be computed easily: if $X_1 = a_i$ is given, $H(X_2)$ is $-\sum_j t_{ij} \log_2(t_{ij})$, so we have

$$H(X_2|X_1) = -\sum_i T_i \sum_j t_{ij} \log_2(t_{ij}) \tag{8.2}$$

What makes this entropy formula particularly important is that for longer sequences average log probability is concentrated on this value. By definition, the probability of any sequence $a = a_{i_1} a_{i_2} \dots a_{i_N}$ of elementary messages is $T_{i_1} \prod_{k=1}^{N-1} t_{i_k i_{k+1}}$, and the probability of a set C containing messages of length N is simply the sum of the probabilities of the individual sequences.

Theorem 8.2.3 (Shannon 1948) For arbitrary small $\varepsilon > 0$ and $\eta > 0$, there is a set $C_{\eta,\varepsilon}$ of messages of length N for sufficiently large N such that $P(a \notin C) < \varepsilon$ and if $a \in C$, then $|\log_2(1/P(a))/N - H| < \eta$.

Proof Let us collect the t_{ij}. If m_{ij} counts the number of times t_{ij} occurred in the product $T_{i_1} \prod_{k=1}^{N-1} t_{i_k i_{k+1}}$, we have

$$P(a) = T_{i_1} \prod_{i,j} t_{ij}^{m_{ij}} \tag{8.3}$$

We define C as containing those and only those sequences a that have positive probability (include no $t_{ij} = 0$) and satisfy $|m_{ij} - NT_i t_{ij}| < N\varepsilon$ for all i, j. For these, the product in (8.3) can be rewritten with exponents $NT_i t_{ij} + N\varepsilon\Theta_{i,j}$ with $|\Theta_{ij}| < 1$. Therefore,

$$\log_2(1/P(a)) = -\log_2(T_{i_1}) - N \sum_{t_{ij} \neq 0} T_i t_{ij} \log_2(t_{ij}) - N\varepsilon \sum_{t_{ij} \neq 0} \Theta_{ij} \log_2(t_{ij}) \tag{8.4}$$

Since the second term is just $-NH$, we have

$$|\log_2(1/P(a))/N - H| < -\log_2(T_{i_1})/N - \varepsilon \sum_{t_{ij} \neq 0} \log_2(t_{ij}).$$

The first term tends to 0 as $N \to \infty$, and the second term can be made less than an arbitrary small η with the appropriate choice of ε. What remains to be seen is that sequences $a \notin C$ with nonzero probability have overall measure $< \varepsilon$. For a nonzero probability a not to belong in C it is sufficient for $|m_{ij} - NT_i t_{ij}| \geq N\varepsilon$ to hold for at least one i, j. Thus we need to calculate $\sum_{t_{ij} \neq 0} P(|m_{ij} - NT_i t_{ij}| \geq N\varepsilon)$. Since this is a finite sum (maximum n^2 terms altogether), take any $t_{ij} \neq 0$ and apply Theorem 8.2.2 to find N large enough for $P(|m_i - NT_i| < N\delta/2) > 1 - \varepsilon$ to hold. By restricting our attention to state a_i, we have a pure Bernoulli experiment whose outcomes (moving to state a_j) satisfy the weak law of large numbers, and thus for any ε and $\delta/2$ we can make $P(|m_{ij}/m_i - t_{ij}| < \delta/2|) > 1 - \varepsilon$. Combining these two, we obtain

$$P(|m_i - NT_i| < N\delta/2)P(|m_{ij} - m_i t_{ij}| < m_i\delta/2) \geq (1-\varepsilon)(1-\varepsilon) \geq 1 - 2\varepsilon$$

By the triangle inequality, $P(|m_{ij} - NT_i t_{ij}| < N\delta) \geq 1 - 2\varepsilon$, so $P(|m_{ij} - NT_i t_{ij}| \geq N\delta) < 2\varepsilon$, and by summing over all i, j, we obtain $P(\overline{C}) < 2n^2\varepsilon$, which can be made as small as desired.

In Section 5.5.2 we defined **hidden Markov models** (HMMs) by weakening the association of states and elementary messages: instead of a single (deterministic) output as in Markov chains, we assign an *output distribution* E_i to each state i. It is assumed that the E_i are independent of time and independent of each other (though possibly identically or very similarly distributed). In the special case $E_i(a_i) = 1, E_i(a_j) = 0 (j \neq i)$, we regain Markov chains, but in the typical case the state cannot be deterministically recovered from the elementary message but remains to some extent hidden, hence the name. Each state i of an HMM can be conceived as a signal process in its own right, with entropy $H(E_i) = H_i$. We do not require the E_i to be discrete. In fact *continuous density* HMMs play an important role in speech recognition, as we shall see in Chapter 9. Although it would be possible to generalize the definition to situations where the underlying Markov chain is also replaced by a continuous process, this makes little sense for our purpose since our goal is to identify the underlying states with linguistic units, which are, by their very nature, discrete (see Section 3.1).

The entropy of the whole process can be computed as a weighted mixture of the output entropies only if each state is final (diagonal transition matrix). In the general case, we have to resort to the original definition $\frac{1}{N}(H(X_1) + H(X_2|X_1) + H(X_3|X_1X_2) + \ldots + H(X_N|X_1X_2\ldots X_{N-1}))$, and we see that $H(X_3|X_1X_2)$ and similar terms can no longer be equated to $H(X_3|X_2)$ since it is the previous *state* of the model not the previous *output* that contributes to the current state and thus indirectly to the current output. Introducing a Markov chain of random *state variables* S_i, we have $P(X_i = a_j) = \sum_{k=1}^{n} S_i(k)E_k(a_j)$ (where the sum ranges over the states). By definition, word entropy will be the limit of

$$\frac{1}{N}H(X_1,\ldots,X_N) = \frac{1}{N}(H(S_1,\ldots,S_N) + H(X_1,\ldots,X_N|S_1,\ldots,S_N) - H(S_1,\ldots,S_N|X_1,\ldots,X_N)) \quad (8.5)$$

We already computed the first term as $H(S_2|S_1)$. The second term is $\frac{1}{N}NH(X|S) = H(X_1|S_1)$, which is generally easy to compute from the underlying Markov process and the output probability distributions. It is only the last term that causes difficulties inasmuch as computing the states from the outputs is a nontrivial task. Under most circumstances, we may be satisfied with pointwise maximum likelihood estimates.

8.3 High-level signal processing

Historically, linguistic pattern recognition systems were heavily slanted towards symbol-manipulation techniques, with the critical pattern recognition step often entirely obscured by the high-level preprocessing techniques that are referred to as *feature detection*. To the extent we can decompose the atomic concatenative units as bundles of distinctive features (see Sections 3.2 and 7.3.2), the simultaneous detection of all features amounts to recognizing the units themselves. In many settings, both linguistic and nonlinguistic, this makes excellent sense since the actual number of distinct units N is considerably larger than the number of binary features used for their feature decomposition, ideally on the order of $\log_2(N)$. Further, some of the well-established features, such as *voicing* in speech and *position* in handwriting recognition, are relatively easy to detect, which gave rise to high hopes that the detection of other features will prove just as unproblematic – in speech recognition, this is known as the Stevens-Blumstein (1981) program.

As we shall see in Chapter 9, low-level signal processing makes good use of the knowledge that phonologists and phoneticians have amassed about speech production and perception. But in high-level processing, engineering practice makes only limited use of the *featural* and *gestural* units proposed in phonology and phonetics: all working systems are based on *(auto)segmental* units. Aside from voicing and a handful of other distinctive features, training good feature detectors proved too hard, and it is only voicing that ends up playing a significant role in speech processing (see Section 9.1). We have little doubt that infants come equipped with such detectors, but research into these is now pursued mainly with the goal of understanding the biological system, as opposed to the goal of building better speech recognition. On the whole, the feature detection problem turned out to be analogous to the problem of flapping wings: a fascinating subject but one with little impact on the design of flying machines.

Therefore, we illustrate high-level signal processing on a simple example from character recognition, that of recognizing the (printed) characters *c, d, e,* and *f*. Only two of these, c and e, are positioned between the normal "n" lines of writing, with d and f, having *ascenders* that extend above the normal top line (*f*, depending on font style, may also have a *descender*). And only two, d and e, have loops that completely surround a white area; c and f leave the plane as a single connected component. Therefore, to recognize these four characters, it is sufficient to detect the features [±ascender] and [±loop], a seemingly trivial task.

Exercise 8.4 Consider the difficulties of extracting either geometrical features such as position or topological features such as connectedness from a grayscale image. Write a list for later comparison with the eventual set of problems that will be discussed in Chapter 9.

First we apply a low-level step of *pixelization*, dividing the line containing the string of characters into a sufficient number of squares, say 30 by 40 for the average character, so that a line with 80 characters is placed into a 2400 by 40 array such that the bottom of this array coincides with the baseline of the print and the top coincides with the horizontal line drawn through the highest points of the ascenders.

Using 8 bit graylevel to describe the amount of black ink found in a pixel, we have devoted some 96 kilobytes to encode the visual representation of 20 bytes of information. To recover the two bits per character that actually interest us, we first need to *segment* the line image into 80 roughly 30 by 40 rectangles, so that each of these contains exactly one character. We do this by considering columns of pixels one by one, adding up the gray values in each to form a *blackness profile*. Those columns where the result is small (zero) are considered dividers, and those where the values are higher than a threshold are considered parts of characters. We obtain an alternating string of divider and character zones, and we consider the segmentation well-established if we have the correct number of character zones (80) and these all have approximately the same width.

Once these preliminarics arc out of the way, the ascender feature can be simply detected by looking at the top five rows of pixels in each *character bounding box*. If these are all white (the sum of grayness is below a low threshold), the character in question has no ascender, otherwise it does. Detecting loops is a more complex computation since we need to find local minima of the grayness function, which involves computing the gradient and determining that the Hessian is positive definite. To compute the gradient, it is actually useful to blur the image, e.g. by averaging grayness over larger neighborhoods (e.g. over a disk of 5–10 pixel radius). Otherwise, completely flat white and black regions would both give zero gradient, and the gradient values near the edges would be numerically unstable.

Visually, such transformations would make the image less legible, as would the ascender transform, which deletes everything but the top five rows of pixels. What this simplified example shows is that the transformations that enhance automatic feature detection may be very different from the ones that enhance human perception – a conclusion that will hold true as long as we focus on the goal of pattern recognition without any attempt to mimic the human perception/recognition process.

Finally, we note here that human speech is intrinsically multimodal: in the typical case, we do not just hear the speakers but also see their mouth, hand gestures, etc. There is clear evidence (McGurk and MacDonald 1976) that the visual cues will significantly interact with the audio cues: the image of a labial sound such as *b* being produced overrides the acoustical cue, so that e.g. the experimental subject will hear *base* even if *vase* was spoken. This *McGurk effect* is already detectable in young infants and is independent of the language being learned by the infant – the effect persists even if the listener does not see the face but just touches it (Fowler and Dekle 1991).

Therefore, it is important to determine the relative contributions of the different channels, a problem that is made very difficult by the fact that the linguistic signal is highly redundant. In telephony, a one second stretch is typically encoded in 8 kilobytes, while in imaging, a one square cm area takes about 56 kilobytes at the 600 dpi resolution common to most printers and scanners. On the average, one second of speech will contain about 10–15 phonemes, each containing no more than 6 bits of information, so the redundancy is about a thousandfold. For the Latin alphabet, one square cm will contain anywhere from three handwritten characters, say 18 bits, to 60 small printed characters (45 bytes), so the redundancy factor is between 1200 and

24000. In fact, these estimates are on the conservative side since they do not take into account the redundancy between adjacent phonemes or characters – we return to the issue of compressing the signal in Chapter 9.

8.4 Document classification

As the number of machine-readable documents grows, finding the ones relevant to a particular query becomes an increasingly important problem. In the ideal case, we would like to have a *question answering* algorithm that would provide the best answer, relative to any collection of documents D, for any (not necessarily well-formed English) query q such as *Who is the CEO of General Motors?* Since D may contain contradictory answers (some of which may have been true at different times, others just plain wrong), it is clear that in general this is not a solvable problem. Question answering, together with unrestricted *machine learning, machine translation, pattern recognition, commonsense reasoning*, etc., belong in the informal class of *AI-complete* problems: a solution to any of them could be leveraged to solve all problems of artificial intelligence.

Exercise 8.5 Define the problems above more formally, and develop Karp-style reduction proofs to show their equivalence.

As Matijasevič (1981) emphasizes, the proof of the unsolvability of a problem is never the final point in our investigations but rather the starting point for tackling more subtle problems. AI-complete problems are so important that finding less general but better solvable formulations is still an important practical goal. Instead of unrestricted machine translation, which would encompass the issue of translating into arbitrary systems of logical calculus, and thus would subsume the whole AI-complete problem of *knowledge representation*, we may restrict attention to translation between natural languages, or even to a given pair of natural languages. Instead of the full question answering problem, we may restrict attention to the more narrow issue of *information extraction*: given a document d (e.g. a news article) and a relation R (e.g. *Is-CEO-Of*), find all pairs of values $\{(x, y) | x \in \textit{Person}, y \in \textit{Company}, (x, y) \in \textit{Is-CEO-Of}\}$ supported by d. This is still rather ambitious, as it requires a more robust approach to parsing the document than we are currently capable of (see Chapter 5), but at least it does away with many thorny problems of knowledge representation and natural language semantics (see Chapter 6). Once the problem is restricted this way, there is no particular reason to believe that it is algorithmically unsolvable, and in fact practical algorithms that run linear in the size of d are available in the public domain (Kornai 1999), though these do not claim to extract all instances, just a large enough percentage to be useful.

Even with linear parsing techniques, syntactic preprocessing of a large collection of documents (such as the web, currently containing about 10^{10} documents) remains impractical, and the problem is typically attacked by means of introducing a crude intermediate classification system T composed of a few hundred to a few thousand *topics*. We assume that T partitions D into largely disjoint subsets $D_t \subset D$ $(t \in T)$ and that queries themselves can be classified for topic(s). The idea is that questions about e.g. current research in computer science are unlikely to be answered by documents discussing the economic conditions prevailing in 19th century Congo, and conversely, questions about slavery, colonial exploitation, or African history are unlikely to be answered by computer science research papers.

Therefore we have two closely related problems: in *query parsing* we try to determine the set of topics relevant to the query, and in *topic detection* we try to determine which topics are discussed in a document. In many practical systems, the two problems are conflated into one by treating queries as (very short) documents in their own right. We thus have the problem of *document classification*: given some documents D, topics T, and some sample of (d, t) pairs from the relation *Is-About* $\subset D \times T$, find the values of *Is-About* for new documents. Since the space of topics is not really structured linguistically (if anything, it is structured by some conceptual organization imposed on encyclopedic knowledge), strictly speaking this is not a linguistic pattern recognition problem, but we discuss it in some detail since the mathematical techniques used are quite relevant to mathematical linguistics as a whole. First, we present some terminology and notation.

We assume a finite set of words $w_1, w_2, \ldots, w_N$ arranged in order of decreasing frequency. N is generally in the range 10^5–10^6 – for words not in this set, we introduce a catch-all *unknown word* w_0. By *general English* we mean a probability distribution G_E that assigns the appropriate frequencies to the w_i either in some large collection of topicless texts or in a corpus that is appropriately representative of all topics. By the (word unigram) probability model of a topic t, we mean a probability distribution G_t that assigns the appropriate frequencies $g_t(w_i)$ to the w_i in a large collection of documents about t. Given a collection C, we call the number of documents that contain w the **document frequency** of the word, denoted $DF(w, C)$, and we call the total number of w tokens its **term frequency** in C, denoted $TF(w, C)$.

Assume that the set of topics $T = \{t_1, t_k, \ldots, t_k\}$ is arranged in order of decreasing probability $Q(T) = q_1, q_2, \ldots, q_k$. Let $\sum_{i=1}^{k} q_i = T \leq 1$, so that a document is topicless with probability $q_0 = 1 - T$. The general English probability of a word w can therefore be computed in topicless documents to be $p_w = G_E(w)$ or as $\sum_{i=1}^{k} q_i g_i(w)$. In practice, it is next to impossible to collect a large set of truly topicless documents, so we estimate p_w based on a collection D that we assume to be representative of the distribution Q of topics. It should be noted that this procedure, while workable, is fraught with difficulties since in general the q_j are not known, and even for very large collections it cannot always be assumed that the proportion of documents falling in topic j estimates q_j well.

As we shall see shortly, within a given topic t, only a few dozen, or perhaps a few hundred, words are truly characteristic (have $g_t(w)$ significantly higher than the background probability $g_E(w)$) and our goal will be to find them. To this end, we need to first estimate G_E. The trivial method is to use the *uncorrected observed frequency* $g_E(w) = TF(w, C)/L(C)$, where $L(C)$ is the **length** of the corpus C (the total number of word tokens in it). While this is obviously very attractive, the numerical values so obtained tend to be highly unstable. For example, the word *with* makes up about 4.44% of a 55 m word sample of the *Wall Street Journal* but

5.00% of a 46 m word sample of the *San Jose Mercury News*. For medium-frequency words, the effect is even more marked. For example, *uniform* appears 7.65 times per million words in the *WSJ* and 18.7 times per million in the *Merc* sample. And for low-frequency words, the straightforward estimate very often comes out as 0, which tends to introduce singularities in models based on the estimates.

The same uncorrected estimate, $g_t(w) = TF(w, D_t)/L(D_t)$, is of course available for G_t, but the problems discussed above are made worse by the fact that any topic-specific collection of documents is likely to be orders of magnitude smaller than our overall corpus. Further, if G_t is a Markov source, the probability of a document containing l_1 instances of w_1, l_2 instances of w_2, etc., will be given by the multinomial formula

$$\binom{l_0 + l_1 + \ldots + l_N}{l_0, l_1, \ldots, l_N} \prod_{i=0}^{N} g_t(w_i)^{l_i} \tag{8.6}$$

which will be zero as long as any of the $g_t(w_i)$ are zero. Therefore, we will *smooth* the probabilities in the topic model by the (uncorrected) probabilities that we obtained for general English since the latter are of necessity positive. Instead of $g_t(w)$ we will therefore use $\alpha g_E(w) + (1-\alpha)g_t(w)$, where α is a small but nonnegligible constant, usually between .1 and .3. Another way of justifying this method is to say that documents are not fully topical but can be expected to contain a small portion α of general English.

Since the probabilities of words can differ by many orders of magnitude (both for general English and for the sublanguage defined by any particular topic), we separate the discussion of the high-, mid-, and low-frequency cases. If a word has approximately constant probability $g_t(w)$ across topics t, we say it is a *function word* of English. Such words are distributed evenly across any sample and will therefore have very low KL divergence. The converse is not true: low KL divergence indicates only that the word is not distinctive for those topics covered in the collection, not that the word is nondistinctive in a larger corpus.

For function words, the estimate $p_w = (1 - T)g_t(w)$ or even simply $g_t(w)$ is reasonable. If a document d has length $l(d) \gg 1/p_w$, we expect the word to appear in d at least once. Let us denote the **size** (number of documents) of a collection C by $S(C)$. If D_l contains only those documents in D that are longer than l, we expect $DF(w, D_l) = S(D_l)$. We can turn this around and use this as a method of discovering function words: a reasonable choice of threshold frequency would be 10^{-4}, and we can say that the function words of English will be those words that appear in all (or a very large proportion of) those documents that have length $\geq 10^5$.

We emphasize that not all words with high observed frequency will meet the test: for example the word *Journal* has about twice the frequency of *when* in the widely used *WSJ* corpora, but it will fail the $DF(w, D_l) = S(D_l)$ test in any other collection, while *when* will pass. The extreme high end of the distribution, words having 0.2% or greater probability, are generally function words, and the first few hundred function words (which go down to the mid-range) collectively account for about half of any corpus (see Section 7.1).

Function words are of course not the only words of general English. In the mid-range and below we will make a distinction between *specific* and *nonspecific* content words. Informally, a content word is nonspecific if it provides little information about the identity of the topic(s) in which it appears. For example, words like *see* or *book* could not be called function words even under the most liberal definition of the term (and there will be many long documents that fail to contain them), but their content is not specific enough: for any topic t, $P(t|w)$ is about the same as the general probability of the topic q_t, or, what is the same by Bayes' rule, $g_t(w)/g_E(w)$ is close to 1.

Exercise 8.6 Assume a large collection of topic-classified data. Define an overall measure of 'closeness to 1' that is independent of the distribution Q of topics (it does not require that the collection be representative of this distribution).

In practice we rarely have access to a large enough collection of topic-classified data, and we have to look at the converse task: what words, if any, are specific to a few topics in the sense that $P(d \in D_t|w \in d) \gg P(d \in D_t)$. This is well measured by the number of documents containing the word. For example *Fourier* appears in only about 200 k documents in a large collection containing over 200 m English documents (see `www.northernlight.com`), while *see* occurs in 42 m and *book* in 29 m. However, in a collection of 13 k documents about digital signal processing, *Fourier* appears 1100 times, so $P(d \in D_t)$ is about $6.5 \cdot 10^{-5}$, while $P(d \in D_t|w)$ is about $5.5 \cdot 10^{-3}$, two orders of magnitude better. In general, words with low DF values, or what is the same, high 1/DF = IDF **inverse document frequency** values, are good candidates for being specific content words. Again, the criterion has to be used with care: it is quite possible that a word has high IDF because of deficiencies in the corpus, not because it is inherently very specific. For example, the word *alternately* has even higher IDF than *Fourier*, yet it is hard to imagine any topic that would call for its use more often than others.

This observation provides strong empirical evidence that the vocabulary of any language cannot be considered finite; for if it was finite, there would be a smallest probability p among the probabilities of the words, and in any random collection of documents with length $\gg 1/p$, we would expect to find no hapaxes at all. Obviously, for hapaxes TF = DF = 1, so to the extent that every document has a topic this could be established deterministically from the hapax in question. In machine-learning terms, this amounts to memorizing the training data, and the general experience is that such methods fail to work well for new data. Overall, we need to balance the TF and IDF factors, and the simplest way of doing this is by the classical TF·IDF formula that looks at the product of these two numbers.

Given a document with word counts l_i and total length n, if we assume the l_i are independent (the 'naive Bayesian' assumption), the log probability quotient that topic t, rather than general English, emitted this document will be given by

$$\sum_{i=0}^{N} l_i \log \frac{\alpha g_E(w_i) + (1-\alpha) g_t(w_i)}{g_E(w_i)}$$

We rearrange this sum in three parts: where $g_E(w_i)$ is significantly larger than $g_t(w_i)$, when it is about the same, and when it is significantly smaller. In the first part, the numerator is dominated by $\alpha g_E(w_i)$, so we have

$$\log(\alpha) \sum_{g_E(w_i) \gg g_t(w_i)} l_i \tag{8.7}$$

which we can think of as the contribution of 'negative evidence', words that are significantly sparser for this topic than for general English. In the second part, the quotient is about 1 and therefore the logs are about 0, so this whole part can be neglected – words that have about the same frequency in the topic as in general English cannot help us distinguish whether the document came from the Markov source associated with the topic or from the one associated with general English. Finally, the part where the probability of the words is significantly higher than the background probability will contribute the 'positive evidence'

$$\sum_{g_E(w_i) \ll g_t(w_i)} l_i \log\left(\alpha + \frac{(1-\alpha)g_t(w_i)}{g_E(w_i)}\right)$$

Since α is a small constant, on the order of .2, while in the interesting cases (such as *Fourier* in DSP vs. in general English) g_t is orders of magnitude larger than g_E, the first term can be neglected and we have, for the positive evidence,

$$\sum_{g_E(w_i) \ll g_t(w_i)} l_i(\log(1-\alpha) + \log(g_t(w_i)) - \log(g_E(w_i)) \tag{8.8}$$

Needless to say, the real interest is not in determining $\log(P(t|d)/P(E|d))$, i.e. whether a document belongs to a particular topic as opposed to general English, but rather in whether it belongs in topic t or topic s. We can compute $\log(P(t|d)/P(s|d))$ as $\log((P(t|d)/P(E|d))/(P(s|d)/P(E|d)))$, and the importance of this step is that we see that the 'negative evidence' given by (8.7) also disappears. Words that are below background probability for topic t will in general also be below background probability for topic s since their instances are concentrated in some other topic u of which they are truly characteristic. The key contribution in distinguishing topics s and t will therefore come from those few words that have significantly higher than background probabilities in at least one of these:

$$\begin{aligned} &\log(P(t|d)/P(s|d)) \\ &= \sum_{g_E(w_i) \ll g_t(w_i)} l_i(\log(1-\alpha) + \log(g_t(w_i)) - \log(g_E(w_i)) \\ &- \sum_{g_E(w_i) \ll g_s(w_i)} l_i(\log(1-\alpha) + \log(g_s(w_i)) - \log(g_E(w_i)) \end{aligned} \tag{8.9}$$

For words w_i that are significant for both topics (such as *Fourier* would be for DSP and for harmonic analysis), the contribution of general English cancels out, and we are left with $\sum l_i \log(g_t(w_i)/g_s(w_i))$. But such words are rare even for closely related topics, and the cases where their probability ratio $g_t(w_i)/g_s(w_i)$ is far from 1 are even rarer, so the bulk of $\log(P(t|d)/P(s|d))$ is contributed by two disjoint sums in (8.9). Even these can be simplified further by noting that in any term $\log(1-\alpha)$ is small compared with $\log(g_s(w_i)) - \log(g_E(w_i))$ since the former is about $-\alpha$ while the latter counts the orders of magnitude in frequency over general English. Thus, if we define the relevance $r(w,t)$ of word w to topic t by $\log(g_s(w_i)) - \log(g_E(w_i))$, we can simply treat this as an additive quantity and for a document d with counts l_i we obtain

$$r(d,t) = \sum l_i r(w,t) \tag{8.10}$$

where the sum is taken over those words w_i whose frequency in documents about t is significantly higher than their background frequency $p_{w_i} = g_E(w_i)$.

What (8.10) defines is the simplest, historically oldest, and best-understood pattern classifier, a *linear machine* where the decision boundaries are simply hyperplanes. As the reasoning above makes clear, linearity is to some extent a matter of choice: certainly the underlying Markovian assumption, that the words are chosen independent of one another, is quite dubious. However, it is a good first-order approximation, and one can extend it to second order, third order, etc., by increasing the Markovian parameter. Once the probabilities of word pairs, word triples, etc., are explicitly modeled, much of the criticism directed at the naive unigram or *bag of words* approach loses its grip.

Of particular importance is the fact that, in topic classification, the models can be *sparse* in the sense of using nonzero coefficients $g_t(w_i)$ only for a few dozen, or perhaps a few hundred, words w_i for a given topic t even though the number of words considered, N, is typically in the hundred thousands to millions (see Kornai 2002). Assuming $k = 10^4$ topics and $N = 10^6$ words, we would need to estimate $kN = 10^{10}$ parameters even for the simplest (unigram) model. This may be (barely) within the limits of our supercomputing ability, but it is definitely beyond the reliability and representativeness of our data. Over the years, this has led to a considerable body of research on *feature selection*, which tries to address the issue by reducing N, and on *hierarchical classification*, which aims at reducing k. We do not attempt to survey this literature here but note that much of it is characterized by an assumption of 'once a feature, always a feature': if a word w_i is found distinctive for topic t, an attempt is made to estimate $g_s(w_i)$ for the whole range of s rather than the one value $g_t(w_i)$ that we really care about.

The fact that high-quality working classifiers can be built using only sparse subsets of the whole potential feature set reflects a deep structural property of the data: at least for the purpose of comparing log emission probabilities across models, the G_t can be approximated by sparse distributions S_t. In fact, this structural property is so strong that it is possible to build classifiers that ignore the differences between the numerical values of $g_s(w_i)$ and $g_t(w_i)$ entirely, replacing both by a uniform estimate $g(w_i)$ based on the IDF (inverse document frequency) of w_i. Traditionally, the l_i multipliers in (8.10) have been known as the term frequency (TF) factor, and such systems, where the classification load is carried entirely by the zero-one decision of using a particular word in a particular topic, are known as TF-IDF classifiers.

8.5 Further reading

The standard pattern recognition handbook is Duda et al. (2000); see also MacKay (2003). These authors approach the problem from a practical standpoint – for a more abstract view, see Devroye et al. (1996) and Hastie et al. (2001). The 'six decibels per bit' rule comes from Bennett (1948). The idea of computing SQNR by assuming uniform distribution with zero mean for each cell comes from Widrow (1960) – for the limits of its applicability, see Sripad and Snyder (1977). In practical applications, analog to digital conversion does not involve circuitry that can quantize to more than 8 bits. Rather, Sigma-Delta conversion (Inose et al. 1962) is used. A-law and μ-law are part of the Consultative Committee for International Telephony and Telegraphy (CCITT) standard G.711 (1972).

The papers and books recommended for Markov processes and HMMs in Section 5.7 approach the subject with linguistic applications in mind. Our treatment follows the the pure mathematical approach taken in Khinchin (1957). It must be admitted that there is still a noticeable gap between the purely mathematically oriented work on the subject such as Cappé et al. (2005) and the central linguistic ideas. While the HMMs used in speech recognition embody the phonemic principle (see Section 3.1), they fall short of full autosegmentalization (Section 3.3) and make no use of the prosodic hierarchy (Section 4.1).

Feature extraction (high-level signal processing) is generally performed through supervised learning, a subject we shall discuss in Chapter 9. The basic literature on speech perception is collected in Miller et al. (1991).

For topic detection experiments, the widely used Reuters Corpus is available at `http://trec.nist.gov/data/reuters/reuters.html`. There is no monographic treatment of the subject (for a survey, see Sebastiani 2002), and the reader is advised to consult the annual SIGIR and TREC proceedings. Classification by linear machine originates with Highleyman (1962), see also Duda et al. (2000 Ch. 5), Haste et al. (2001 Ch. 4), Devroye et al. (1996 Ch. 4).

9

Speech and handwriting

Conceptually, the techniques of linguistic pattern recognition are largely independent of the medium, but overall performance is influenced by the preprocessing to such an extent that until a few years ago the pattern recognition step was generally viewed as a small appendix to the main body of signal processing knowledge. To this day, it remains impossible to build a serious system without paying close attention to preprocessing, and deep algorithmic work on the recognizer will often yield smaller gains than seemingly more superficial changes to the front end. In Section 9.1, we introduce a speech coding method, linear prediction, that has played an important role in practical application since the 1970s. We extend the discussion of quantization started in Section 8.1 from scalars to vectors and discuss the Fourier transform-based (homomorphic) techniques that currently dominate the field.

These techniques, in spite of their analytic sophistication, are still low-level inasmuch as the signal can still be reconstructed, often without perceptually noticeable loss from the encoding, yet they suffice to decrease the bitrate by several orders of magnitude. As we shall see, the bitrate provides a surprisingly good measure of our understanding of the nature of speech: the more we know, the better we can compress the signal. This observation extends well beyond low-level signal processing in that incorporating deeper knowledge about the speech signal leads to further gains in compression. In Section 9.2, we discuss how a central component of the linguistic theory of speech, the phonemic principle introduced in Section 3.1, can be leveraged to yield further compression gains in HMMs.

The recognition of handwritten or printed text by computer is referred to as *optical character recognition* (OCR). When the input device is a digitizer tablet that transmits the signal in real time (as in pen-based computers and personal digital assistants) or includes timing information together with pen position (as in signature capture), we speak of *dynamic* recognition. When the input device is a still camera or a scanner, which captures the position of digital ink on the page but not the order in which the ink was laid down, we speak of *static* or *image-based* OCR. The difficulties encountered in dynamic OCR are largely similar to those found in speech recognition: the stream of position/pen pressure values output by the digitizer tablet is analogous to the stream of speech signal vectors output by the audio processing front end, and the same kinds of low-level signal processing and pattern recognition techniques are widely employed for both. In Section 9.3, we will deal primarily with static OCR, emphasizing those aspects of the problem that have no counterpart in the recognition of spoken or signed language.

9.1 Low-level speech processing

A two-way infinite sequence $s = \ldots s_{-2}, s_{-1}, s_0, s_1, s_2, \ldots$ will be called a (discrete) **signal**, and its generating function $\sum_{n=-\infty}^{\infty} s_n z^{-n}$ will be called its **z transform** **Z**(s). Signals will also be denoted by $\{s_n\}$. If the signal is bounded (a condition we can always enforce by clipping it; see Section 8.1) and satisfies further conditions that are generally met, the z transform will be absolutely convergent on a disk of positive radius and the signal can be uniquely reconstructed from it. A **filter** (sometimes called a *system*) is a mapping from signals to signals: we are particularly interested in the case where the mapping is both linear and time-invariant (invariant under the shift operator S). The signal u defined by $u_0 = 1, u_n = 0\ (n \neq 0)$ is called the unit **impulse**, and the image $\{h_n\}$ of this signal under some filter F is called the **impulse response** $h = F(u)$ of the filter F. As long as h_n is absolute convergent, h completely characterizes any linear and time-invariant filter. To see this, consider any arbitrary input x and write it as $\sum_{m=-\infty}^{\infty} x_m u_0^m$, where u_0^m is u_0 shifted by m. Since F is linear and time-invariant, we obtain

$$F(x)_n = \sum_{m=-\infty}^{\infty} x_m h_{n-m} = \sum_{m=-\infty}^{\infty} x_{n-m} h_m \tag{9.1}$$

This sum, which will always converge for x bounded, determines the output uniquely just from the values of h. A linear and time-invariant filter is called **causal** if $h_n = 0$ for $n < 0$ and **stable** if bounded input always produces bounded output – it is trivial to see that absolute convergence of h_n is both necessary and sufficient for stability. In what follows, we will be chiefly concerned with filters that are linear, time-invariant, causal, and stable and will omit these qualifications.

In speech processing applications, we are particularly concerned with the output of a filter F with impulse response h when the input is a sampled complex sine wave $x = \{e^{i\omega n}\}$. Using (9.1), we obtain

$$y_n = F(x)_n = \sum_{m=-\infty}^{\infty} e^{i\omega(n-m)} h_m = e^{i\omega n} \sum_{m=-\infty}^{\infty} h_m e^{-i\omega m} \tag{9.2}$$

The term outside the sum is x_n. The function $\sum_{m=-\infty}^{\infty} h_m e^{-i\omega m}$ will be denoted $H(e^{i\omega})$ and called the **frequency response** or **transfer function** of F. With this notation, (9.2) becomes the more perspicuous

$$y = H(e^{i\omega})x \tag{9.3}$$

where equality among signals means termwise equality. In general, we define the **frequency spectrum F**(x) of a signal x as

$$X(e^{i\omega}) = \sum_{m=-\infty}^{\infty} x_m e^{-i\omega m} \tag{9.4}$$

so that the frequency response is just the frequency spectrum of the impulse response. Since (9.4) is a Fourier series with coefficients x, we can make good use of the orthogonality of $e^{i\omega n}$ and $e^{i\omega m}$ on the $-\pi \le \omega \le \pi$ interval and find x from its frequency spectrum by the well-known Fourier inversion formula

$$x_n = \frac{1}{2\pi} \int_{-\pi}^{\pi} X(e^{i\omega}) e^{i\omega n} d\omega \tag{9.5}$$

Now if $y = F(x)$, and we denote the frequency spectrum of x, y, and h (the impulse response of F) by X, Y, and H, respectively, we have, by applying (9.5) to y,

$$y_n = \frac{1}{2\pi} \int_{-\pi}^{\pi} Y(e^{i\omega}) e^{i\omega n} d\omega \tag{9.6}$$

Since (9.5) holds for all n, we can apply F to the whole series, obtaining

$$y_n = \frac{1}{2\pi} \int_{-\pi}^{\pi} X(e^{i\omega}) F(e^{i\omega n}) d\omega \tag{9.7}$$

By applying (9.3) to the series $e^{i\omega n}$, the last term in this integral, $F(e^{i\omega n})$, can be expressed as $H(e^{i\omega})e^{i\omega n}$. Comparing this to (9.6) yields

$$\frac{1}{2\pi} \int_{-\pi}^{\pi} X(e^{i\omega}) H(e^{i\omega}) e^{i\omega n} d\omega = y_n = \frac{1}{2\pi} \int_{-\pi}^{\pi} Y(e^{i\omega}) e^{i\omega n} d\omega \tag{9.8}$$

which, by the uniqueness of Fourier coefficients, leads to

$$X(e^{i\omega}) H(e^{i\omega}) = Y(e^{i\omega}) \tag{9.9}$$

Thus, the frequency spectrum of the output is obtained by multiplying the frequency spectrum of the input with the frequency response of the filter.

The frequency spectrum and the z transform are both obtained from a signal by using the terms as coefficients in a series of complex functions: the frequency spectrum is the z transform with $z = e^{i\omega}$ (i.e. investigated over the unit circle). Taking the z transform of both sides of (9.1), it is trivially seen that (9.9) is valid for X, Y, H z transforms (rather than frequency spectra) of x, y, h. In general, we make little distinction between **Z** and **F** and for any signal $s, t, \ldots, x, y, z$ will follow the engineering convention of using uppercase $S, T, \ldots, X, Y, Z$ to denote its z transform and frequency spectrum alike. The use of lowercase letters indicates, in engineering parlance, signals and operations in the *time domain*, where the independent variable is time, and the use of uppercase letters refers to the *frequency domain*, where the independent variable is frequency. The two domains are connected by the transforms and their inverses.

While we are chiefly interested in discrete (digital) signals, it is clear that the speech waveform is inherently continuous (analog), and one could replace (9.4) by a continuous version (distinguished by a subscript C):

$$X_C(i\Omega) = \int_{-\infty}^{\infty} x(t)e^{-i\Omega t}\,dt \tag{9.10}$$

If we measure the continuous signal $x(t)$ at time intervals T apart (this is called using a **sampling frequency** $\omega_s = 2\pi/T$), by the inverse Fourier transform (which of course remains applicable in the continuous case), we obtain

$$x(nT) = \frac{1}{2\pi}\int_{-\infty}^{\infty} X_C(i\Omega)e^{i\Omega nT}\,d\Omega \tag{9.11}$$

We shall now subdivide the integration domain to intervals of length $2\pi/T$ starting at $(2m-1)\pi/T$ and obtain the Fourier expansion

$$x(nT) = \frac{T}{2\pi}\int_{-\pi/T}^{\pi/T} \frac{1}{T}\sum_{m=-\infty}^{\infty} X_C(i(\omega + 2\pi m/T))e^{i\omega nT}\,d\omega \tag{9.12}$$

where we substituted $\omega + 2\pi m/T$ for Ω in the mth interval. By (9.5) the discrete signal $\{x(nT)\}$ satisfies

$$x(nT) = \frac{T}{2\pi}\int_{-\pi/T}^{\pi/T} X(e^{i\omega T})e^{i\omega nT}\,d\omega \tag{9.13}$$

Comparing (9.12) with (9.13) yields

$$\frac{T}{2\pi}\int_{-\pi/T}^{\pi/T} \frac{1}{T}\sum_{m=-\infty}^{\infty} X_C(i(\omega + 2\pi m/T))e^{i\omega nT}\,d\omega = \frac{T}{2\pi}\int_{-\pi/T}^{\pi/T} X(e^{i\omega T})e^{i\omega nT}\,d\omega \tag{9.14}$$

and by the uniqueness of Fourier coefficients we have

$$\frac{1}{T}\sum_{m=-\infty}^{\infty} X_C(i(\omega + 2\pi m/T)) = X(e^{i\omega T}) \tag{9.15}$$

If $x(t)$ is **bandlimited** in the sense that $X_C(i\omega) = 0$ for any $|\omega| \geq \omega_c$ and we restrict ourselves to sampling frequencies $\omega_s = 2\pi/T > 2\omega_c$, within the interval $|\omega_c| < \omega_s/2$ all but the central term of (9.15) will be zero and there we have

$$\frac{1}{T}X_C(i\omega) = X(e^{i\omega T}) \tag{9.16}$$

In other words, the continuous frequency spectrum (Fourier transform) X_C is fully determined by the discrete frequency spectrum X, even though though X will have *sidebands* ($k\omega_s$ translates of the central band) while X_C will be, as we assumed, identically zero outside the critical band. Since the Fourier transform uniquely determines the function, and the discrete spectrum was computed entirely on the basis of the sampled values, these values uniquely determine the original continuous function, and we have the following theorem.

Theorem 9.1.1 Sampling theorem. If a continuous signal $x(t)$ is bandlimited with cutoff frequency ω_c, sampling it at any frequency $\omega_s = 2\pi/T > 2\omega_c$ or higher provides a discrete signal $\{x(nT)\}$ from which $x(t)$ can be fully reconstructed.

Discussion As is well-known, the upper frequency threshold of human hearing is about 20 kHz, and human speech actually contains only a negligible amount of energy above 12 kHz. The 44.1 kHz sampling rate of audio CDs was set with the sampling theorem in mind since analog HiFi equipment was designed to operate in the 20 Hz – 20 kHz range and the intention was to preserve all high fidelity audio in digital format. Telephone speech, sampled at 8 kHz, is perfectly understandable, in spite of the fact that much of the frequency range below 350 Hz is also severely attenuated. In speech research, a sampling rate of 20 kHz, and in speech communications a 10 kHz sampling rate, is common. In what follows, we concentrate on the digital case, since Theorem 9.1.1 guarantees that nothing of importance is lost that way.

In Section 8.1, we discussed log PCM speech coding, which, by means of fitting the quantization steps to the long-term amplitude distribution of speech, achieves toll quality transmission at 64 kbps. To go beyond this limit, we need to exploit short-term regularities. Perhaps the simplest idea is to use **delta coding**, transmitting the difference between two adjacent samples rather than the samples themselves. Since the differences are generally smaller than the values, it turns out we can save about 1 bit per sample and still provide speech quality equivalent to the original (see Jayant and Noll 1984). *Linear predictive coding* (LPC) extends the basic idea of delta coding by considering not only the previous sample but the previous p samples to be available for predicting the next sample. Thus we shall consider the general problem of obtaining the signal s from some unknown input signal u such that

$$s_n = -\sum_{k=1}^{p} a_k s_{n-k} + G \sum_{l=0}^{q} b_l u_{n-l} \tag{9.17}$$

i.e. as a linear combination of the past $q+1$ inputs and the past p outputs. (We assume $b_0 = 1$ and separate out a *gain factor* G for putting the equations in a more intuitive form.) With these conventions, the transfer function can be written in terms of z transforms as

$$H(z) = \frac{S(z)}{U(z)} = G\frac{1+\sum_{l=1}^{q} b_l z^{-l}}{1+\sum_{k=1}^{p} a_k z^{-k}} \tag{9.18}$$

i.e. as a Padé approximation. Since the zeros of the denominator appear as poles in the whole expression, in signal processing (9.18) is known as the *pole-zero* model, with the key case where $b_l = 0$ for $l \geq 1$ called the *all-pole* model – the number of terms p is called the *order* of the model.

Engineers call a signal *stationary* if its statistical characteristics, as summarized in the frequency spectrum, stay constant over long periods of time. The vibrations produced by rotating machinery or the spattering of rain are good examples. We say a signal $X(t)$ is **stationary in the wide sense** if its expectation $E(X(t))$ is constant and the correlation function $E(X(t)X(s))$ depends only on the difference $t - s$. This is less strict than Definition 8.2.1 because we do not demand full invariance under a single sample shift (and thus by transitivity over the entire timeline) but use a weaker notion of 'approximate' invariance instead. The situation is further complicated by the fact that speech is rarely stationary, even in this wide sense, over more than a few pitch periods – this is referred to as the signal being *quasistationary*. At the high end (infants, sopranos), the glottal pulses can follow each other as fast as every 2 ms, for adult male speech 6–12 ms is typical, while at the low end (basso profondo) 22 ms or even longer pitch periods are found.

To take the quasistationary nature of the signal into account, speech processing generally relies on the use of *windowing* techniques, computing spectra on the basis of samples within a 20 ms stretch. Typically such a window will contain more than a full pitch period and thus allow for very good reconstruction of the signal. Because of edge effects (produced when the pitch period is close to the window size), rectangular windows are rarely used. Rather, the signal within the window is multiplied with a windowing function such as the *Hamming window*, $w_n = 0.54 - 0.46\cos(2\pi n/(N-1))$. In speech processing, given a 20 kHz sampling rate and 20 ms windows, N is about 400, depending on how the edges are treated. Windows are generally overlapped 50% so that each sample appears in two successive windows. In other words, analysis proceeds at a *frame rate* of 100 Hz.

Aside from computing windowed functions, pointwise multiplication of two signals is rarely called for. A more important operation is the **convolution** $s \circ t$ of two signals s and t given by $(s \circ t)_k = \sum_{n=0}^{N-1} s_n t_{k-n}$ in the discrete case and by $\int s(\tau)t(x-\tau)d\tau$ in the continuous case. When the signal has only finite support (e.g. because of windowing), we can consider the indexes mod N in performing the summation, a method known as *circular convolution*, or we can take values of s and t outside the range $0, \ldots, N$ to be zeros, a method known as *linear convolution*. Unless explicitly stated otherwise, in what follows we assume circular convolution for finite signals. For the infinite case, (9.1) asserts that the output of a filter F is simply the convolution of the input and the impulse response of F, and (9.9) asserts that convolution in the time domain amounts to multiplication in the frequency domain.

The practical design of filters, being concerned mostly with the frequency response, generally proceeds in the frequency domain. Of particular interest are *highpass* filters, whose response is a step function (0 below the frequency cutoff and 1 above), *lowpass* filters (1 below the cutoff and 0 above), and *bandpass* filters (0 below the bottom and above the top of the band, 1 inside the band). Since these ideal pass characteristics can be approximated by both analog and digital circuitry at reasonable cost, it is common for signal processing algorithms to be implemented using filters as building blocks, and much of the literature on speech production, processing, and perception is actually presented in these terms. One tool used particularly often is a *filter bank* composed of several filters, each sensitive only in a specified *subband* of the region of interest. In *mel* filtering, triangular transfer function filters are used in an overlapping sequence that follows the classical mel subjective pitch scale (Stevens and Volkman 1940, Beranek 1949).

Exercise 9.1 Given capacitors, resistors, and coils of arbitrary precision, design high- low- and bandpass filters that pass pure sinusoidal voltage signals within a prescribed error factor relative to the ideal filter characteristics.

After windowing, we are faced with a very different data reduction problem: instead of a fast (20 kHz) sequence of (16 bit) *scalars*, we need to compress a much slower (0.1 kHz) sequence of *vectors*, where a single vector accounts for the contents of a whole frame. We begin by replacing the signal within a window by some appropriate function of it such as its **discrete Fourier transform** (DFT), given by

$$S_k = \sum_{n=0}^{N-1} s_n e^{-\frac{2\pi i}{N}kn} \tag{9.19}$$

Here S_k is the amplitude of the signal at frequency $2\pi k/N$, and it is common to use notation like $S(\omega)$ in analogy with the continuous case even though, strictly speaking, we are now concerned only with frequencies that are a multiple of $2\pi/N$. Of special note is the discrete version of Parseval's theorem:

$$\sum_{n=0}^{N-1} |s_n|^2 = \frac{1}{N} \sum_{n=0}^{N-1} |S_n|^2 \tag{9.20}$$

As in the continuous case, Parseval's theorem is interpreted as saying that the total energy of the signal (defined there by $\int_{-\infty}^{\infty} s^2(t)dt$) can also be obtained by integrating the energy in the frequency domain. The contribution of a frequency range (also known as a frequency *band* in signal processing) $a \leq \omega \leq b$ is given by $\int_a^b S(\omega)d\omega + \int_{-b}^{-a} S(\omega)d\omega$ – the first term is known as the contribution of the *positive frequency* and the second as that of the *negative frequency*. In analogy with the continuous case, the squared absolute values of the DFT coefficients are called the **energy spectrum** and are denoted $P(\omega)$ (with implicit time normalization, the term *power spectrum* is also often used).

From a given energy spectrum, we can recover the moduli of the DFT coefficients by taking square roots, but we lose their argument (known in this context as the *phase*), so we cannot fully reconstruct the original signal. However, hearers are relatively insensitive to the distinction between waveforms inverted from full spectra and from modulus information alone (at least for sound presented by loudspeakers in a room with normal acoustics – for sounds presented directly through earphones, discarding the phase information is more problematic; see Klatt 1987) and for many, if not all, purposes in speech processing, energy spectra are sufficient. This is not to say that the phase is irrelevant (to the contrary, providing phase continuity is important to the perceived naturalness of synthesized speech) or that it contains no information (see Paliwal and Atal (2003) for recognition based on phase information alone), but the situation is somewhat analogous to speech, which contains enough information in the 150 Hz – 1.5 kHz band to be nearly perfectly intelligible but also contains enough information in the 1.5–15 kHz band to be equally intelligible.

The DFT has several important properties that make it especially well-suited for the analysis of speech. First, note that (9.19) is actually a *linear* transform of the inputs (with complex coefficients, to be sure, but still linear). The inverse DFT (IDFT) is therefore also a linear transform and one that is practically identical to the DFT:

$$s_k = \frac{1}{N} \sum_{n=0}^{N-1} S_n e^{\frac{2\pi i}{N} kn} \tag{9.21}$$

i.e. only the sign in the exponents and the normalization factor $1/N$ are different. There are, significantly, *fast Fourier transform* (FFT) algorithms that compute the DFT or the IDFT in $O(N \log(N))$ steps instead of the expected N^2 steps.

Since in (linear) acoustics an echo is computed as a convolution of the original signal with a function representing the objects on which the sound is reflected, it is natural to model speech as an acoustic source (either the glottis, or, in the case of unvoiced sounds, frication noise generated at some constriction in the vocal tract) getting filtered by the shape of the vocal tract downstream from the source. This is known as the *source-filter model of speech production* (Fant 1960). As convolution corresponds to multiplication of generating functions or DFTs, the next natural step is to take logarithms and investigate the additive version of the transform, especially as signal processing offers very effective filtering techniques for separating signals whose energy lies in separate frequency ranges.

Experience shows that before taking the logarithm it is advantageous to rescale the energy spectrum by using $\Omega = 6 \log(\omega/1200\pi + \sqrt{1 + (\omega/1200\pi)^2})$, which converts the frequency variable ω given in radians/sec into *bark* units Ω (Schroeder 1977), as the bark scale (Zwicker et al. 1957) models important features of human hearing better than the linear (physical) or logarithmic (musical) frequency scale would. Essentially the same step, nonlinear warping of frequencies, can be accomplished by more complex signal processing in the time domain using filter banks arranged on the mel scale (Davis and Mermelstein 1980). Further perceptually motivated signal processing steps, such as preemphasis of the data to model the different sensitivities of human hearing at different frequencies, or amplitude compression to model the Weber-Fechner law, are used in various schemes, but we do not follow this line of development here as it contributes little to speech recognition beyond making it more noise-robust (Hermansky 1990, Hermansky et al. 1991, Shannon and Paliwal 2003).

If we now take the log of the (mel- or bark-scaled) energy spectral coefficients, what was a multiplicative relationship (9.9) between a *source* signal (a glottal pulse for voiced sounds or fricative noise for unvoiced sounds) and a *filter* given by the shape of the vocal tract becomes an additive relation. The log energy spectrum itself, being a finite sequence of (real) values, can be considered a finite signal, amenable to analysis by DFT or by IDFT (the two differ only in circular order and a constant multiplier). We call the IDFT of the log of the energy spectrum the **cepstrum** of the original signal and call its independent variable the *quefrency* in order to avoid collision with the standard notion of frequency. For example, if we find a cepstral peak at quefrency 166, this means recurrence at every 166 samples, which at a sampling rate of 20 kHz amounts to a real frequency of 120 Hz (Noll 1964,1967).

As we are particularly interested in the case where the original signal or the energy spectrum is mel- or bark-wrapped, we note here that in mel scaling, a great deal of economy can be achieved by keeping only a few (generally 12–16) filter bank outputs, so that the dimension of the cepstral parameter space is kept at 12–16. In the bark case, we achieve the same effect by *downsampling*; i.e. keeping only a few (generally 16–20) points in the energy spectrum. To fully assess the savings entailed by these steps would require an analysis of the quantization losses. Clearly, 32 bit resolution on the cepstral parameters is far more than what we need. We postpone this step but note that 18 32-bit parameters for 100 frames per second would mean 57.6 kbps coding, not a particularly significant improvement over the 64 kbps log PCM scheme discussed earlier.

The savings will come from the source-filter approach: the glottal pulse can be modeled by an impulse function (in many cases, more realistic models are desirable – source modeling is a major research topic on its own) and frication can be modeled by white noise. Obviously, neither the impulse nor the white noise needs to be transmitted. All that is required is some encoding of the filter shape plus one bit per frame to encode whether the frame is voiced or unvoiced – in voiced frames, another 8–10 bits are used to convey the fundamental frequency F0. As we shall see, the relevant information about the transfer function of the filter can be transmitted in 40–80 bits, and frame rates as low as 40–50 Hz are sufficient, yielding toll or communications quality speech coding at 16 kbps or lower. The main advantage of the cepstral representation is that spectral characteristics of the source and the filter are largely separated in the quefrency domain and can be extracted (or suppressed) by bandpass (resp. notch) filtering (called *liftering* in this domain).

Let us now turn to the issue of modeling a signal, be it an actual time domain signal or a series of spectral or cepstral coefficients, in the form (9.17), using the all-pole model. We will for the moment ignore the input signal u_n and look for a least squares error solution to $e_n = s_n + \sum_{k=1}^{p} a_k s_{n-k}$. By setting the partial derivatives of $\sum_n e_n^2$ to zero, we obtain the set of normal equations

$$\sum_{k=1}^{p} a_k \sum_n s_{n-k} s_{n-i} = -\sum_n s_n s_{n-i} \tag{9.22}$$

For N finite, the expressions $\sum_{n=0}^{N-1} s_{n-k} s_{n-i}$ are known as the *covariances* and will be collected in the covariance matrix ϕ_{ik}, which is symmetric. If N is infinite, only the differences $i - k$ matter, and the same expression is called the *autocorrelation* and is denoted by $R(i - k)$. Either way, we have p linear equations in p unknowns, the LPC coefficients, and may take advantage of the special form of the covariance or autocorrelation matrix to solve the problem relatively quickly.

Exercise 9.2 Assuming that the autocorrelation coefficients $R(j)$ have already been computed (e.g. from estimating them in a fixed size window), find an algorithm to solve equations (9.22) in $O(p^2)$ steps.

If we assume that the signal values s_i are samples of random variables X_i, it is the expected value $E(e_n^2)$ of the squared error that we wish to minimize, and again by setting the partial derivatives to zero we obtain

$$\sum_{k=1}^{p} a_k E(s_{n-k}s_{n-i}) = -E(s_n s_{n-i}) \tag{9.23}$$

For a stationary process, $E(s_{n-k}s_{n-i}) = R(k-i)$ holds, and speech is generally considered quasistationary to the extent that this remains a reasonable approximation if R is estimated on a window comprising at most a few pitch periods.

So far, we have ignored the input signal u in (9.17), but it is clear that the equation can hold with error $e_n = 0$ in an all-pole model only if $Gu_n = e_n$. We cannot take advantage of this observation point by point (to introduce a term that corrects for the prediction error would require knowing the error in advance), but we can use it in the average in the following sense: if we want the energy of the actual signal to be the same as the energy of the predicted signal, the total energy of the input signal must equal the total energy of the error signal, $\sum_n e_n^2 = \sum_n (s_n + \sum_{k=1}^{p} a_k s_{n-k}^2)$. Using (9.22), we can see this to be

$$E_p = \sum_n s_n^2 + \sum_{k=1}^{p} a_k \sum_n s_n s_{n-k} \tag{9.24}$$

or, using the autocorrelations $R(i)$, just $E_p = R(0) + \sum_{k=1}^{p} a_k R(k)$. If the input signal u is the unit impulse, its energy will be G^2, which must be set equal to the energy E_p of the error signal that we just computed. Since the $R(i)$ must be computed anyway if the autocorrelation method is used to determine the a_i, the gain G now can also be computed from

$$G^2 = R(0) + \sum_{k=1}^{p} a_k R(k) \tag{9.25}$$

to provide a complete characterization of the transfer function $H(z)$ in (9.18) as long as no zeros, just poles (zeros in the denominator), are used.

Exercise 9.3 If the samples u_n are uncorrelated (white noise), show that the autocorrelations $\hat{R}(i)$ of the output signal $\hat{s}_n = -\sum_{k=1}^{p} a_k \hat{s}_{n-k} + Gu_n$ are the same as the autocorrelations $R(i)$ of the original signal as long as G is set so as to preserve the total energy $\hat{R}(0) = R(0)$. Does (9.25) remain true in this case?

The considerations above provide justification only for the LPC coding of voiced (glottal pulse source) and unvoiced (white noise source) signals, but it turns out that all-pole models are applied with noticeable success to signals such as cepstra that have characteristics very different from those of the raw speech signal. So far, we have decomposed the original signal into voiced and unvoiced frames and devoted 8–10 bits per frame to encoding the pitch (F0) of voiced frames. What remains is transmitting the LPC coefficients a_i and the gain G. Since these are sensitive to quantization noise, the predictors a_i are often replaced by *reflection coefficients* k_i, which can be computed essentially by the same recursive procedure that is used in solving (9.22). Taking $p = 10$ and using 10 bits per coefficient, 50 frames/sec, we obtain cell grade speech coding at about one-tenth the bitrate of log PCM.

To make further progress, it makes sense to model the properties of the LPC (reflection or predictor) coefficients jointly. A **vector quantizer** (VQ) is an algorithm that maps a sequence $\mathbf{x}_i$ of (real or complex) vectors on binary sequences $\gamma(\mathbf{x}_i)$ suitable for data transmission. It is generally assumed that the range of γ is finite: elements of the range are called **channel symbols**. For each channel symbol c, the decoder recovers a fixed vector $\hat{\mathbf{x}}_c$ from a table M called the **codebook** or the **reproduction alphabet**. Since the coding is generally lossy (the exception would be the rare case when all inputs exactly match some codebook vector), we do not insist that the dimension of the output match the dimension of the input. When they do, the error (quantization loss) can be measured by the average (expected) distortion introduced by the VQ scheme, again measured in decibels:

$$\text{SQNR} = 10 \log_{10} \frac{E(\|\mathbf{x}\|)}{E(d(\mathbf{x}, \hat{\mathbf{x}}))} \tag{9.26}$$

In general, there is no reason to believe that Euclidean distance is ideally suited to measuring the actual distortion caused by a VQ system. For example, if the vectors to be transmitted are vowel formant frequencies, as in the Peterson-Barney data discussed in Section 8.1, it is well-known that the perceptual effects of perturbing F1, F2, and F3 are markedly different (see Flanagan 1955), and a distortion measure that is invariant under permutation of the vector components is structurally incapable of describing this situation well.

If we know the probability distribution P over some space X of the vectors to be transmitted and know the ideal distance measure d, the problem reduces to a task of *unsupervised clustering*: find codebook vectors $\hat{\mathbf{x}}_1, \ldots, \hat{\mathbf{x}}_n$ such that SQNR is maximized. Since the numerator of (9.26) is given, the task is to minimize the denominator, which, assuming the $\mathbf{x}$ inputs are drawn randomly according to P, is given by

$$\int_X d(\mathbf{x}, \hat{\mathbf{x}}) dP(\mathbf{x}) \tag{9.27}$$

a quantity known as the **reconstruction error** of the VG system defined by γ and the vectors $\hat{\mathbf{x}}_i$. Clearly, the larger the codebook, the more we can decrease the reconstruction error, but this is offset by an increased number of bits required to transmit the VQ codes. According to the MDL principle (see Section 7.3.1), we need to optimize the cost (in bits) of transmitting the channel symbols plus the cost of encoding $\mathbf{x}$ given $\hat{\mathbf{x}}_i$. In practical systems where the inputs are N-dimensional vectors of 32-bit reals, the cost of transmitting the residuals $\mathbf{x} - \mathbf{x}_i$ would be overwhelming compared with the $\log_2(|M|)$ bits required to transmit the codes. Without knowing much about the probability distribution P, we would have to dedicate $32N$ bits to transmitting a residual, and in practice codebooks over $|M| > 2^{32}$ make little sense as they would take up too much memory (in fact, typical codebook sizes are in the 2^{12}–2^{20} range).

If the size m of the codebook (also known as the *number of levels* in analogy to the scalar case) is set in advance, many clustering algorithms are applicable. The most commonly used one is Lloyd's algorithm, also known as the LBG algorithm, which uses a random sample of signals $\mathbf{s}_k$ $(1 \le k \le r \gg m)$. First we pick, either based on some knowledge about P or randomly, cluster seeds $\mathbf{x}_i^{(0)}$ $(1 \le i \le m)$ and consider for each $\mathbf{s}_i$ the $\mathbf{x}_j^{(0)}$ that is closest to it as measured by distance d: the index j (given in binary) is defined as $\gamma^{(0)}(\mathbf{s}_i)$. In step $k+1$, we take the (Euclidean) centroid of all samples in the inverse image of $\gamma^{(k)}(i)$, and use these as the reconstruction vector $\mathbf{x}_i^{(k+1)}$ transmitted by the label $\gamma^{(k+1)}(i)$. We stop when the recognition error no longer improves (which can happen without having reached a global optimum; see Gray and Karnin 1982). With carefully controlled VQ techniques, taking full advantage of the HMM structure, communications quality speech coding at 100–400 bps, approaching the phonemic bitrate is possible (Picone and Doddington 1989). Concatenation techniques, which use the same Viterbi search as HMMs, can improve this to toll quality without increasing the bitrate (Lee and Cox 2001).

If the final goal is not just compression but recognition of the compressed signal, the situation changes quite markedly in that *supervised clustering* techniques now become available. Here we assume that we have *labeled* (also known as *truthed*) vectors $\mathbf{s}_k$ whose recognized (truth) value is $t(\mathbf{s}_k)$ taken from a finite set $t_1, \dots, t_l$. Ideally, we would want to set $m = l$, so that the quantization provides exactly as many levels as we would ultimately need to distinguish in the classification stage, but in practice this is not always feasible. For example, if the labels correspond to phonemes, there may be very distinct signals (corresponding e.g. to trilled vs. tapped r sounds) that occupy very distinct regions of the signal space, so that clustering them together would result in a centroid that is part of neither the trilled nor the tapped region of signals. (English makes no phonological distinction between these two clusters, but in languages like Spanish the distinction is phonological: consider *perro* 'dog' vs. *pero* 'but'.) Assuming $m \ge l$ still makes sense because transmitting differently labeled $\mathbf{s}$ values by the same code would doom to failure whatever recognition algorithm we may want to use for the reconstructed signal.

In the supervised case, any distortion that leaves the truth value intact can be neglected, so the distance $d(\mathbf{s}, \hat{\mathbf{s}})$ between the original and the reconstructed signal should be set as 0 if they have the same label and as 1 if they do not. As long as there is no absolute neutralization (i.e. no indistinguishable signals can carry different labels, see Section 8.1), the inverse images of the labels $t_1, \dots, t_l$ partition the signal space X into disjoint sets $X_1, \dots, X_l$ (plus possibly a remainder set X_0 containing nonlabelable 'junk' data), and it is natural to take the centroids of X_i as the codebook vectors. When the X_i are very far from perfect spheres in Euclidean space, it may make a great deal of sense to approximate them by the union of many spheres and simply take the labeled samples as the center of each sphere. Such systems are known as **nearest neighbor classifiers** since for any incoming signal $\mathbf{x}$ they find the nearest (in Euclidean distance) labeled sample $\mathbf{s}_i$ and assign it the same label $t(\mathbf{s}_i)$.

The ideal metric d that would yield 0 (resp. 1) between two samples exactly when they have the same (resp. different) labels is only known to us in artificially created examples. On real data we must work with some approximation. In the case of speech signals, it is reasonable to assume that if two signals have very little difference in their energy spectra, they are more likely to be tokens of the same type than when their energy spectra indicate audible differences at some frequency. We therefore investigate the case where d measures the total energy of the quantization error – for this to be meaningful, we assume the original signals are all normalized (amplitude scaled) to have the same energy $\sum_i s_i^2 = 1$ and that the gain of the all-pole filters used to encode the data is set in accordance with (9.25).

Suppose the task is to recognize utterances of roughly equal duration (N samples) from a closed set (e.g. isolated digits), with each of the utterance types $t_1, \ldots, t_l$ having equal probability $1/l$. We have a number of labeled sample utterances, but our goal is not to identify them (since no two utterances are perfectly identical, the sample could just be memorized) but rather to identify hitherto unseen signals taken from the same subset X of $\mathbb{R}^N$, $X = \bigcup_{i=0}^{l} X_l$, where $P(X_0) = 0, P(X_i) = 1/l$ $(1 \leq i \leq l)$. Our goal is to create codebook vectors $\mathbf{c}_i$ in p-dimensional space, $p \ll N$ for each of the l clusters X_i so that we get a nearest neighbor classifier. For any incoming signal s, we compute $d(s, \mathbf{c}_i)$ for $1 \leq i \leq l$ and select the i for which this value is minimal. We want the quantization error to be relatively small when the LPC model of s is close to one of the $\mathbf{c}_i$, and we are not much worried about the fringe cases when it is far from each $\mathbf{c}_i$ since their probability is low.

Let us pick a single cluster center $\mathbf{c}$, given by an all-pole filter with predictor parameters $a_1, \ldots, a_p$ and gain G. If we use white noise for input to this filter, we should obtain unvoiced versions (akin to whispering but without decrease in signal energy) of the original signals. The log probability of obtaining any particular signal s from $\mathbf{c}$ is given approximately by

$$\log P(s|\mathbf{c}) = \frac{N}{2}\left(\log 2\pi G^2 + \frac{1}{G}\mathbf{c}R_s\mathbf{c}^T\right) \tag{9.28}$$

Here R_s is the autocorrelation matrix of s, which is symmetric and positive semidefinite. Instead of a true distance (symmetrical and satisfying the triangle inequality) we only have a weaker type of d called a *divergence*, which will be small if the signal s is close to the model $\mathbf{c}$ and large if it is not. Equation (9.28), known as the **Itakura-Saito divergence**, is a special case of the Kullback-Leibler divergence (see Section 7.2). If we apply Lloyd's algorithm with the Itakura-Saito divergence, what we do in effect is average (in the Euclidean sense) the R_x matrices for each class X_i and compute the model $\mathbf{c}_i$ on the basis of these average matrices using the autocorrelation version of (9.22).

9.2 Phonemes as hidden units

Conceptually we can distinguish two main clusters of phenomena in the study of speech: *phonological* and *phonetic*. On the phonological side, we find discrete mental units defined in terms of contrast, where change from one unit to the other results in change of meaning. A good example is tone, where typically only two levels, high and low, are available – some languages have three, but languages such as Mandarin

Chinese that distinguish many 'tones' are really distinguishing many tonal configurations (sequences of high and low tones, see Section 3.3). On the phonetic side we have continuously variable physical parameters like pitch (the frequency F0 with which the vocal folds open and close) that can be changed by any small amount without affecting the meaning. Almost all phenomena we discussed in Chapters 3 and 4 fit squarely in the phonological cluster, while almost everything about the signals discussed so far indicates continuity, and discretization by sampling or VQ does not alter this picture since the number of discrete levels used in these steps is vastly larger than the number of discrete units. For example, in discretizing pitch, 256–1024 levels (8–10 bits) are common, while for tone we would generally need only 2–4 levels (1 or 2 bits).

The conceptual distinction is matched by reliance on different sorts of evidence: phonology views the human apparatus for speech production and perception as a legitimate instrument of data collection and relies almost exclusively on data (judgments concerning the grammaticality and well-formedness of certain forms) that phoneticians regard as subjective, while phonetics prefers to consider objective data such as speech waveforms. Yet the distinction between phonological and phonetic is by no means clear-cut, and the theory of lexical phonology and morphology (LPM, see Kiparsky 1982) distinguishes between two classes of phonological rules, *lexical* and *postlexical*, of which only the lexical class has clearly and unambiguously phonological character – the postlexical class shares many key features with purely phonetic rules. Here the distinctions, as summarized in Kaisse and Hargus (1993), are drawn as follows:

	Lexical	Postlexical
(a)	word-bounded	not word-bounded
(b)	access to word-internal structure assigned at same level	access to phrase structure only
(c)	precede all postlexical rules	follow all lexical rules
(d)	cyclic	apply once
(e)	disjunctively ordered with respect to other lexical rules	conjunctively ordered with respect to lexical rules
(f)	apply in derived environments	apply across the board
(g)	structure-preserving	not structure-preserving
(h)	apply to lexical categories only	apply to all categories
(i)	may have exceptions	automatic
(j)	not transferred to a second language	transferable to second language
(k)	outputs subject to lexical diffusion	subject to neogrammarian sound change
(l)	apply categorically	may have gradient outputs

While the criteria *(a–l)* are only near-truths, they are sufficient for classifying almost any rule as clearly lexical or postlexical. This is particularly striking when processes that historically start out as phonetic get *phonologized.* Not only will such rules change character from gradual (continuous) to discrete *(l)*, they will also begin to affect different elements of the lexicon differently *(k)*, acquire exceptions *(i)* and morphological conditions *(h)*, and begin to participate in the phonological rule system in ways phonetic processes never do *(c–f)*.

A good example is provided by phonetic *coarticulation*, a process that refers both to the local smoothing of articulatory trajectories and to longer-range interactions that can be observed e.g. between two vowels separated by a consonant (Öhman 1966) whenever some speech organs move into position before the phoneme that will require this comes up or stay in position afterwards. When the effect gets phonologized, it can operate over very long, in principle unbounded, ranges – a famous example is the *ruki* rule in Sanskrit, which triggers retroflexation of *s* after phonemes in the ruki class no matter how many nonruki consonants and vowels intervene. In reference to their phonetic origin, such rules are known in phonology as *anticipatory* and *perseveratory* rules of assimilation, irrespective of their range.

In Chapters 3 and 4 we presented all the discrete units generally agreed upon in phonology: features (autosegments), phonemes, syllables, and words. (We also presented some that are less widely used, such as moras, feet, and cola – here we will simplify the discussion by concentrating on the better-known units.) Of these, only words have a clear relationship to meanings. All others are motivated by the mechanics of speech production and are meaningless in themselves. Even for words, the appropriate phonological notion, the *prosodic word*, does not entirely coincide with the grammatical (syntactic) notion of wordhood (see Section 4.1), but the two are close enough to say that we can separate words from one another on the basis of meaning most of the time.

How can the discrete and meaningless units used in phonology be realized in, and recovered from, the undifferentiated continuous data provided by acoustic signals? What we call the *naturalistic* approach is to trace the causal chain, to the extent feasible, from the brain through the movement of articulators and the resulting air pressure changes. Within the brain, we assume some kind of combinatorical mechanism capable of computing the phonological representation of an utterance from pieces stored in the mental lexicon. This representation in turn serves as a source of complex nerve impulse patterns driving the articulators (Halle 1983), with the final output determined by the acoustics of the vocal tract.

Note that the combinatorical mechanism does not necessarily operate left to right. In particular, intermediate representations, procedures, and structures are generally viewed as having no theoretical status whatsoever, comparable to the scratch paper that holds the intermediate results in long division. This view is shared by context-sensitive rule-based phonology (Chomsky and Halle 1968), finite-state approaches (Koskenniemi 1983), and optimality theory (Prince and Smolensky 1993). Only multistratal theories, such as LPM, treat the output of the individual levels as real, and even then there is no promise of left to right computation. In particular, there is no *Greibach normal form* (see Ex. 2.4) that would force outputting of a segment for each unit of computation effort.

To the extent that speech production and speech perception rely on the same discrete phonological representation, tracing the causal chain on the decoder side implies that the acoustic signal is perceived in terms of the same articulator movement patterns as were used in producing the signal. This is known as the *motor theory of speech perception* (Liberman 1957). To quote Liberman and Mattingly (1989:491), speech perception

> processes the acoustic signal so as to recover the coarticulated gestures that produced it. These gestures are the primitives that the mechanisms of speech production translate into actual articulator movements, and they are also the primitives that the specialized mechanisms of speech perception recover from the signal.

Tracing the causal chain this way goes a long way toward explaining what the phonological representations, so painstakingly built by the linguist, are good for (besides accounting for the linguistic data of course). If the representations can be recast in terms of articulatory gestures, and moreover if indeed these gestures provide the key to speech perception, a wealth of extralinguistic evidence, ranging from X-ray microbeam tracing of the articulators (Fujimura et al. 1973) to perception studies of formant location (Klein et al. 1970), can be brought to bear on the description of these representations.

There are two problems left unanswered by following the causal chain. First, the gap between the discrete and the continuous is left unbridged: even if we identify phonological representations with gestural scores, these are continuous at best for timing parameters and the main gestural subcomponents (such as opening or closing the lips, raising or lowering the velum, etc.) remain discrete. Whatever we may do, we still need to recover the discrete articulatory configurations from a continuum of signals. Second, there is an important range of phenomena from sign language to handwriting that raises the same technical issues, but this time without the benefit of a complex (and arguably genetically set) mechanism between the representation and the perceived signal.

It is at this point that the modern theory of speech recognition parts with the naturalistic theory: if there is a need to create a mapping from discrete elements to continuous realizations in any case, there does not seem to be a significant advantage in creating an intermediate representation that is tightly coupled to a complex mechanism that is specific to the physiology of the vocal tract. As a case in point, let us consider the theory of distinctive features (see Section 3.2). A rather detailed qualitative description of the articulatory and acoustic correlates of distinctive features was available as early as in Jakobson et al. (1952). Nearly three decades later, Stevens and Blumstein (1981) still had not found a way of turning this into a quantitative description that could be used to automatically detect features (see e.g. Remez 1979), and to this day research in this area has failed to reveal a set of reliable acoustic cues for phonological features of the sort envisioned in Cherry, Halle, and Jakobson (1953) and Cherry (1956).

Thus the naturalistic model that interposes a gestural layer between the mental representations and the acoustic signal has been replaced by a simpler and more direct view of the mental lexicon that is assumed to store highly specific acoustic engrams recorded during the language acquisition process: these engrams can be directly used as lookup keys into a mental database that will contain syntactic, semantic, morphological, and other nonacoustic information about the form in question (Klatt 1980). Under this view, surface forms are just acoustic signals, while underlying forms could contain detailed articulatory plans for the production of the form, together with links to semantic, syntactic, and morphological information stored in various formats.

The relationship between psychological units of linguistic processing and their physical realizations is many to many, both with different phonological representations corresponding to the same utterance (neutralization; see Section 8.1) and with the same representation having many realizations, and many conditioning factors, ranging from the physical differences among speakers sharing the same competence to the amount of distortion tolerated in the realization process. While phonology generally works with idealized data that preserve dialectally and grammatically conditioned variation but suppress variation within the speech of an individual and across individuals sharing the same dialect/sociolect, for the moment we lump all sources of variation together, and defer the issue of how to separate these out.

The units that we shall take as basic are the phonemes, which are instrumental in describing such a broad range of phenomena that their psychological reality can hardly be disputed. A subjective, but nevertheless important factor is that most researchers are convinced that they are in fact communicating using sentences, words, syllables, and phonemes. A great deal of the reluctance of speech engineers to accept distinctive features can no doubt be attributed to the fact that for features this subjective aspect is missing: no amount of introspection reveals the featural composition of vowels, and to the extent introspection works (e.g. with place of articulation) it is yielding results that are not easily expressible in terms of distinctive features unless a more complex structure (feature geometry, see Section 3.2) is invoked.

A less subjective argument in favor of certain linguistic units can be made on the basis of particular systems of writing. To the extent that a morpheme-, mora-, syllable-, or phoneme-based writing system can be easily acquired and consistently used by any speaker of the language, the psychological reality of the units forming the basis of the system becomes hard to deny. Distinctive features fare slightly better under this argument, given sound-writing systems such as Bell's (1867) Visible Speech or Sweet's (1881) Sound Notation, but to make the point more forcefully, the ease of use and portability of such writing systems to other languages needs to be demonstrated. For now, the most portable system we have, the *International Phonetic Alphabet* (IPA), is alphabetic, though the idea that its organization should reflect the featural composition of sounds is no longer in doubt (see Halle and Ladefoged 1988).

The simplest direct model with phonemic units would be one where the mapping is one to one, storing a single signal template with each phoneme. To account for variation, we need to use a probability model of templates instead, leaving open the possibility, corresponding to neutralization, that the same template may have nonzero probability in the distribution associated to more than one phonemic unit. We thus obtain a hidden Markov model, where the hidden units are phonemic, and the emission probabilities model the acoustic realization of phonemes. Transition probabilities can be set in accordance with the phonotactics of the language being modeled or, if a description of the lexically stored phoneme strings (words) is available, in accordance with this description.

Since the phonemic units are concatenative, the output signals should also be, meaning either that we *smooth* the abrupt change between the end of one signal (coming from one hidden state) and the beginning of the next signal (coming from another hidden state) or we adjust the emissions so that only smoothly fitting signals can follow each other. While concatenation and smoothing continues to play an important role in speech synthesis systems (see Klatt 1987, van Santen et al. 1997), in speech recognition the second option has proven more fruitful: instead of directly modeling phonemes, we model phonemes *in context*. For example, instead of a single model for the i in *mint*, *hint*, *lint*, and *tint* we build as many separate models as there can bc phonemes preceding the i. If we do the same with the l phoneme, building as many models as there can be phonemes following it (so that different ls are used in *lee*, *lie*, *low*, *lieu*, etc.), we can be reasonably certain that the appropriate l model (one that is based on the context $\_i$), when followed by the appropriate i model (one that is based on the context $l\_$), will contain only signals that can be concatenated without much need for smoothing.

HMMs in which the hidden units are phonemes in two-sided phoneme contexts are called *triphone* models (the name is somewhat misleading in that the units are single phones, restricted to particular contexts, rather than sequences of three phones) and will contain, if there were n phonemes, no more than n^3 hidden units, and possibly significantly fewer if phonotactic regularities rule out many cases of phoneme b appearing in context a_c. State of the art systems extend this method to *quinphones* (phonemes in the context of two phonemes on each side) and beyond, using cross-word contexts where necessary.

Starting with Bloomfield, a great deal of effort in mathematical linguistics has been devoted to defining models that explicate the relation between the low-level (continuous, phonetic, meaning-preserving) and the high-level (discrete, phonological, meaning-changing) items and processes involved in speech. But there are some persistent difficulties that could not be solved without a full appreciation of the variability of the system. A triphone or quinphone model will account for a great deal of this variability, but even a cursory look at Fig. 8.1 makes it evident that other sources, in particular the identity of the speaker, will still contribute significant variability once contextual effects are factored out. Also, it is clear that steady-state vowels are more of an idealization than typical speech samples: major spectral shifts occur every 5–10 milliseconds, and windows that contain no such shifts are a rarity. This problem is to some extent mitigated by adding *delta* (first derivative) and *delta delta* (second derivative) features to the basic feature set (Furui 1986), since this will re-emphasize the spectral shifts that the original features may dampen.

The most natural probabilistic model of emission is a normal distribution, where a single sample acts as the mean μ, and all samples are assigned probabilities in accordance with the density function

$$\frac{1}{(2\pi)^{N/2}|R|^{1/2}} \exp\left[-\frac{1}{2}(x-\mu)R^{-1}(x-\mu)^T\right] \tag{9.29}$$

where R is the N-dimensional covariance matrix that determines the distribution. In the simplest case, $N = 1$; i.e. there is a single measurable parameter that characterizes every sample. Ideally, this is what we would like to see e.g. for tone, which is determined by a single parameter F0. But when we measure F0 for H (phonologically high tone) and L (phonologically low tone) syllables, we obtain two distributions that are nearly identical, with neither the means μ_H and μ_L nor the variances σ_H and σ_L showing marked differences. The reason for this is that tonal languages show a steady, cumulative lowering of F0 called *downdrift*, which obscures the differences between H and L tones on average. At any given position, in particular sentence-initially, the differences between H and L produced by the same speaker are perceptible, but averaging speakers and positions together blurs the distinction to such an extent that separation of H and L by unsupervised clustering becomes impossible. This is not to say that the distinction cannot be captured statistically, e.g. by focusing on the phrase-initial portion of the data, but rather to emphasize that our current techniques are often insufficient for the automatic discovery of structure from the raw data: unless we *know* about downdrift, there is no reason to inspect the phrase-initial portion of the data separately.

This is not to say that unsupervised, or minimally supervised, clustering techniques have no chance of obtaining the correct classes, at least if the data are presented clearly. For example, if we restrict attention to steady-state vowels with unambiguous pronunciation and a homogeneous set of speakers (adult males), the approximate (F1, F2, F3) centroids for the ten vowels measured by Peterson and Barney (1952) can be found using just the fact that there are exactly ten clusters to be built (Kornai 1998a). But to accommodate cases of major allophonic variation, such as trilled vs. tapped r, distinct Gaussians must be assigned to a single phoneme model. In this case, we talk about *mixture models* because the density function describing the distribution of data points belonging to a single phoneme is the mixture (weighted sum) of ordinary Gaussians. By using a large number of mixture components, we can achieve any desired fit with the training data. In the limiting case, we can fit a very narrow Gaussian to each data point and thereby achieve a perfect fit.

The number of Gaussians that can be justified is limited both by the MDL principle (see Section 7.3) and by the availability of training data. If we are to model n^3 hidden units (triphones), each with N parameters for the mean and $N(N+1)/2$ for variances and covariances, using m mixture components will require a total of $O(n^3N^2m/2)$ parameters. With typical values like $n = 50$, $N = 40$, $m = 10$, this would mean 10^9 parameters. Of the many strategies used for reducing this number, we single out the use of *diagonal* models where only the variances are kept and the covariances are all set to zero. This will reduce the parameter space by a factor of $N/2$ at minimal cost since the covariances refer to different dimensions of a highly compressed (mel cepstral) representation whose individual components should already be almost entirely uncorrelated. Another method is to use only a limited number of Gaussians and share these (but not the mixture weights) across different hidden units. This is called a *tied mixture* model (Bellegarda and Nahamoo 1990). Tying the parameters and reducing the number of Gaussians is particularly important in those cases where not all phonotactically permissible triphones occur in the training data.

Another important approach to reducing variation is based on *speaker adaptation*. The Peterson-Barney data already show clearly the effects of having men, women, and children in the sample, and indeed training separate models for men and women (Nishimura and Sugawara 1988) is now common. If there are separate mixture components for men and women, it makes eminent sense to deploy *selection* strategies that use some short initial segment of speech to determine which component the data fits best and afterwards suppress the mixtures belonging in the other component or components.

Speaker adaptation can also work by noting the characteristic differences between the speech of the current speaker and those whose speech was used to train the model, and employ some transformation T^{-1} to the incoming speech or, alternately, applying T to the model data, as a means of achieving a better fit between the two. The former method, e.g. by the normalization of cepstral means, is used more when the variation is due to the environment (background noises, echoes, channel distortion), while the latter is used chiefly to control variability across speakers and to some extent across dialects (especially for nonnative speakers).

Adapting the variances (and covariances, if full covariance models are used) is less important than adapting the means, for which generally T is chosen as an affine transform $\mathbf{x} \mapsto \mathbf{x}A + \mathbf{b}$, and the new means are computed by maximum likelihood linear regression (MMLR; see Leggetter and Woodland 1995). A more naturalistic, but not particularly more successful, method is vocal tract length normalization (VTLN, see Wakita 1977), where the objective is to compensate for the normal biological variation in the length of the vocal tract. Altogether, the naturalistic model remains a source of inspiration for the more abstract direct approach, but the success of the latter is no longer tied to progress in articulatory or perceptual research.

9.3 Handwriting and machine print

In dynamic OCR, we can be reasonably certain that the input the system receives is writing, but in image-based OCR the first task is *page decomposition*, the separation of linguistic material from photos, line drawings, and other nonlinguistic information. A further challenge is that we often find different scripts, such as Kanji and Kana, or Cyrillic and Latin, in the same running text.

The input is generally a scanned image, increasingly available in very high resolution (600 dpi or more) and in full (4 bytes per pixel) color. Uncompressed, such an image of a regular letter-size page would take up over 130 MB. The naturalistic program would suggest applying algorithms of *early vision* such as edge detection, computation of optical flow, lightness, albedo, etc., to derive representations more suitable for page decomposition and character recognition. However, practical systems take the opposite tack and generally begin with a variety of data reduction steps, such as *downsampling*, typically to fax resolution (200 dpi horizontal, 100 dpi vertical) and *binarization*; i.e. replacing color or grayscale pixel values by binary (black and white) values. After these steps, a typical uncompressed image will be over 1.8 MB, still too large for fax transmission of multipage documents.

Here we describe the **G3** (CCITT Group 3) standard in some detail because it illustrates not only the basic ideas of Huffman coding (see Section 7.1) but also the distance between theoretical constructs and practical standards. Each line is defined as having 1728 binary pixels (thus somewhat exceeding the 8.5 inch page width used in the United States at 200 dpi) that are *run-length encoded* (RLE). In RLE, instead of transmitting the 0s and 1s, the length of the alternating runs of 0s and 1s get transmitted. In the G3 standard, length is viewed as a base 64 number. For shorter runs, only the last digit (called *terminating length* in this system) gets transmitted, but for longer runs, the preceding digit (called the *make-up*) is also used. Since the length distribution of white and black runs differs considerably, two separate codebooks are used. Terminating length and make-up codes jointly have the prefix-free property both for white and black, but not for the union of the two codebooks. The following terminating length and make-up codes are used:

term. length	white code	black code	term. length	white code	black code	make up	white	black
0	00110101	0000110111	32	00011011	000001101010			
1	000111	010	33	00010010	000001101011	64	11011	0000001111
2	0111	11	34	00010011	000011010010	128	10010	000011001000
3	1000	10	35	00010100	000011010011	192	010111	000011001001
4	1011	011	36	00010101	000011010100	256	0110111	000001011011
5	1100	0011	37	00010110	000011010101	320	00110110	000000110011
6	1110	0010	38	00010111	000011010110	384	00110111	000000110100
7	1111	00011	39	00101000	000011010111	448	01100100	000000110101
8	10011	000101	40	00101001	000001101100	512	01100101	0000001101100
9	10100	000100	41	00101010	000001101101	576	01101000	0000001101101
10	00111	0000100	42	00101011	000011011010	640	01100111	0000001001010
11	01000	0000101	43	00101100	000011011011	704	011001100	0000001001011
12	001000	0000111	44	00101101	000001010100	768	011001101	0000001001100
13	000011	00000100	45	00000100	000001010101	832	011010010	0000001001101
14	110100	00000111	46	00000101	000001010110	896	011010011	0000001110010
15	110101	000011000	47	00001010	000001010111	960	011010100	0000001110011
16	101010	0000010111	48	00001011	000001100100	1024	011010101	0000001110100
17	101011	0000011000	49	01010010	000001100101	1088	011010110	0000001110101
18	0100111	0000001000	50	01010011	000001010010	1152	011010111	0000001110110
19	0001100	00001100111	51	01010100	000001010011	1216	011011000	0000001110111
20	0001000	00001101000	52	01010101	000000100100	1280	011011001	0000001010010
21	0010111	00001101100	53	00100100	000000110111	1344	011011010	0000001010011
22	0000011	00000110111	54	00100101	000000111000	1408	011011011	0000001010100
23	0000100	00000101000	55	01011000	000000100111	1472	010011000	0000001010101
24	0101000	00000010111	56	01011001	000000101000	1536	010011001	0000001011010
25	0101011	00000011000	57	01011010	000001011000	1600	010011010	0000001011011
26	0010011	000011001010	58	01011011	000001011001	1664	011000	0000001100100
27	0100100	000011001011	59	01001010	000000101011	1728	010011011	0000001100101
28	0011000	000011001100	60	01001011	000000101100			
29	00000010	000011001101	61	00110010	000001011010			
30	00000011	000001101000	62	00110011	000001100110			
31	00011010	000001101001	63	00110100	000001100111			

In order to ensure that the receiver maintains color synchronization, all lines must begin with a white run length code word (if the actual scanning line begins with a black run, a white run length of zero will be sent). For each page, first end of line (EOL, 000000000001) is sent, followed by variable-length line codes, each terminated by EOL, with six EOLs at the end of the page. On the average, G3 compression reduces the size of the image by a factor of 20.

Exercise 9.4 Research the CCITT Group 4 (G4) standard, which also exploits some of the redundancy between successive lines of the scanned image and thereby improves compression by a factor of 2. Research JBIG and JBIG2, which generally improve G4 compression by another factor of 4.

Since the black and white runs give a good indication of the rough position of content elements, the first step of page decomposition is often performed on RLE data, which is generally sufficient for establishing the local horizontal and vertical directions and for the appropriate grouping of titles, headers, footers, and other material set in a font different from the main body of the text. Adaptation to the directions inherent in the page is called *deskewing*, and again it can take the form of transforming (rotating or shearing) the image or transforming the models. The tasks of deskewing and page decomposition are somewhat intertwined because the simplest page decomposition methods work best when the image is not skewed. Unlike in speech recognition, where models that incorporate an explicit segmentation step have long been replaced by models that integrate the segmentation and the recognition step in a single HMM search, in OCR there is still very often a series of segmentation steps, first for *text zones* (i.e. rectangular windows that contain text only), then for *lines*, and finally for *characters*.

The search for text zones can proceed top-down or bottom-up. In the top-down approach, we first count the black pixels in each row (column), obtaining a column (row) of blackness counts known as the *projection profiles* (see Wang and Srihari 1989). These are generally sufficient for finding the headers and footers. Once these are separated out, the vertical profile can be used to separate text columns, and on each column horizontal profiles can be used to separate the text into lines. Besides sensitivity to skew, a big drawback of the method is that it presumes a regular, rectangular arrangement of the page, and more fancy typography, such as text flowing around a circular drawing, will cause problems. In the bottom-up approach, we begin with the smallest elements, *connected components*, and gradually organize them or their *bounding boxes* (the smallest rectangle entirely containing them) into larger structures. By searching for the dominant peak in a histogram of vectors connecting each component to its nearest neighbor, the skew of the document can be reliably detected (Hashizume et al. 1986, O'Gorman 1993). Both the identification of connected components and the finding of nearest neighbors are computationally very expensive steps, and once we are willing to spend the resources, other skew-insensitive methods of segmenting text from images, such as the use of Gabor filters (Jain and Bhattacharjee 1992) are also available.

Besides deskewing, there are several other normalization steps performed on the entire image as needed; for example, xeroxing or scanning thick, bound documents introduces *perspective distortion* at the edge of the page (see Kanungo et al. 1993), and the vibrations of the scanner can cause blurring. Detection and removal of *speckle noise* (also known as *salt-and-pepper noise* because noise can take the form of unwanted white pixels, not just black) is also best performed on the basis of estimating the noise parameters globally. Other normalization steps, in particular removing distortions in the horizontal baseline caused by a loose paper forwarding mechanism (common in mechanical typewriters and low-quality scanners/faxes), are better performed line by line.

To the extent that small skew (generally within one degree) and simple rectangular page layout are valid assumptions for the vast majority of holdings in digital libraries, the less expensive top-down algorithms remain viable and have the advantage that finding text lines and characters can generally proceed by the same steps. For machine-printed input and for *handprint* (block letters, as opposed to cursive writing), these steps reduce the problem to that of *isolated character recognition.* Here the dominant technology is template matching, typically by neural nets or other trainable algorithms. Since direct matching of templates at the bitmap level, the method used in the first commercially available OCR system in 1955, works well only for fixed fonts known in advance, attention turned early on to deriving a suitable set of features so that variants of the same character will map to close points in feature space.

For isolated characters, the first step is generally *size* normalization; i.e. rescaling the image to a standardized window. Since this window is typically much smaller than the original (it can be as small as 5 by 7 for Latin, Cyrillic, and similar alphabets, 16 by 20 for Oriental characters), the new pixels correspond to larger zones of the original bounding box. The new pixel values are set to the average blackness of each zone, so that the rescaled image will be grayscale (4–8 bits) even if the original was binary (Bokser 1992). Because characters can vary from the extremely narrow to the extremely broad, the aspect ratio of the original bounding box is generally kept as a separate feature, together with the position of the bounding box relative to the baseline so as to indicate the presence or absence of ascenders and descenders.

Besides absolute size, we also wish to normalize *stroke width* in handwritten characters and *font weight* in machine print. An elegant, and widely used, technology for this is *mathematical morphology* (MM), which is based on the dual operations of *erosion* and *dilation.* We begin with a fixed set $B \subset \mathbb{R}^2$ called the *structuring element*, typically a disk or square about the origin. The **reflection** of B, denoted $\hat{B}$, is defined as $\{-\mathbf{x}|\mathbf{x} \in B\}$ – for the typical structuring elements, symmetrical about the origin, we have $\hat{B} = B$. The **translation** $B_{\mathbf{x}}$ of B by vector $\mathbf{x}$ is defined as $\{\mathbf{b}+\mathbf{x}|\mathbf{b} \in B\}$. The **dilation** $A \oplus B$ of A by B is defined as $\{x|\hat{B}_{\mathbf{x}} \cap A \neq \emptyset\}$, and the **erosion** $A \ominus B$ of A by B is defined as $\{x|B_{\mathbf{x}} \subset A\}$. Two MM operations defined on top of these are the **opening** $A \circ B$ of A by B given by $(A \ominus B) \oplus B$ and the **closing** $A \bullet B$ of A by B given by $(A \oplus B) \ominus B$. Finally, the **boundary** ∂A of A according to B is defined as $A \setminus (A \ominus B)$.

What makes MM particularly useful for image processing is that all the operations above remain meaningful if the sets A and B are composed of pixels (elements of a tiling of $\mathbb{R}^2$). For example, if the structuring element B is chosen to be slightly smaller than the regular printed dot (both periods at sentence end and dots over *i* s), the operation $(A \circ B) \bullet B$ will filter out all salt-and-pepper noise below this size while leaving the information-carrying symbols largely intact. By iterated erosion, we can also obtain the **skeleton** of an image, defined as the set of those points in the image that are equidistant from at least two points of the boundary. While in principle the skeleton should be an ideal replacement of characters with greater stroke width, in practice skeletonization and all forms of thinning are quite sensitive to noise, even after despeckling. Noise is also a persistent problem for approaches based on *chain codes* that express a simply connected two-dimensional shape in terms of its one-dimensional boundary.

In **vertex chain coding**, we apply the same principle to triangular, rectangular, and hexagonal tilings: we pick any vertex of the polygon that bounds the object and count the number of pixels that are connected to the boundary at that vertex. The total chain code is obtained by reading off these numbers sequentially following the polygon counterclockwise. Several chain codes, all cyclic permutations of the same string, could be obtained, depending on the vertex at which we start. For uniqueness, we pick the one that is minimal when interpreted as a number. This will make the code rotation invariant.

Chain codes offer a relatively compact description of simply connected binary images, and efficient algorithms exist to compute many important properties of an image (such as its skew; see Kapogiannopoulos and Kalouptsidis 2002) based on chain codes for all connected components. However, for isolated character recognition, chain codes are very brittle: many characters have holes, and features such as 'being n times connected' (and in general all Betti numbers and other topological invariants) are greatly affected by noise. In the analysis of handwriting, it is a particularly attractive idea (already present in Eden 1961) to build a naturalistic model along the same lines we were all taught in first grade: t is a straight line down, curve to the right, cross near the top. Noise stands in the way of such *structural decomposition* approaches to a remarkable degree, and simple, robust features such as the *height contour*, which is obtained from a line by computing the highest and the lowest black pixel in each column of pixels, turn out to be more helpful in OCR even though they clearly ignore structural features of the original such as loops.

Therefore it is particularly important to look for noise-robust features that preserve as many of the desirable invariance properties of chain codes as feasible. After size normalization, the character image fits in a fixed domain and can be thought of as a function $f(x, y)$ with value zero outside e.g. the unit circle (which is mathematically more convenient than a rectangular bounding box). A typical normalization/feature extraction step is computing the **central moments** μ_{pq} defined in the usual manner by computing the x and y directional means $\overline{x} = \iint x f(x, y) dx dy / \iint f(x, y) dx dy$ and $\overline{y} = \iint y f(x, y) dx dy / \iint f(x, y) dx dy$ and defining

$$\mu_{pq} = \iint (x - \overline{x})^p (y - \overline{y})^q f(x, y) d(x - \overline{x}) d(y - \overline{y}) \tag{9.30}$$

Clearly, the central moments are invariant under arbitrary translation. Since the moment-generating function and the two-dimensional Fourier transform of f are essentially the same complex function viewed on the real and the imaginary axes, it comes as little surprise that Fourier techniques, both discrete and continuous, play the same pivotal role in the two-dimensional (handwriting signal) case that they have played in the one-dimensional (speech signal) case.

To obtain features that are also invariant under rotation, we can express f in terms of any orthogonal basis where rotation invariance is easily captured. One standard method is to define the n, m radial polynomials $R_{nm}(\rho)$, where n is the *order* of the polynomial, m is the *winding number*, $n - |m|$ is assumed even, $|m| < n$ by

$$R_{nm}(\rho) = \sum_{s=0}^{n-|m|/2} (-1)^s \frac{(n-s)!\rho^{n-2s}}{s!\left(\frac{n+|m|}{2} - s\right)!\left(\frac{n-|m|}{2} - s\right)!} \tag{9.31}$$

and define the n, m **Zernike basis function** $V_{nm}(\rho, \theta)$ as $V_{nm}(\rho, \theta) = R_{nm}(\rho)e^{im\theta}$. These functions are orthogonal on the unit disk, with

$$\iint_{x^2+y^2 \leq 1} \overline{V_{nm}(x, y)V_{pq}(x, y)dxdy} = \frac{\pi}{n+1}\delta_{np}\delta_{mq} \tag{9.32}$$

The **Zernike moment** A_{nm} of a function f (already assumed size and translation normalized) is given by

$$A_{nm} = \iint_{x^2+y^2 \leq 1} f(x, y)\overline{V_{nm}(\rho, \theta)}dxdy \tag{9.33}$$

If $g(\rho, \theta)$ is obtained from $f(\rho, \theta)$ by rotation with angle α (i.e. $g(\rho, \theta) = f(\rho, \theta - \alpha)$), the n, m-th Zernike moment of g is $A_{nm}e^{-im\alpha}$, where A_{nm} was the n, m-th Zernike moment of f. Thus, the absolute values $|A_{nm}|$ are rotation invariant. Notice that fully rotation-invariant features are not ideally suited for OCR – for example, 6 and 9 would get confused. A more important goal is the normalization of *slant*, both for the recognition of handwriting and for machine print that contains italicized portions. Projection profiles, taken in multiple directions, give a good indication of writing slant and are often used.

Altogether, the computation of features for isolated characters often involves a mixture of normalization and feature extraction steps that may end up producing more data, as measured in bits, than were present in the original image. This is particularly true in cases such as Zernike moments or other series expansion techniques (Fourier, wavelet, and similar techniques are often used), which could lead to a large number of terms limited only by two-dimensional versions of Theorem 9.1.1. In general, the goal is not simply data reduction but rather finding the features that provide good clustering of the data for VQ and other techniques the way Itakura-Saito divergence does for speech signals.

Typically, data reduction is accomplished by **Karhunen-Loève transformation,** also known as **principal component analysis** (PCA). We begin with a set of (real or complex) feature vectors $\mathbf{x}_i$ $(1 \le i \le n)$ with some high dimension N and seek to find the projection P to d-dimensional space that retains as much of the variance in the data as possible. To this end, we first compute the (empirical) mean of the data set $\overline{\mathbf{x}} = \frac{1}{n}\sum_{i=1}^{n} \mathbf{x}_i$ and subtract it from all feature vectors (in practical applications, the variances are also often normalized). The covariance matrix of the (now zero mean) vectors is symmetric (or Hermitian, if complex features are used) and thus has orthogonal eigenvectors. These are ordered according to decreasing size of the eigenvalues, and only the first d are kept. P projects the (zero mean) data on the subspace spanned by these eigenvectors.

If the task is recognition, it is at least in principle possible for important information to be lost in PCA since the variability according to some critical feature may be low. To account for this possibility, sometimes linear discriminant analysis (LDA; see Fisher 1936, 1937) is used, but the improvement over PCA is generally slight when LDA is just a preprocessing stage for some more complex recognition strategy such as nearest neighbor classification. The advantage of LDA is in the fact that it will derive a robust classifier in its own right and the training process, which uses only the first and second moments of the distributions of $\mathbf{x}_i$, is very fast.

Exercise 9.5†† Develop a handprint classifier using the NIST isolated character database available at `http://www.nist.gov/srd/nistsd19.htm`. Compare your results with the state of the art.

Most classification methods that work well for machine print and handprint run into serious problems when applied to cursive writing because in the character identification stage it is very hard to recover from errors made in the segmentation stage. The filter- and projection-based page decomposition methods discussed here generalize reasonably well to cursive writing as far as segmentation into text blocks and lines is concerned, but segmenting a line into separate words and a word into separate characters based on these and similar methods is prone to very significant errors. For cursive writing, the segmentation problem must be confronted the same way as in speech recognition. To quote Halle and Stevens (1962:156),

> The analysis procedure that has enjoyed the widest acceptance postulates that the listener first segments the utterance and then identifies the individual segments with particular phonemes. *No analysis scheme based on this principle has ever been successfully implemented.* This failure is understandable in the light of the preceding account of speech production, where it was observed that segments of an utterance do not in general stand in a one-to-one relation with the phonemes. The problem, therefore, is to devise a procedure which will transform the continuously-changing speech signal into a discrete output without depending crucially on segmentation.

To this day, we do not have successful early segmentation, and not for lack of trying. Until the advent of HMMs, there were many systems based on the segmentation-classification-identification pipeline, but none of them achieved performance at the desired level. Today, many of the design features deemed necessary by the prescient Halle-Stevens work, such as the use of generative language models for the lexicon and larger utterances or the pruning of alternatives by multiple passes, are built into HMMs, but the main strength of these systems comes from two principal sources: first, that the recognition algorithm explores segmentation and classification alternatives in parallel, and second, that the system is *trainable*. Parallelism means that HMMs are capable of considering all segmentation alternatives (hypothesizing every new frame as potentially beginning a new phoneme) without unduly burdening a separate phoneme-level recognizer. Perhaps ironically, the best segmentation results are the ones obtained in the course of HMM recognition: rather than furnishing an essential first step of analysis, segment boundaries arise as a by-product of the full analysis.

9.4 Further reading

Viewed from a contemporary perspective, the great bulk of (digital) signal processing knowledge rests on a clean, elegant foundation of 19th century complex function theory and should be very accessible to mathematicians once the terminological and notational gap between mathematics and electrical engineering is crossed.[1] Yet the rational reconstruction of the fundamentals presented here has little in common with the actual historical development, which is better understood through major textbooks such as Flanagan (1972) (for analog) and Rabiner and Schafer (1978) (for early digital). In particular, Theorem 9.1.1 goes back to Whittaker (1915) but has been rediscovered many times, most notably by Shannon and Weaver (1949).

In the theory of time series analysis, Padé approximation is known as the *autoregressive moving average* (ARMA) model, with the all-pole case referred to as *autoregressive* (AR) and the all-zero case as *moving average* (MA); see e.g. Box and Jenkins (1970). In digital signal processing, there are a wide variety of windowing functions in use (see e.g. Oppenheim and Schafer 1999), but for speech little improvement, if any, results from replacing the standard Hamming window by other windowing functions. The importance of the fast Fourier transform can hardly be overstated – see Brigham (1988) for a textbook devoted entirely to this subject. Even though harmonic analysis is a natural framework for dealing with speech, algorithms that relied on actually computing Fourier coefficients were considered impractical before the modern rediscovery of the FFT (Cooley and Tukey 1965). We mention here that the FFT was already known to Gauss (see Heideman et al. 1984).

The standard introduction to homomorphic speech processing is Schafer and Rabiner (1975), but the presentation here follows more closely the logic of Makhoul (1975). Cepstra, and the attendant syllable-reversal terminology, were introduced in Bogert et al. (1963); see also Childers et al. (1977). Mel-cepstral features have effectively replaced direct (time domain) LPC features, but linear prediction, either applied in the quefrency domain or directly (as in the GSM 6.10 standard), remains a standard data compression method in modern audio transmission.

The savings effected by the fast algorithm of Exercise 9.2 are trivial by today's standards since p is generally on the order of 10^1, so p^3 gives less than 10^5 instructions per second, while the chips embedded in contemporary cell phones are

[1] In particular, in engineering works we often find the imaginary unit i denoted by j.

increasingly capable of hundreds of MIPS. Note, however, that in the image domain a variety of operations, such as adaptive binarization (setting the binarization threshold according to local conditions; see Kamel and Zhao 1993), are still quite expensive.

For an overview of unsupervised clustering, see Everitt (1980) and Duda et al. (2000 Ch. 10), and for the supervised case, see Anderberg (1973). For Itakura-Saito divergence, see Itakura (1975), and for its relation to Bregman divergences, see McAulay (1984), Wei and Gibson (2000), and Banerjee at al (2005). For optimal decorrelation of cepstral features, see Demuynck et al. (1998). Tying techniques, which are critical for training high-quality HMMs, are discussed further in Jelinek (1997 Ch. 10). For speaker adaptation, see Richard Stern's survey article on robust speech recognition in Cole (1997, with a new edition planned for 2007).

Although somewhat dated, both Bunke and Wang (1997) and O'Gorman and Kasturi (1995) offer excellent introductions to the many specialized topics related to OCR. For an overview of the early history, see Mori et al. (1992). For adaptive thresholding, see Sezgin and Sankur (2004). For page segmentation, see Antonacopoulos et al. (2005). Mathematical morphology was invented by Matheron and Serra (see Serra 1982); for various generalizations, see Maragos et al. (1996). For a comparison of skeletalization and thinning algorithms see Lee et al. (1991) and Lam et al. (1992). Chain codes were introduced by Freeman (1961). The vertex chain code presented here is from Bribiesca (1999).

Moment normalization originated with Hu (1962). Zernike polynomials arise in the study of wavefronts for optical systems with a central axis (Zernike 1934) and are widely used in opthalmology to this day. Their use in character recognition was first proposed in Khotanzad and Hong (1990). For an overview of feature extraction methods for isolated character recognition, see Trier et al. (1996). PCA and LDA are basic tools in pattern recognition; see e.g. Duda et al. (2001 Sec. 3.8). For the use of MMLR in dynamic handwriting recognition, see Senior and Nathan (1997), and for image-based handwriting recognition, see Vinciarelli and Bengio (2002). The state of the art in handwriting recognition is closely tracked by the International Workshop on Frontiers of Handwriting Recognition (IWFHR).

Early systems incorporating explicit rule-based segmentation steps are discussed in Makhoul (2006) from a historical perspective. In linguistic pattern recognition, trainability (called *adaptive learning* at the time) was first explored in early OCR work (see Highleyman 1962, Munson 1968), but the practical importance and far-reaching theoretical impact of trainable models remained something of a trade secret to speech and OCR until the early 1990s, when Brown et al. (1990) demonstrated the use of trainable models in machine translation. This is not to say that theoretical linguistics ignored the matter entirely, and certainly the single most influential work of the period, Chomsky (1965), was very explicit about the need for grammatical models to be learnable. Until the 1990s, theoretical linguistics focused on the relationship of learnability and child language development (see e.g. Pinker 1984, 2nd revised ed. 1996), mostly from the perspective of stimulus poverty, and it took the clear victory of trainable models over handcrafted rule systems in what was viewed as a core semantic competence, translation, to bring the pure symbol-manipulation and the statistical approaches together again (Pereira 2000).

10

Simplicity

Writing a textbook is a process that forces upon the author a neutral stance since it is clear that the student is best served by impartial analysis. Yet textbook authors can feel as compelled to push their own intellectual agenda as authors of research papers, and for this final chapter I will drop the impersonal *we* together with its implied neutrality and present a view of the overall situation in mathematical linguistics that makes no claims to being the standard view or even the view of a well-definable minority.

If one had to select a single theme running through this book, it would no doubt be the emphasis on regular (finite state) structures. The classic results from this field originate with the work of Mealy (1955) on electric circuits and Kleene (1956) on nerve nets with discrete sets of inputs and outputs as opposed to the continuous inputs and outputs of perceptrons and contemporary neural nets. For the textbook writer, it was reasonable to assume that at least a streamlined version of this material would be familiar to the intended readership, as it is generally regarded as an essential part of the core computer science/discrete mathematics curriculum, and it was also reasonable to give short shrift to some of the more modern developments, especially to the use of (k-tape) finite state transducers in phonology/morphology, as these are amply covered in more specialized volumes such as Roche and Schabes (1997) and Kornai (1999).

Yet I feel that, without this chapter, some of the impact of this work would be lost on the reader. On the one hand, the methods pioneered by Mealy, Kleene, Schützenberger, Krohn, Rhodes, Eilenberg, Angluin, and a host of others remain incredibly versatile and highly generalizable as mathematical techniques. For example, much of the work on regular generalizations of FSA, both to transducers and to k-strings, is nothing but a rehash of the basic techniques relying on the identification of the congruence classes (elements of the syntactic monoid) with automata states: if the (right)congruence has finite index, finite automata are available. But before turning to the issue of *why* these techniques are so highly generalizable, it should be noted that there is an important aspect of these methods that has, to some extent, been suppressed by the general turning of the tide against cybernetics, systems theory, and artificial intelligence, namely the direct appeal to neurological models originating with Kleene.

As we all know, dealing with natural language is hard. It is hard from the standpoint of the child, who must spend many years acquiring a language (compare this time span to that required for the acquisition of motor skills such as eating solids, walking, or swimming), it is hard for the adult language learner, it is hard for the scientist who attempts to model the relevant phenomena, and it is hard for the engineer who attempts to build systems that deal with natural language input or output. These tasks are so hard that Turing (1950) could rightly make fluent conversation in natural language the centerpiece of his test for intelligence. In more than half a century of work toward this goal, we have largely penetrated the outermost layer, speech production and perception, and we have done so by relying on learning algorithms specific to the Markovian nature of the signal. In spite of notable advances in the automatic acquisition of morphology, in this next layer we are still at a stage where speech recognition was forty years ago, with handcrafted models quite comparable, and in some cases superior, to machine-learned models.

Returning briefly to the issue raised at the beginning of this book: what do we have when a model is 70%, 95%, or even 99.99% correct? Isn't a single contradiction or empirically false prediction enough to render a theory invalid? The answer to the first question is clearly positive: contradictions must be managed carefully in a metatheory (multivalued logic) and cannot pervade the theory itself. But the answer to the second question is negative: a single wrong prediction, or even a great number of wrong predictions, does *not* render the theory useless or invalid, a methodological point that has been repeatedly urged by Chomsky. To quantify this better, let us try to derive at least some rough estimates on the information content of linguistic theory.

There is a substantive body of information stored in the lexicon, which is clearly irreducible in the sense that universal grammar will have little to say about the idiosyncrasies of particular words or set phrases: a conservative estimate would be the size of an abridged dictionary, 20,000–30,000 words, each requiring a few hundred unpredictable bits to encode their morphological, syntactic, and semantic aspects, for a total of about, say, a megabyte. Knowing that speech compression is already within a factor of five of the phonemic bitrate, we can conclude that the entire lexicon, including phonetic/phonological information, is unlikely to exceed a couple of megabytes.

Traditional wisdom, largely confirmed by the practical experience of adult language learners, says that to know a language is to know the words, and in fact, from the middle ages until the 19th century, syntactic theory was viewed as a small appendix to the main body of lexical and morphological knowledge. Perhaps a more realistic estimate can be gathered from construction grammar, which relies on a few dozen generic rules as opposed to thousands of specific constructions, or from Pāṇini, whose grammar is about 10% the size of his lexicons, the *dhātupāṭha* (about 2,000 verbal roots) and the *gaṇapāṭha* (about 20,000 nominal stems), even though he attempts neither detailed semantic analysis of content words nor exhaustive coverage in the lexicon. Be that as it may, a great deal of modern syntax seems to investigate the recursive behavior of a few dozen constructions, likely encodable in less than

10 kilobytes, while the bulk of the training/learning effort is clearly directed at the nonrecursive (list-like) megabytes of data one needs to memorize in order to master the language.

As the reader who went through this book knows, mathematical linguistics is not yet a unified theory: there are many ideas and fragments of theories, but there is nothing we could call a full theory of the domain. This is, perhaps, a reflection on the state of linguistics itself, which is blessed with many great ideas but few that have found good use beyond the immediate range of phenomena for which they were developed. Notions such as the *phoneme* or the *paradigm* have had a tremendous impact on anthropology, sociology, and literary theory, and some of the best work that predates the still fashionable hodge-podge of 'critical' thought in the humanities clearly has its inspiration in these and similar notions of structural linguistics. If anything, language stands as a good example of a system "lacking a clear central hierarchy or organizing principle and embodying extreme complexity, contradiction, ambiguity, diversity, interconnectedness, and intereferentiality" – the very definition of the postmodern intellectual state according to Wikipedia. In these pages, in order to gain clarity and to make a rigorous analysis possible, I have systematically chosen the simplest versions of the problems, and even these no doubt embody extreme complexity.

Is there, then, a single thread that binds mathematical linguistics together? This book makes the extended argument that there is, and it is the attempt to *find fast algorithms.* As all practicing engineers know, polynomial algorithms do not scale: for large problems only linear algorithms work. For example, Gaussian elimination, or any other $O(N^3)$ method, must be replaced for large but sparse matrices by rotation techniques that scale linearly or nearly so with the amount of (nonzero) data. Since finite state methods are by their nature linear, improving the state of the art requires a realignment of focus from the algebraically complex to the algebraically simple, in particular to the noncounting finite state realm. The work is by no means done: there is surprisingly much that we do not know about this domain, especially about the weighted versions of the models. For example, it is not known whether noncounting languages are pac learnable (under mild noise conditions). As Theorem 7.3.4 shows, there is little reason to believe that noncounting is the end of the story: to get a good characterization of natural language syntax we need to establish further restrictions so as to enable iitl learning. That said, noncounting still provides a valuable upper bound on the complexity of the problem.

The same drastic realignment of focus is called for in the case of semantics. Far too great attention has been lavished on complex systems with incredibly powerful deduction, to the detriment of investigating simpler, decidable calculi. There are a massive amount of facts for which a well-developed theory of semantics must account, and in this domain we do not even have the beginnings of a good automatic acquisition strategy (Solomonoff-Levin universal learning is unlikely to scale). Yet it is clear that the standard approach of linguistics, ascribing as much complexity to universal grammar (and, by implication, to the genetic makeup of humans) as can be amortized over the set of possible languages, is not very promising for semantics.

The only words that can possibly yield to an approach from this direction belong in the narrow class of physical sensations and emotional experiences.

In sum, it is not enough to stand on the shoulders of giants – we must also face in a better direction.

10.1 Previous reading

The best ideas in this book should be credited to people other than the author even if no reference could be dug up. In particular, the heavy emphasis on noncounting languages originates with an apocryphal remark by John von Neumann: *The brain does not use the language of mathematics*. The idea that VBTOs are first class entities is from William Lawvere, and the notion that linking is variable binding is from lectures by Manfred Bierwisch. The concluding sentence follows from a bon mot of Alan Kay.

References

Peter Aczel. 1988. *Non-Well-Founded Sets*. CSLI, Stanford, CA.

Alfred V. Aho. 1968. Indexed grammars – an extension of context-free grammars. *Journal of the ACM*, 15(4):647–671.

Judith Aissen and David Perlmutter. 1983. Clause reduction in Spanish. In David Perlmutter, editor, *Studies In Relational Grammar*, volume 1, pages 360–403. University of Chicago Press.

Kazimierz Ajdukiewicz. 1935. Die syntaktische konnexität. *Studia Philosophica*, 1:1–27.

Robin Allott. 1995. Sound symbolism. In Udo L. Figge, editor, *Language In the Würm Glaciation*, pages 15–38. Brockmeyer, Bochum.

Michael R. Anderberg. 1973. *Cluster Analysis For Applications*. Academic Press.

John R. Anderson and Gordon H. Bower. 1973. *Human Associative Memory*. Winston, Washington, DC.

Stephen R. Anderson. 1982. Where is morphology? *Linguistic Inquiry*, 13: 571–612.

Stephen R. Anderson. 1985. *Phonology In the Twentieth Century: Theories of Rules and Theories of Representations*. University of Chicago Press.

Stephen R. Anderson. 1992. *A-Morphous Morphology*. Cambridge University Press.

Stephen R. Anderson. 2000. Reflections on 'On the phonetic rules of Russian'. *Folia Linguistica*, 34(1–2):11–27.

Dana Angluin. 1980. Inductive inference of formal languages from positive data. *Information and Control*, 21:46–62.

Dana Angluin. 1981. A note on the number of queries needed to identify regular languages. *Information and Control*, 51:76–87.

Dana Angluin. 1982. Inference of reversible languages. *Journal of the ACM*, 29: 741–765.

Dana Angluin. 1987. Learning regular sets from queries and counterexamples. *Information and Computation*, 75:87–106.

Dana Angluin. 1992. Computational learning theory: Survey and selected bibliography. In *Proceedings of STOC 92*, pages 351–369.

Dana Angluin and Carl H. Smith. 1983. Inductive inference: Theory and methods. *ACM Computing Surveys*, 15(3):237–269.

Martin Anthony, Norman Biggs, and John Shawe-Taylor. 1990. The learnability of formal concepts. In *Proceedings of the 3rd COLT*, pages 246–257.

Apostolos Antonacopoulos, Basilios Gatos, and David Bridson. 2005. ICDAR2005 page segmentation competition. In *Proceedings of the 8th ICDAR*, pages 75–79, Seoul, South Korea. IEEE Computer Society Press.

G. Aldo Antonelli. 1999. A directly cautious theory of defeasible consequence for default logic via the notion of general extension. *Artificial Intelligence*, 109 (1–2):71–109.

Anthony Aristar. 1999. Typology and the Saussurean dichotomy. In E. Polomé and C.F. Justus, editors, *Language Change and Typological Variation: In Honor of Winfred P. Lehmann*. Journal of Indo-European Studies Monograph.

Mark Aronoff. 1976. *Word Formation in Generative Grammar*. MIT Press.

Mark Aronoff. 1985. Orthography and linguistic theory: The syntactic basis of masoretic Hebrew punctuation. *Language*, 61(1):28–72.

John L. Austin. 1962. *How to do things with Words*. Clarendon Press, Oxford.

Rainer Bäuerle and Max J. Cresswell. 1989. Propositional attitudes. In F. Guenthner and D.M. Gabbay, editors, *Handbook of Philosophical Logic*, volume IV, pages 491–512. Reidel, Dordrecht.

Emmon Bach. 1980. In defense of passive. *Linguistics and Philosophy*, 3:297–341.

Emmon Bach. 1981. Discontinuous constituents in generalized categorial grammars. In *Proceedings of the NELS*, volume II, pages 1–12.

Mark Baltin and Anthony Kroch. 1989. *Alternative Conceptions of Phrase Structure*. University of Chicago Press.

Arindam Banerjee, Srujana Merugu, Inderjit S. Dhillon, and Joydeep Ghosh. 2005. Clustering with Bregman divergences. *Journal of Machine Learning Research*, 6:1705–1749.

Yehoshua Bar-Hillel. 1953. A quasi-arithmetical notation for syntactic description. *Language*, 29:47–58.

Yehoshua Bar-Hillel. 1964. *Language and Information*. Addison-Wesley.

Yehoshua Bar-Hillel, Chaim Gaifman, and Eli Shamir. 1960. On categorial and phrase structure grammars. *Bulletin of the Research Council of Israel*, 9F:1–16.

Chris Barker. 2005. *Variable-free semantics in 2005*. ms, NYU.

Chris Barker and Pauline Jacobson, editors. 2007. *Direct Compositionality*. Oxford University Press.

Jon Barwise and John Etchemendy. 1987. *The Liar: An Essay in Truth and Circularity*. Oxford University Press.

Jon Barwise and John Perry. 1983. *Situations and Attitudes*. MIT Press.

Edwin L. Battistella. 1996. *The Logic of Markedness*. Oxford University Press.

Leonard E. Baum and Ted Petrie. 1966. Statistical inference for probabilistic functions of finite state Markov chains. *Annals of Mathematical Statistics*, 37:1554–1563.

Leonard E. Baum, Ted Petrie, George Soules, and Norman Weiss. 1970. A maximization technique occurring in the statistical analysis of probabilistic functions of Markov chains. *Annals of Mathematical Statistics*, 41(1):164–171.

Joffroy Beauquier and Loÿs Thimonier. 1986. Formal languages and Bernoulli processes. In J. Demetrovics, G.O.H. Katona, and A. Salomaa, editors, *Algebra, Combinatorics, and Logic in Computer Science*, pages 97–112. North-Holland, Amsterdam.

A. Melville Bell. 1867. *Visible Speech*. Simpkin and Marshall, London.

Jerome R. Bellegarda and David Nahamoo. 1990. Tied mixture continuous parameter modeling for speech recognition. *IEEE Transactions on Acoustics, Speech, and Signal Processing*, 38(12):2033–2045.

Nuel D. Belnap. 1977. How a computer should think. In G. Ryle, editor, *Contemporary Aspects of Philosophy*, pages 30–56. Oriel Press, Newcastle upon Tyne.

Nuel D. Belnap and L.E. Szabó. 1996. Branching space-time analysis of the GHZ theorem. *Foundations of Physics*, 26(8):989–1002.

W.R. Bennett. 1948. Spectra of quantized signals. *Bell System Technical Journal*, 27:446–472.

Leo L. Beranek. 1949. *Acoustic Measurements*. Wiley, New York.

Roger L. Berger. 1966. *The Undecidability of the Domino Problem*, volume 66. Memoires of the American Mathematical Society, Providence, RI.

Anne Bergeron and Sylvie Hamel. 2004. From cascade decompositions to bit-vector algorithms. *Theoretical Computer Science*, 313:3–16.

George Berkeley. 1734. *The Analyst, or a Discourse addressed to an Infidel Mathematician*. Reprinted in W. Ewald, editor: From Kant to Hilbert: A Source Book in the Foundations of Mathematics, Oxford University Press 1966.

Jean Berko. 1958. The child's learning of English morphology. *Word*, 14:150–177.

Jean Berstel. 1973. Sur la densité asymptotique de langages formels. In M. Nivat, editor, *Automata, Languages, and Programming*, pages 345–358. North-Holland.

Manfred Bierwisch. 1988. *Thematic grids as the interface between syntax and semantics*. Lectures, Stanford, CA.

Steven Bird and T. Mark Ellison. 1994. One-level phonology: Autosegmental representations and rules as finite automata. *Computational Linguistics*, 20(1):55–90.

Garrett Birkhoff. 1940. *Lattice theory*. American Mathematical Society Colloquium Publications, Providence, RI.

James P. Blevins. 2003. Stems and paradigms. *Language*, 79(4):737–767.

Juliette Blevins. 1995. The syllable in phonological theory. In John Goldsmith, editor, *Handbook of Phonology*, pages 206–244. Blackwell, Oxford.

Juliette Blevins and Sheldon P. Harrison. 1999. Trimoraic feet in Gilbertese. *Oceanic Linguistics*, 38:203–230.

Bernard Bloch and George L. Trager. 1942. *Outline of Linguistic Analysis*. Linguistic Society of America, Baltimore.

Leonard Bloomfield. 1926. A set of postulates for the science of language. *Language*, 2:153–164.

Leonard Bloomfield. 1933. *Language*. George Allen and Unwin, London.

Anselm Blumer, Andrzej Ehrenfeucht, David Haussler, and Manfred K. Warmuth. 1989. Learnability and the Vapnik-Chervonenkis dimension. *Journal of the ACM*, 36(4):929–965.

Bruce P. Bogert, Michael J.R. Healy, and John W. Tukey. 1963. The quefrency alanysis of time series for echoes: cepstrum, pseudo-autocovariance, cross-cepstrum, and saphe cracking. In M. Rosenblatt, editor, *Proceedings of the Symposium on Time Series Analysis*, pages 209–243. Wiley.

Mindy Bokser. 1992. Omnidocument technologies. *Proceedings of the IEEE*, 80(7):1066–1078.

George Box and Gwilym Jenkins. 1970. *Time Series Analysis: Forecasting and Control*. Holden-Day, San Francisco.

Otto Böhtlingk. 1964. *Pāṇini's Grammatik*. Georg Olms, Hildesheim.

Michael Böttner. 2001. Peirce grammar. *Grammars*, 4(1):1–19.

Ruth M. Brend and Kenneth L. Pike. 1976. *Tagmemics, volume 2: Theoretical Discussion*. Trends in Linguistics 2. Mouton, The Hague.

Joan Bresnan. 1982. Control and complementation. In Joan Bresnan, editor, *The mental representation of grammatical relations*, pages 282–390. MIT Press.

Ernesto Bribiesca. 1999. A new chain code. *Pattern Recognition*, 32(2):235–251.

E. Oran Brigham. 1988. *The Fast Fourier Transform and its Applications*. Prentice-Hall.

Ellen Broselow. 1995. The skeletal tier and moras. In John Goldsmith, editor, *Handbook of Phonology*, pages 175–205. Blackwell, Oxford.

Peter Brown, John Cocke, Stephen Della Pietra, Vincent J. Della Pietra, Fredrick Jelinek, John D. Lafferty, Robert L. Mercer, and Paul S. Roossin. 1990. A statistical approach to machine translation. *Computational Linguistics*, 16:79–85.

Herman E. Buiskool. 1939. *The Tripadi. Being an abridged English recast of Purvatrasiddham (an analytical-synthetical inquiry into the system of the last three chapters of Panini's Astadhyayi)*. E.J. Brill, Leiden.

Horst Bunke and Patrick S.P. Wang, editors. 1997. *Handbook of Character Recognition and Document Image Analysis*. World Scientific, Singapore.

Oliver Cappé, Eric Moulines, and Tobias Rydén. 2005. *Inference in Hidden Markov Models*. Springer.

Guy Carden. 1983. The non-finite = state-ness of the word formation component. *Linguistic Inquiry*, 14(3):537–541.

George Cardona. 1965. On Pāṇini's morphophonemic principles. *Language*, 41: 225–237.

George Cardona. 1969. Studies in Indian grammarians I: The method of description reflected in the Śiva-sūtras. *Transactions of the American Philosophical Society*, 59(1):3–48.

George Cardona. 1970. Some principles of Pāṇini's grammar. *Journal of Indian Philosophy*, 1:40–74.

George Cardona. 1976. Some features of Pāṇinian derivations. In H. Parret, editor, *History of Linguistic Thought and Contemporary Linguistics*. de Gruyter, Berlin/New York.

George Cardona. 1988. *Pāṇini: his Work and its Traditions*, volume I. Motilal Banarsidass, New Delhi.

Greg Carlson and Francis J. Pelletier, editors. 1995. *The Generic Book*. University of Chicago Press.

Robert Carpenter and Glynn Morrill. 2005. Switch graphs for parsing type logical grammars. In *Proceedings of the International Workshop on Parsing Technology*, Vancouver. IWPT05.

Claudia Casadio, Philip J. Scott, and Robert A.G. Seely, editors. 2005. *Language and Grammar: Studies in Mathematical Linguistics and Natural Language*. CSLI, Stanford CA.

Henrietta Cedergren and David Sankoff. 1974. Variable rules: performance as a statistical reflection of competence. *Language*, 50:335–355.

Gregory J. Chaitin. 1982. Gödel's theorem and information. *International Journal of Theoretical Physics*, 22:941–954.

David G. Champernowne. 1952. The graduation of income distributions. *Econometrica*, 20:591–615.

David G. Champernowne. 1953. A model of income distribution. *Economic Journal*, 63:318–351.

David G. Champernowne. 1973. *The Distribution of Income*. Cambridge University Press.

Bruce Chandler and Wilhelm Magnus. 1982. *The History of Combinatorial Group Theory: A Case Study in the History of Ideas*. Springer.

Yuen Ren Chao. 1961. Graphic and phonetic aspects of linguistic and mathematical symbols. In R. Jakobson, editor, *Structure of Language and its Mathematical Aspects*, pages 69–82. American Mathematical Society, Providence, RI.

Eugene Charniak. 1993. *Statistical Language Learning*. MIT Press.

Colin Cherry. 1956. Roman Jakobson's distinctive features as the normal coordinates of a language. In Morris Halle, editor, *For Roman Jakobson*. Mouton, The Hague.

Colin Cherry, Morris Halle, and Roman Jakobson. 1953. Toward the logical description of languages in their phonemic aspect. *Language*, 29:34–46.

Donald G. Childers, David P. Skinner, and Robert C. Kemerait. 1977. The cepstrum: A guide to processing. *Proceedings of the IEEE*, 65(10):1428–1443.

Noam Chomsky. 1956. Three models for the description of language. *IRE Transactions on Information Theory*, 2:113–124.

Noam Chomsky. 1957. *Syntactic Structures*. Mouton, The Hague.

Noam Chomsky. 1959. On certain formal properties of grammars. *Information and Control*, 2:137–167.

Noam Chomsky. 1961. On the notion 'rule of grammar'. In R. Jakobson, editor, *Structure of Language and its Mathematical Aspects*, pages 6–24. American Mathematical Society, Providence, RI.

Noam Chomsky. 1965. *Aspects of the Theory of Syntax*. MIT Press.

Noam Chomsky. 1967. Degrees of grammaticalness. In J.A. Fodor and J.J. Katz, editors, *The Structure of Language*, pages 384–389. Prentice-Hall.

Noam Chomsky. 1970. Remarks on nominalization. In R. Jacobs and P. Rosenbaum, editors, *Readings in English Transformational Grammar*, pages 184–221. Blaisdell, Waltham, MA.

Noam Chomsky. 1980. *Rules and Representations*. Columbia University Press.

Noam Chomsky. 1981. *Lectures on Government and Binding*. Foris, Dordrecht.

Noam Chomsky. 1995. *The Minimalist Program*. MIT Press.

Noam Chomsky and Morris Halle. 1965a. Some controversial questions in phonological theory. *Journal of Linguistics*, 1:97–138.

Noam Chomsky and Morris Halle. 1968. *The Sound Pattern of English*. Harper and Row, New York.

Noam Chomsky and Howard Lasnik. 1993. Principles and parameters theory. In J. Jacobs, editor, *Syntax: An International Handbook of Contemporary Research*, volume 1, pages 505–569. de Gruyter, Berlin.

Noam Chomsky and Marcel Paul Schützenberger. 1963. The algebraic theory of context-free languages. In P. Braffort and D. Hirschberg, editors, *Computer Programming and Formal Systems*, pages 118–161. North-Holland, Amsterdam.

Alexander Clark and Rémy Eyraud. 2005. Identification in the limit of substitutable context-free languages. In S. Jain, H.U. Simon, and E.Tomita, editors, *Proceedings of the 16th International Conference on Algorithmic Learning Theory (ALT 2005)*, pages 283–296. Springer.

Robin Clark. 1994. Kolmogorov complexity and the information content of parameters. IRCS Report 94-17, Institute for Research in Cognitive Science, University of Pennsylvania.

George N. Clements. 1977. Vowel harmony in nonlinear generative phonology. In W.U. Dressler and O.E. Pfeiffer, editors, *Phonologica 1976*, pages 111–119. Institut fur Sprachwissenschaft der Universität Innsbruck.

George N. Clements. 1985. The problem of transfer in nonlinear phonology. *Cornell Working Papers in Linguistics*, 5:38–73.

George N. Clements and Kevin C. Ford. 1979. Kikuyu tone shift and its synchronic consequences. *Linguistic Inquiry*, 10:179–210.

George N. Clements and S. Jay Keyser. 1983. *CV Phonology: A Generative Theory of the Syllable*. MIT Press.

Alan Cobham. 1969. On the base-dependence of sets of numbers recognizable by finite automata. *Mathematical Systems Theory*, 3:186–192.

Ronald A. Cole, editor. 1997. *Survey of the State of the Art in Human Language Technology*. Cambridge University Press.

N.E. Collinge. 1985. *The Laws of Indo-European*. Benjamins.

Bernard Comrie, editor. 1990. *The World's Major Languages*. Oxford University Press.

Confucius. 1979. *The Analects*. Penguin, Harmondsworth.

James W. Cooley and John W. Tukey. 1965. An algorithm for the machine calculation of complex Fourier series. *Mathematics of Computation*, 19:297–301.

Robin Cooper. 1975. *Montague's Semantic Theory and Transformational Syntax*. PhD thesis, University of Massachusetts, Amherst.

John Corcoran, William Hatcher, and John Herring. 1972. Variable binding term operators. *Zeitschrift für mathematische Logik und Grundlagen der Mathematik*, 18:177–182.

Michael A. Covington. 1984. *Syntactic Theory in the High Middle Ages*. Cambridge University Press.

David R. Cox and H. David Miller. 1965. *The Theory of Stochastic Processes*. Methuen, London.

Max J. Cresswell. 1985. *Structured Meanings*. MIT Press.

Haskell B. Curry. 1961. Some logical aspects of grammatical structure. In R. Jakobson, editor, *Structure of Language and its Mathematical Aspects*, pages 56–68. American Mathematical Society, Providence, RI.

Haskell B. Curry and Robert Feys. 1958. *Combinatory Logic I*. North-Holland, Amsterdam.

James Cussens and Stephen G. Pulman. 2001. Grammar learning using inductive logic programming. *Oxford University Working Papers in Linguistics, Philology and Phonetics*, 6:31–45.

Steven B. Davis and Paul Mermelstein. 1980. Comparison of parametric representations for monosyllabic word recognition in continuously spoken sentences. *IEEE Transactions on Acoustics, Speech, and Signal Processing*, 28(4):357–366.

Max W. Dehn. 1912. Über unendlich diskontinuierliche Gruppen. *Mathematische Annalen*, 71:116–144.

François Dell and Mohamed Elmedlaoui. 1985. Syllabic consonants and syllabification in Imdlawn Tashlhiyt Berber. *Journal of African Languages and Linguistics*, 7:105–130.

François Dell and Mohamed Elmedlaoui. 1988. Syllabic consonants in Berber: some new evidence. *Journal of African Languages and Linguistics*, 10:1–17.

Hans den Besten. 1985. The ergative hypothesis and free word order in Dutch and German. In J. Toman, editor, *Studies in German Grammar*, pages 23–64. Foris, Dordrecht.

Jim des Rivières and Hector J. Levesque. 1986. The consistency of syntactical treatments of knowledge. In M. Vardi, editor, *Theoretical aspects of reasoning about knowledge*, pages 115–130. Morgan Kaufmann, Los Altos, CA.

Kris Demuynck, Jacques Duchateau, Dirk Van Compernolle, and Patrick Wambacq. 1998. Improved feature decorrelation for HMM-based speech recognition. In *Proceedings of the ICSLP*, volume VII, pages 2907–2910, Sydney, Australia.

Keith Devlin. 1991. *Logic and Information*. Cambridge University Press.

Keith Devlin. 2001. *The mathematics of information*. ESSLLI Lecture Notes, Helsinki.

Luc Devroye, László Győrfi, and Gábor Lugosi. 1996. *A Probabilistic Theory of Pattern Recognition*. Springer.

Robert M.W. Dixon. 1994. *Ergativity*. Cambridge University Press.

David Dowty, Robert Wall, and Stanley Peters. 1981. *Introduction to Montague Semantics*. Reidel, Dordrecht.

David Dowty. 1979. *Word Meaning and Montague Grammar*. Reidel, Dordrecht.

David Dowty. 1985. On recent analyses of the semantics of control. *Linguistics and Philosophy*, 8(3):291–331.

David Dowty. 1991. Thematic proto-roles and argument selection. *Language*, 67:547–619.

Erich Drach. 1937. *Grundgedanken der deutschen Satzlehre*. Diesterweg, Frankfurt.

Fred Dretske. 1981. *Knowledge and the Flow of Information*. MIT Press.

Richard O. Duda, Peter E. Hart, and David G. Stork. 2000. *Pattern Classification*. Wiley.

Mark Durie. 1987. Grammatical relations in Acehnese. *Studies in Language*, 11(2):365–399.

Umberto Eco. 1995. *The Search for the Perfect Language*. Blackwell, Oxford.

Murray Eden. 1961. On the formalization of handwriting. In R. Jakobson, editor, *Structure of Language and its Mathematical Aspects*, pages 83–88. American Mathematical Society, Providence, RI.

Andrzej Ehrenfeucht, David Haussler, Michael Kearns, and Leslie Valiant. 1989. A general lower bound on the number of examples needed for learning. *Information and Computation*, 82:247–251.

Samuel Eilenberg. 1974. *Automata, Languages, and Machines*, volume A. Academic Press.

P.D. Eimas, E.R. Siqueland, P. Jusczyk, and J. Vigorito. 1971. Speech perception in infants. *Science*, 171:303–306.

Jason Eisner. 2001. Expectation semirings: Flexible EM for learning finite-state transducers. In G. van Noord, editor, *ESSLLI Workshop on Finite-State Methods in Natural Language Processing*, Helsinki.

Clarence A. Ellis. 1969. *Probabilistic Languages and Automata*. PhD thesis, University of Illinois, Urbana.

Miklós Erdélyi-Szabó, László Kálmán, and Ágnes Kurucz. 2007. Towards a natural language semantics without functors and operands. *Journal of Logic, Language and Information*.

J.R. Estoup. 1916. *Gammes Stenographiques*. Institut Stenographique de France, Paris.

Brian Everitt. 1980. *Cluster Analysis*. Halsted Press, New York.

Gunnar Fant. 1960. *Acoustic Theory of Speech Production*. Mouton, The Hague.

Gilles Fauconnier. 1985. *Mental Spaces*. MIT Press.

Fibonacci. 1202. *Liber Abaci*. Translated by L.E. Siegler. Springer (2002).

Charles Fillmore. 1968. The case for case. In E. Bach and R. Harms, editors, *Universals in Linguistic Theory*, pages 1–90. Holt and Rinehart, New York.

Charles Fillmore. 1977. The case for case reopened. In P. Cole and J.M. Sadock, editors, *Grammatical Relations*, pages 59–82. Academic Press.

Charles Fillmore and Paul Kay. 1997. *Berkeley Construction Grammar*. http://www.icsi.berkeley.edu/~kay/bcg/ConGram.html.

Kit Fine. 1985. *Reasoning with Arbitrary Objects*. Blackwell, Oxford.

John Rupert Firth. 1948. Sounds and Prosodies. *Transactions of the Philological Society*, 46:127–152.

Ronald A. Fisher. 1936. The use of multiple measurements in taxonomic problems. *Annals of Eugenics*, 7:179–188.

Ronald A. Fisher. 1937. The statistical utilization of multiple measurements. *Annals of Eugenics*, 8:376–385.

James Flanagan. 1955. Difference limen for vowel formant frequency. *Journal of the Acoustical Society of America*, 27:613–617.

James Flanagan. 1972. *Speech Analysis, Synthesis and Perception*. Springer.

Daniel P. Flickinger. 1987. *Lexical Rules in the Hierarchical Lexicon*. PhD Thesis, Stanford University.

Christophe Costa Florencio. 2003. *Learning Categorial Grammars*. PhD Thesis, Universiteit Utrecht.

William A. Foley and Robert van Valin. 1984. *Functional Syntax and Universal Grammar*. Cambridge University Press.

Carol A. Fowler and Dawn J. Dekle. 1991. Listening with eye and hand: Crossmodal contributions to speech perception. *Journal of Experimental Psychology: Human Perception and Performance*, 17:816–828.

Herbert Freeman. 1961. On the encoding of arbitrary geometric configurations. *IRE Transactions on Electronic Computers*, 10:260–268.

Gottlob Frege. 1879. *Begriffsschrift: eine der arithmetischen nachgebildete Formelsprache des reinen Denkens*. L. Nebert, Halle.

Osamu Fujimura, S. Kiritani, and H. Ishida. 1973. Computer controlled radiography for observation of movements of articulatory and other human organs. *Computer Biological Medicine*, 3:371–384.

Sadaoki Furui. 1986. Speaker-independent isolated word recognition using dynamic features of speech spectrum. *IEEE Transactions on Acoustics, Speech, and Signal Processing*, 34(1):52–59.

Péter Gács. 2003. *Lecture notes on descriptional complexity and randomness*. ms, Boston University.

Chaim Gaifman. 1965. Dependency systems and phrase-structure systems. *Information and Control*, 8:304–337.

L.T.F. Gamut. 1991. *Logic, Language, and Meaning*. University of Chicago Press.

Gerald Gazdar. 1981. Unbounded dependencies and coordinate structure. *Linguistic Inquiry*, 12:155–184.

Gerald Gazdar, Ewan Klein, Geoffrey K. Pullum, and Ivan A. Sag. 1985. *Generalized Phrase Structure Grammar*. Blackwell, Oxford.

Matthew L. Ginsberg. 1986. Multi-valued logics. In Matthew L. Ginsberg, editor, *Readings in Non-monotonic Reasoning*, pages 251–255. Morgan Kaufmann, San Mateo, CA.

Matthew L. Ginsberg, editor. 1986a. *Readings in Non-monotonic Reasoning*. Morgan Kauffman, San Mateo, CA.

Abraham Ginzburg. 1968. *Algebraic Theory of Automata*. Academic Press.

Henry A. Gleason. 1961. Genetic relationship among languages. In R. Jakobson, editor, *Structure of Language and its Mathematical Aspects*, pages 179–189. American Mathematical Society, Providence, RI.

Cliff Goddard. 2002. The search for the shared semantic core of all languages. In Cliff Goddard and Anna Wierzbicka, editors, *Meaning and Universal Grammar – Theory and Empirical Findings*, volume 1, pages 5–40. Benjamins.

Rob W.N. Goedemans, Harry G. van der Hulst, and Ellis M. Visch. 1996. *Stress Patterns of the World. Part 1: Background*. Holland Academic Graphics, The Hague.

E. Mark Gold. 1967. Language identification in the limit. *Information and Control*, 10:447–474.

E. Mark Gold. 1978. Complexity of automaton identification from given data. *Information and Control*, 37(3):302–320.

John A. Goldsmith. 1976. *Autosegmental Phonology*. PhD thesis MIT.

John A. Goldsmith. 1990. *Autosegmental and Metrical Phonology*. Blackwell, Cambridge, MA.

John Goldsmith and Gary Larson. 1990a. Local modeling and syllabification. In M. Noske, K. Deaton, and M. Ziolkowski, editors, *CLS 26 Parasession on the Syllable in Phonetics and Phonology*, pages 129–142. Chicago Linguistic Society.

Raymond G. Gordon, Jr. 2005. *Ethnologue: Languages of the World*. SIL International, Dallas, TX.

Robert M. Gray and E.D. Karnin. 1982. Multiple local optima in vector quantizers. *IEEE Transactions on Information Theory*, 28(2):256–261.

Joseph H. Greenberg. 1963. Some universals of grammar with particular reference to the order of meaningful elements. In J.H. Greenberg, editor, *Universals of Human Language*, pages 73–113. MIT Press.

Barbara B. Greene and Gerald M. Rubin. 1971. *Automatic grammatical tagging of English*. Department of Linguistics, Brown University, Providence, RI, ms.

Maurice Gross. 1972. *Mathematical Models in Linguistics*. Prentice-Hall.

Jozef Gruska. 1997. *Foundations of Computing*. Thomson International Computer Press.

Pierre Guiraud. 1954. *Les charactères statistiques du vocabulaire*. Presses Universitaires de France, Paris.

Kenneth Hale. 1983. Warlpiri and the grammar of non-configurational languages. *Natural Language and Linguistic Theory*, 1:5–47.

Morris Halle. 1961. On the role of simplicity in linguistic descriptions. In R. Jakobson, editor, *Structure of Language and its Mathematical Aspects*, pages 89–94. American Mathematical Society, Providence, RI.

Morris Halle. 1964. On the bases of phonology. In J.A. Fodor and J. Katz, editors, *The Structure of Language*, pages 324–333. Prentice-Hall.

Morris Halle. 1983a. Distinctive features and their articulatory implementation. *Natural Language and Linguistic Theory*, 1:91–107.

Morris Halle and George N. Clements. 1983. *Problem Book in Phonology*. MIT Press.

Morris Halle and Samuel Jay Keyser. 1971. *English Stress: Its Growth and Its Role in Verse*. Harper and Row, New York.

Morris Halle and Peter Ladefoged. 1988. Some major features of the international phonetic alphabet. *Language*, 64:577–582.

Morris Halle and Ken Stevens. 1962. Speech recognition: A model and a program for research. *IRE Transactions on Information Theory*, 49:155–159.

Michael Hammond. 1987. Hungarian cola. *Phonology Yearbook*, 4:267–269.

Michael Hammond. 1995. Metrical phonology. *Annual Review of Anthropology*, 24:313–342.

Zellig Harris. 1951. *Methods in Structural Linguistics*. University of Chicago Press.

Zellig Harris. 1957. Coocurence and transformation in liguistic structure. *Language*, 33:283–340.

Randy Allen Harris. 1995. *The Linguistics Wars*. Oxford University Press.

Michael A. Harrison. 1978. *Introduction to Formal Language Theory*. Addison-Wesley.

Ralph V.L. Hartley. 1928. Transmission of information. *Bell System Technical Journal*, 7:535–563.

Akihide Hashizume, Pen-Shu Yeh, and Azriel Rosenfeld. 1986. A method of detecting the orientation of aligned components. *Pattern Recognition Letters*, 4:125–132.

Trevor Hastie, Robert Tibshirani, and Jerome Friedman. 2001. *Elements of Statistical Learning: Data Mining, Inference, and Prediction*. Springer.

John A. Hawkins. 1983. *World Order Universals*. Academic Press.

Patrick J. Hayes. 1978. *The Naive Physics Manifesto*. Institut Dalle Molle, Geneva.

Bruce Hayes. 1980. *A Metrical Theory of Stress Rules*. PhD thesis, MIT.

Bruce Hayes. 1995. *Metrical Stress Theory*. University of Chicago Press.

Harold S. Heaps. 1978. *Information Retrieval – Computational and Theoretical Aspects*. Academic Press.

Jeffrey Heath. 1987. *Ablaut and Ambiguity: Phonology of a Moroccan Arabic Dialect*. State University of New York Press, Albany.

Michael T. Heideman, Don H. Johnson, and C. Sidney Burrus. 1984. Gauss and the history of the FFT. *IEEE Acoustics, Speech, and Signal Processing Magazine*, 1:14–21.

Irene Heim. 1982. *The Semantics of Definite and Indefinite Noun Phrases*. PhD thesis, University of Massachusetts, Amherst, MA.

Steve J. Heims. 1991. *The Cybernetics Group, 1946–1953*. MIT Press.

Leon Henkin. 1950. Completeness in the theory of types. *Journal of Symbolic Logic*, 15:81–91.

Jacques Herbrand. 1930. *Recherches sur la théorie de la demonstration*. PhD thesis, Paris.

Gustav Herdan. 1960. *Type-Token Mathematics*. Mouton, The Hague.

Gustav Herdan. 1966. *The Advanced Theory of Language as Choice and Chance*. Springer.

Hynek Hermansky. 1990. Perceptual linear predictive (PLP) analysis of speech. *Journal of the Acoustical Society of America*, 87(4):1738–1752.

Hynek Hermansky, Nelson Morgan, Aruna Bayya, and Phil Kohn. 1991. RASTA-PLP speech analysis. *US West Advanced Technologies TR-91-069*.

Susan R. Hertz. 1982. From text-to-speech with SRS. *Journal of the Acoustical Society of America*, 72:1155–1170.

Wilbur H. Highleyman. 1962. Linear decision functions with application to pattern recognition. *Proceedings of the IRE*, 50:1501–1514.

Henry Hiz. 1961. Congrammaticality, batteries of transformations and grammatical categories. In R. Jakobson, editor, *Structure of Language and its Mathematical Aspects*, pages 43–50. American Mathematical Society, Providence, RI.

Louis Hjelmslev. 1961. *Prolegomena to a Theory of Language*. University of Wisconsin Press, Madison.

Robert D. Hoberman. 1987. Emphasis (pharyngealization) as an autosegmental harmony feature. In A. Bosch, B. Need, and E. Schiller, editors, *Parasession on Autosegmental and Metrical Phonology*, volume 23, pages 167–181. CLS.

Charles Hockett. 1960. Logical considerations in the study of animal communication. In W.E. Lanyon and W.N. Tavolga, editors, *Animal Sounds and Communications*, pages 392–430. American Institute of Biological Sciences, Washington, DC.

Fred W. Householder. 1965. On some recent claims in phonological theory. *Journal of Linguistics*, 1:13–34.

Fred W. Householder. 1966. Phonological theory: a brief comment. *Journal of Linguistics*, 2:99–100.

Ming-Kuei Hu. 1962. Visual pattern recognition by moment invariants. *IRE Transactions on Information Theory*, 49:179–187.

Geoffrey J. Huck and John A. Goldsmith. 1995. *Ideology and Linguistics Theory: Noam Chomsky and the Deep Structure Debates*. Routledge, London.

Richard A. Hudson. 1990. *English Word Grammar*. Blackwell, Oxford.

M. Sharon Hunnicutt. 1976. Phonological rules for a text to speech system. *American Journal of Computational Linguistics*, Microfiche 57:1–72.

James R. Hurford. 1975. *The Linguistic Theory of Numerals*. Cambridge University Press.

Rini Huybregts. 1976. Overlapping dependencies in Dutch. *Utrecht Working Papers in Linguistics*, 1:224–265.

Larry M. Hyman. 1975. *Phonology: Theory and Analysis*. Holt, Rinehart, Winston, New York.

Larry M. Hyman. 1982. The representation of nasality in Gokana. In H. van der Hulst and N. Smith, editors, *The Structure of Phonological Representation*, pages 111–130. Foris, Dordrecht.

William J. Idsardi. 2006. Misplaced optimism. *Rutgers Optimality Archive*, 840.

Neil Immerman. 1988. Nondeterministic space is closed under complementation. *SIAM Journal of Computing*, 17(5):935–938.

H. Inose, Y. Yasuda, and J. Murakami. 1962. A telemetering system by code manipulation – delta-sigma modulation. *IRE Transactions on Space Electronics and Telemetry*, 8:204–209.

Fumitada Itakura. 1975. Minimum prediction residual principle applied to speech recognition. *IEEE Transactions on Acoustics, Speech, and Signal Processing*, 23:67–72.

Ray S. Jackendoff. 1972. *Semantic Interpretation in Generative Grammar*. MIT Press.

Ray S. Jackendoff. 1977. *X-bar Syntax: A Study of Phrase Structure*. MIT Press.

Ray S. Jackendoff. 1983. *Semantics and Cognition*. MIT Press.

Ray S. Jackendoff. 1990. *Semantic Structures*. MIT Press.

Pauline Jacobson. 1990. Raising as function composition. *Linguistics and Philosophy*, 13(4):423–475.

Pauline Jacobson. 1999. Towards a variable-free semantics. *Linguistics and Philosophy*, 22:117–184.

Anil K. Jain and Sushil Bhattacharjee. 1992. Text segmentation using Gabor filters for automatic document processing. *Machine Vision and Applications*, 5: 169–184.

Roman Jakobson. 1936. *Beitrag zur allgemeinen Kasuslehre*. Travaux du Cercle linguistique de Prague.

Roman Jakobson. 1941. *Kindersprache, Aphasie, und allgemeine Lautgesetze*. Uppsala Universitets Arsskrift, Uppsala.

Roman Jakobson. 1957. Mufaxxama – the emphatic phonemes in Arabic. In E. Pulgrim, editor, *Studies Presented to Joshua Whatmough*. Mouton, The Hague.

Roman Jakobson, editor. 1961. *Structure of Language and its Mathematical Aspects*. American Mathematical Society, Providence, RI.

Roman Jakobson, Gunnar Fant, and Morris Halle. 1952. *Preliminaries to Speech Analysis: The Distinctive Features and Their Correlates*. MIT Press.

Theo M.V. Janssen. 1997. Compositionality. In J. van Benthem, editor, *Handbook of Logic and Language*, pages 414–474. Elsevier.

Nikil S. Jayant and Peter Noll. 1984. *Digital Coding of Waveforms: Principles and Applications to Speech and Video*. Prentice-Hall.

Frederick Jelinek. 1997. *Statistical Methods for Speech Recognition*. MIT Press.

Otto Jespersen. 1897. *Fonetik*. Det Schubotheske Forlag, Copenhagen.

Otto Jespersen. 1924. *The Philosophy of Grammar*. George Allen and Unwin, London.

Stig Johansson, Eric Atwell, Roger Garside, and Geoffrey Leech. 1986. *The Tagged LOB Corpus: Users' Manual*. ICAME: The Norwegian Computing Centre for the Humanities.

Ch. Douglas Johnson. 1970. *Formal aspects of phonological representation*. PhD thesis, UC Berkeley.

Neil D. Jones. 1966. *A survey of formal language theory*. University of Western Ontario Computer Science Tech Report 3, London, ON.

S.D. Joshi and Saroja Bhate. 1984. *Fundamentals of Anuvrtti*. Poona University Press.

Aravind K. Joshi. 2003. Tree adjoining grammars. In R. Mitkov, editor, *Handbook of Computational Linguistics*, pages 483–500. Oxford University Press.

Daniel Jurafsky and James H. Martin. 2000. *Speech and Language Processing*. Prentice-Hall.

Daniel Kahn. 1976. *Syllable-based Generalizations in English Phonology*. PhD thesis, MIT.

Ellen Kaisse and Sharon Hargus. 1993. Introduction. In Ellen Kaisse and Sharon Hargus, editors, *Studies in Lexical Phonology*, pages 1–19. Academic Press.

Mohamed Kamel and Aiguo Zhao. 1993. Extraction of binary character/graphics images from grayscale document images. *Graphical Models and Image Processing*, 55:203–217.

Hans Kamp. 1981. A theory of truth and semantic representation. In J.A.G. Groenendijk, T.M.V. Jansen, and M.B.J. Stokhof, editors, *Formal Methods in the Study of Language*, pages 277–322. Mathematisch Centrum, Amsterdam.

Makoto Kanazawa. 1996. Identification in the limit of categorial grammars. *Journal of Logic, Language and Information*, 5(2):115–155.

Tapas Kanungo, Robert M. Haralick, and Ishin T. Phillips. 1993. Global and local document degradation models. In *Proceedings of the 2nd ICDAR*, pages 730–734.

Ronald M. Kaplan and Martin Kay. 1994. Regular models of phonological rule systems. *Computational Linguistics*, 20:331–378.

David Kaplan. 1978. On the logic of demonstratives. *Journal of Philosophical Logic*, 8:81–98.

David Kaplan. 1989. Demonstratives. In J. Almog, J. Perry, and H. Wettstein, editors, *Themes from Kaplan*, pages 481–563. Oxford University Press.

Gerge S. Kapogiannopoulos and Nicholas Kalouptsidis. 2002. A fast high precision algorithm for the estimation of skew angle using moments. In M. H. Hamza, editor, *Proceedings of the Signal Processing, Pattern Recognition, and Applications (SPPRA 2002)*, pages 275–279, Calgary, AB. Acta Press.

Richard M. Karp. 1972. Reducibility among combinatorial problems. In R. Miller and J.W. Thatcher, editors, *Complexity of Computer Computations*, pages 85–104. Plenum Press, New York.

Lauri Karttunen. 1995. The replace operator. In *Proceedings of ACL95*, pages 16–23.

Paul Kay. 2002. An informal sketch of a formal architecture for construction grammar. *Grammars*, 5:1–19.

Paul Kay and Chad K. McDaniel. 1979. On the logic of variable rules. *Language in Society*, 8:151–187.

Michael Kearns and Leslie Valiant. 1989. Cryptographic limitations on learning Boolean formulae and finite automata. In *Proceedings of the 21st Annual ACM Symposium on Theory of Computing*, pages 433–444.

Aleksandr I. Khinchin. 1957. *Mathematical Foundations of Information Theory*. Dover, New York.

Alireza Khotanzad and Yaw Hua Hong. 1990. Invariant image recognition by Zernike moments. *IEEE Transactions on Pattern Analysis and Machine Intelligence*, 12(5):489–497.

Paul Kiparsky. 1968. Linguistic universals and linguistic change. In E. Bach and R. Harms, editors, *Universals in Linguistic Theory*, pages 171–202. Holt, New York.

Paul Kiparsky. 1979. *Pāṇini as a Variationist*. MIT Press and Poona University Press, Cambridge and Poona.

Paul Kiparsky. 1982. From cyclic phonology to lexical phonology. In H. van der Hulst and N. Smith, editors, *The structure of phonological representations, I*, pages 131–175. Foris, Dordrecht.

Paul Kiparsky. 1982a. The ordering of rules in Pāṇini's grammar. In *Some Theoretical Problems in Pāṇini's Grammar*, Post-graduate and Research Department Series, pages 77–121. Bhandarkar Oriental Research Institute, Poona.

Paul Kiparsky. 1987. *Morphosyntax*. ms, Stanford University.

Paul Kiparsky. 2002. *On the Architecture of Pāṇini's grammar*. ms, Stanford University.

Paul Kiparsky. 2006. *Paradigms and opacity*. CSLI and University of Chicago Press.

Georg Anton Kiraz. 2001. *Computational Nonlinear Morphology with Emphasis on Semitic Languages*. Cambridge University Press.

Charles Kisseberth. 1970. On the functional unity of phonological rules. *Linguistic Inquiry*, 1:291–306.

Thomas Klammer and Muriel R. Schulz. 1996. *Analyzing English Grammar*. Allyn and Bacon, Boston.

Dennis H. Klatt. 1980. SCRIBER and LAFS: Two new approaches to speech analysis. In Wayne Lea, editor, *Trends in Speech Recognition*, pages 529–555. Prentice-Hall.

Dennis H. Klatt. 1987. Review of text-to-speech conversion for English. *Journal of the Acoustical Society of America*, 82(3):737–793.

Stephen C. Kleene. 1956. Representation of events in nerve nets and finite automata. In C. Shannon and J. McCarthy, editors, *Automata Studies*, pages 3–41. Princeton University Press.

W. Klein, Reinier Plomp, and Louis C.W. Pols. 1970. Vowel spectra, vowel spaces, and vowel identification. *Journal of the Acoustical Society of America*, 48: 999–1009.

Donald E. Knuth. 1971. *The Art of Computer Programming*. Addison-Wesley.

Martha Kolln. 1994. *Understanding English Grammar*. MacMillan, New York.

Andrei N. Kolmogorov. 1965. Three approaches to the quantitative definition of information. *Problems of Information Transmission*, 1:1–7.

András Kornai. 1985. Natural languages and the Chomsky hierarchy. In Margaret King, editor, *Proceedings of the 2nd European ACL Conference*, pages 1–7.

András Kornai. 1987. Finite state semantics. In U. Klenk, P. Scherber, and M. Thaller, editors, *Computerlinguistik und philologische Datenverarbeitung*, pages 59–70. Georg Olms, Hildesheim.

András Kornai. 1995. *Formal Phonology*. Garland Publishing, New York.

András Kornai. 1996. Analytic models in phonology. In J. Durand and B. Laks, editors, *Current Trends in Phonology: Models and Methods*, volume 2, pages 395–418. CNRS, ESRI, Paris X, Paris.

András Kornai. 1998. Quantitative comparison of languages. *Grammars*, 1(2): 155–165.

András Kornai, editor. 1999. *Extended Finite State Models of Language*. Cambridge University Press.

András Kornai. 1999a. Zipf's law outside the middle range. In J. Rogers, editor, *Proceedings of the Sixth Meeting on Mathematics of Language*, pages 347–356. University of Central Florida.

András Kornai. 2002. How many words are there? *Glottometrics*, 2(4):61–86.

András Kornai and Geoffrey K. Pullum. 1990. The X-bar theory of phrase structure. *Language*, 66:24–50.

András Kornai and Zsolt Tuza. 1992. Narrowness, pathwidth, and their application in natural language processing. *Discrete Applied Mathematics*, 36:87–92.

Frederik H.H. Kortlandt. 1973. The identification of phonemic units. *La Linguistique*, 9(2):119–130.

Kimmo Koskenniemi. 1983. Two-level model for morphological analysis. In *Proceedings of IJCAI-83*, pages 683–685.

C.H.A. Koster. 1970. Two-level grammars. *Seminar on Automata Theory and Mathematical Linguistics*, 11.

Marcus Kracht. 2003. *The Mathematics of Language*. Mouton de Gruyter, Berlin.

Marcus Kracht. 2007. *Counter-free languages are not learnable*. ms, UCLA.

Angelika Kratzer. 1995. Stage level and individual level predicates. In G. Carlson and F.J. Pelletier, editors, *The Generic Book*. University of Chicago Press.

Saul A. Kripke. 1972. Naming and necessity. In D. Davidson, editor, *Semantics of Natural Language*, pages 253–355. D. Reidel, Dordrecht.

Saul A. Kripke. 1975. Outline of a theory of truth. *Journal of Philosophy*, 72: 690–716.

Anthony Kroch. 1994. Morphosyntactic variation. In K. Beals, editor, *Proceedings of the CLS 30 Parasession on Variation and Linguistic Theory*, volume 2, pages 180–201. Chicago Linguistic Society.

Kenneth Krohn and John Rhodes. 1965. Algebraic theory of machines. I. Prime decomposition theorem for finite semigroups and machines. *Transactions of the American Mathematical Society*, 116:450–464.

Henry Kucera and Nelson Francis, editors. 1967. *Computational Analysis of Present-Day American English*. Brown University Press, Providence, RI.

Patrica K. Kuhl and James D. Miller. 1975. Speech perception by the chinchilla: voiced-voiceless distinction in alveolar plosive consonants. *Science*, 190:69–72.

Marco Kuhlmann. 2005. *Mildly Context-Sensitive Dependency Languages*. Programming Systems Lab, Saarland University.

Werner Kuich. 1970. On the entropy of context-free languages. *Information and Control*, 16:173–200.

Werner Kuich and Arto Salomaa. 1986. *Semirings, Automata, Languages*. Springer.

William Labov. 1966. *The social stratification of English in New York City*. Center for Applied Linguistics, Washington, DC.

William Labov. 1981. Resolving the neogrammarian controversy. *Language*, 57:267–308.

William Labov. 1994. *Principles of Linguistic Change: Internal Factors*. Blackwell, Oxford.

Louisa Lam, Seong-Whan Lee, and Ching Y. Suen. 1992. Thinning methodologies – a comprehensive survey. *IEEE Transactions on Pattern Analysis and Machine Intelligence*, 14(9):869–885.

Joachim Lambek. 1958. The mathematics of sentence structure. *American Mathematical Monthly*, 65:154–170.

Joachim Lambek. 1961. On the calculus of syntactic types. In R. Jakobson, editor, *Structure of Language and its Mathematical Aspects*, pages 166–178. American Mathematical Society, Providence, RI.

Joachim Lambek. 2004. A computational approach to English grammar. *Syntax*.

Ronald Langacker. 1982. Space grammar, analyzability, and the English passive. *Language*, 58:22–80.

Ronald Langacker. 1987. *Foundations of Cognitive Grammar*, volume 1. Stanford University Press.

Ronald Langacker. 1991. *Foundations of Cognitive Grammar*, volume 2. Stanford University Press.

D. Terence Langendoen and Paul M. Postal. 1984. *The vastness of natural language*. Blackwell, Oxford.

Shalom Lappin. 1996. Generalized quantifiers, exception phrases, and logicality. *Journal of Semantics*, 13:197–220.

William Leben. 1973. *Suprasegmental phonology*. PhD thesis, MIT.

Ki-Seung Lee and Richard V. Cox. 2001. A very low bit rate speech coder based on a recognition/synthesis paradigm. *IEEE Transactions on Speech and Audio Processing*, 9(5):482–491.

Seong-Whan Lee, Louisa Lam, and Chin Y. Suen. 1991. Performance evaluation of skeletonization algorithms for document image processing. In *Proceedings of ICDAR'91*, pages 272–282.

Geoffrey Leech, Roger Garside, and Michael Bryant. 1994. CLAWS4: the tagging of the British National Corpus. In *Proceedings of the 15th International Conference on Computational Linguistics*, pages 622–628, Kyoto, Japan.

Karel de Leeuw, Edward F. Moore, Claude E. Shannon, and N. Shapiro. 1956. Computability by probabilistic machines. In C.E. Shannon and J. McCarthy, editors, *Automata studies*, pages 185–212. Princeton University Press.

Christopher J. Leggetter and Philip C. Woodland. 1995. Maximum likelihood linear regression for speaker adaptation of continuous density Hidden Markov Models. *Computer Speech and Language*, 9:171–185.

Ilse Lehiste. 1970. *Suprasegmentals*. MIT Press.

Willem J.M. Levelt. 1974. *Formal Grammars in Linguistics and Psycholinguistics*, volume 1–3. Mouton, The Hague.

Beth Levin. 1993. *English Verb Classes and Alternations: A Preliminary Investigation*. University of Chicago Press.

Beth Levin and Malka Rappaport Hovav. 2005. *Argument Realization*. Cambridge University Press.

Juliette Levin. 1985. *A Metrical Theory of Syllabicity*. PhD thesis, MIT.

Ming Li and Paul Vitányi. 1995. A new approach to formal language theory by kolmogorov complexity. *SIAM J. Comput*, 24(2):398–419.

Ming Li and Paul Vitányi. 1997. *An Introduction to Kolmogorov Complexity and Its Applications*. Springer.

Wentian Li. 1992. Random texts exhibit Zipf's-law-like word frequency distribution. *IEEE Transactions on Information Theory*, 38(6):1842–1845.

Alvin M. Liberman. 1957. Some results of research on speech perception. *Journal of the Acoustical Society of America*, 29:117–123.

Alvin M. Liberman and Ignatius G. Mattingly. 1989. A specialization for speech perception. *Science*, 243:489–494.

Mark Y. Liberman. 1975. *The Intonation System of English*. PhD thesis, MIT.

Scott K. Liddell and Robert E. Johnson. 1989. American Sign Language: The phonological base. *Sign Language Studies*, 64:195–277.

Elliott H. Lieb and Jakob Yngvason. 2003. The entropy of classical thermodynamics. In A. Greven, G. Keller, and G. Warnecke, editors, *Entropy*, pages 147–196. Princeton University Press.

Rochelle Lieber. 1980. *On the Organization of the Lexicon*. PhD Thesis, MIT.

John Lyons. 1968. *Introduction to Theoretical Linguistics*. Cambridge University Press.

David J.C. MacKay. 2003. *Information Theory, Inference, and Learning Algorithms*. Cambridge University Press.

John Makhoul. 1975. Linear prediction: A tutorial review. *Proceedings of the IEEE*, 63(4):561–580.

John Makhoul. 2006. Speech processing at BBN. *Annals of the History of Computing*, 28(1):32–45.

Oded Maler and Amir Pnueli. 1994. On the cascaded decomposition of automata, its complexity, and its application to logic. *ms*.

Alexis Manaster-Ramer. 1986. Copying in natural languages, context-freeness, and queue grammars. In *Proceedings of the 24th Meeting of the Association for Computational Linguistics*, pages 85–89.

Alexis Manaster-Ramer, editor. 1987. *Mathematics of Language*. Benjamins.

Benoit Mandelbrot. 1952. An informational theory of the structure of language based upon the theory of the statistical matching of messages and coding. In

W. Jackson, editor, *Second Symposium on Information Theory*, pages 486–500. Butterworths, London.

Benoit Mandelbrot. 1959. A note on a class of skew distribution functions. Analysis and critique of a paper by H.A. Simon. *Information and Control*, 2:90–99.

Benoit Mandelbrot. 1961a. Final note on a class of skew distribution functions: Analysis and critique of a model due to Herbert A. Simon. *Information and Control*, 4:198–216.

Benoit Mandelbrot. 1961b. On the theory of word frequencies and on related Markovian models of discourse. In R. Jakobson, editor, *Structure of Language and its Mathematical Aspects*, pages 190–219. American Mathematical Society, Providence, RI.

Benoit Mandelbrot. 1961c. Post scriptum to 'final note'. *Information and Control*, 4:300–304.

Christopher Manning and Hinrich Schütze. 1999. *Foundations of Statistical Natural Language Processing*. MIT Press.

Petros Maragos, Ronald W. Schafer, and Muhammad Akmal Butt, editors. 1996. *Mathematical Morphology and Its Applications to Image and Signal Processing*. Kluwer.

Andrei A. Markov. 1912. Answer to P. A. Nekrasov (in Russian). *Mat. Sbornik*, 28:215–227.

André Martinet. 1957. Arbitraire linguistique et double articulation. *Cahiers Ferdinand de Saussure*, 15:105–116.

Yuri Matijasevic. 1981. What should we do having proved a decision problem to be unsolvable? In Andrei P. Ershov and Donald E. Knuth, editors, *Proceedings on Algorithms in Modern Mathematics and Computer Science*, pages 441–448. Springer.

Robert J. McAulay. 1984. Maximum likelihood spectral estimation and its application to narrow-band speech coding. *IEEE Transactions on Acoustics, Speech and Signal Processing*, 32(2):243–251.

John J. McCarthy. 1979. *Formal Problems in Semitic Phonology and Morphology*. PhD thesis, MIT.

John J. McCarthy. 1988. Feature geometry and dependency: A review. *Phonetica*, 45(2–4):84–108.

John J. McCarthy. 2003. OT constraints are categorical. *Phonology*, 20:75–138.

James D. McCawley. 1968. *The Phonological Component of a Grammar of Japanese*. Mouton, The Hague.

James McCloskey. 1979. *Transformational syntax and model-theoretic semantics: a case study in Modern Irish*. Reidel, Dordrecht.

Michael C. McCord. 1990. Slot grammar: A system for simpler construction of practical natural language grammars. In R. Studer, editor, *International Symposium on Natural Language and Logic*, pages 118–145. Springer.

Harry McGurk and John MacDonald. 1976. Hearing lips and seeing voices. *Nature*, 264:746–748.

James B. McMillan. 1977. A controversial consonant. *American Speech*, 52 (1–2):84–97.

Robert McNaughton and Seymour Papert. 1968. The syntactic monoid of a regular event. In M. Arbib, editor, *Algebraic Theory of Machines, Languages, and Semigroups*, pages 297–312. Academic Press.

Robert McNaughton and Seymour Papert. 1971. *Counter-Free Automata*. MIT Press.

George H. Mealy. 1955. A method for synthesizing sequential circuits. *Bell System Technical Journal*, 34:1045–1079.

Igor A. Mel'cuk. 1988. *Dependency Syntax: Theory and Practice*. SUNY Press, Albany, NY.

Igor A. Mel'cuk. 1993–2000. *Cours de morphologie générale*, volume 1–5. Les Presses de l'Université de Montréal.

Jason Merchant. 2001. *The Syntax of Silence: Sluicing, Islands, and the Theory of Ellipsis*. Oxford University Press.

Albert R. Meyer. 1969. A note on star-free events. *Journal of the ACM*, 16(2): 220–225.

Andrei Mikheev. 2002. Periods, capitalized words, etc. *Computational Linguistics*, 28(3):289–318.

George A. Miller. 1957. Some effects of intermittent silence. *American Journal of Psychology*, 70:311–314.

George A. Miller and Noam Chomsky. 1963. Finitary models of language users. In R.D. Luce, R.R. Bush, and E. Galanter, editors, *Handbook of Mathematical Psychology*, pages 419–491. Wiley.

Joanne L. Miller, Raymond D. Kent, and Bishnu S. Atal. 1991. *Papers in Speech Communication: Speech Perception*. Acoustical Society of America, New York.

Philip H. Miller. 1999. *Strong Generative Capacity*. CSLI, Stanford, CA.

Marvin Minsky. 1975. A framework for representing knowledge. In P.H. Winston, editor, *The Psychology of Computer Vision*, pages 211–277. McGraw-Hill.

Marvin Minsky and Seymour Papert. 1966. Unrecognizable sets of numbers. *Journal of the ACM*, 13(2):281–286.

Michael Mitzenmacher. 2004. A brief history of generative models for power law and lognormal distributions. *Internet Mathematics*, 1(2):226–251.

Mehryar Mohri, Fernando C.N. Pereira, and Michael Riley. 1996. Weighted automata in text and speech processing. In András Kornai, editor, *Proceedings of the ECAI-96 Workshop on Extended Finite State Models of Language*, pages 46–50, Budapest. John von Neumann Computer Society.

Friederike Moltmann. 1995. Exception phrases and polyadic quantification. *Linguistics and Philosophy*, 18:223–280.

Richard Montague. 1963. Syntactical treatments of modality, with corollaries on reflexion principles and finite axiomatizability. *Acta Philosophica Fennica*, 16:153–167.

Richard Montague. 1970a. English as a formal language. In R. Thomason, editor, *Formal Philosophy*, volume 1974, pages 188–221. Yale University Press.

Richard Montague. 1970b. Universal grammar. *Theoria*, 36:373–398.

Richard Montague. 1973. The proper treatment of quantification in ordinary English. In R. Thomason, editor, *Formal Philosophy*, pages 247–270. Yale University Press.

Edit A. Moravcsik and Jessica Wirth, editors. 1980. *Current Approaches to Syntax*. Academic Press.

Shunji Mori, Ching Y. Suen, and Kazuhiko Yamamoto. 1992. Historical review of OCR research and development. *Proceedings of the IEEE*, 80:1029–1058.

Andrzej Mostowski. 1957. On a generalization of quantifiers. *Fundamenta Mathematicae*, 44:12–36.

John H. Munson. 1968. Experiments in the recognition of hand-printed text: part I. In *Proceedings of Fall Joint Computer Conference 1125–1138*, Washington, DC. Thompson Books.

John Myhill. 1957. Finite automata and the representation of events. WADC TR-57-624, Wright Patterson Air Force Base, Ohio, USA.

Trygve Nagell, Atle Selberg, Sigmund Selberg, and Knut Thalberg, editors. 1977. *Selected mathematical papers of Axel Thue*. Universitetsforlaget, Oslo.

Anil Nerode. 1958. Linear automaton transformations. *Proceedings of the American Mathematical Society*, 9(4):541–544.

Marina Nespor and Irene Vogel. 1989. On clashes and lapses. *Phonology*, 7:69–116.

Joel Nevis. 1988. *Finnish Particle Clitics and General Clitic Theory*. Garland Press, New York.

Frederick J. Newmeyer. 1980. *Linguistic Theory in America*. Academic Press.

Eugene Nida. 1949. *Morphology: The Descriptive Analysis of Words*. University of Michigan, Ann Arbor.

Masafumi Nishimura and Kazuhide Sugawara. 1988. Speaker adaptation method for HMM-based speech recognition. In *Proceedings of ICASSP88*, pages 207–210.

A. Michael Noll. 1964. Short-time spectrum and cepstrum techniques for vocal-pitch detection. *Journal of the Acoustical Society of America*, 36(2):296–302.

A. Michael Noll. 1967. Cepstrum pitch determination. *Journal of the Acoustical Society of America*, 41(2):293–309.

Harry Nyquist. 1924. Certain factors affecting telegraph speed. *Bell System Technical Journal*, 3:324–346.

David Odden. 1995. Tone: African languages. In J. Goldsmith, editor, *Handbook of Phonology*, pages 445–475. Blackwell.

Richard T. Oehrle. 2000. Context-sensitive node admissibility. *Grammars*, 3:275–293.

Anthony G. Oettinger. 1961. Automatic syntactic analysis and the pushdown store. In R. Jakobson, editor, *Structure of Language and its Mathematical Aspects*, pages 104–129. American Mathematical Society, Providence, RI.

Lawrence O'Gorman. 1993. The document spectrum for page layout analysis. *IEEE Transactions on Pattern Analysis and Machine Intelligence*, 15(11):1162–1173.

Lawrence O'Gorman and Rangachar Kasturi, editors. 1995. *Document Image Analysis*. IEEE Computer Society Press, New York.

Sven E.G. Öhman. 1966. Coarticulation in VCV utterances: spectographic measurements. *Journal of the Acoustical Society of America*, 39:151–168.

Almerindo Ojeda. 1988. A linear precedence account of cross-serial dependencies. *Linguistics and Philosophy*, 11:457–492.

Almerindo Ojeda. 2006a. *Are intensions necessary? Sense as the construction of reference*. ms, UC Davis.

Almerindo Ojeda. 2006b. Discontinuous constituents. In *Elsevier Encyclopedia of Languages and Linguistics*. Elsevier.

Alan V. Oppenheim and Ronald W. Schafer. 1999. *Discrete-Time Signal Processing*. Prentice-Hall.

Nicholas Ostler. 1979. *Case-Linking: a Theory of Case and Verb Diathesis Applied to Classical Sanskrit*. PhD thesis, MIT.

Jaye Padgett. 2002. Feature classes in phonology. *Language*, 78(1):81–110.

Kuldip K. Paliwal and Bishnu S. Atal. 2003. Frequency-related representation of speech. In *Proceedings of Eurospeech 2003*, pages 65–68.

Vilfredo Pareto. 1897. *Cours d'economie politique*. Rouge, Lausanne.

Terence Parsons. 1970. Some problems concerning the logic of grammatical modifiers. *Synthese*, 21(3–4):320–334.

Barbara Partee, Alice ter Meulen, and Robert Wall. 1990. *Mathematical methods in linguistics*. Kluwer.

Francis J. Pelletier and Nicholas Asher. 1997. Generics and defaults. In J. van Benthem and A. ter Meulen, editors, *Handbook of Logic and Language*, pages 1125–1177. North-Holland.

Mati Pentus. 1997. Product-free Lambek calculus and context-free grammars. *Journal of Symbolic Logic*, 62:648–660.

Fernando Pereira. 2000a. The hedgehog and the fox: Language technology and the knowledge of language. In *Fall 2000 Seminar Series*. Johns Hopkins University The Center for Language and Speech Processing, Baltimore.

David M. Perlmutter. 1983. *Studies in Relational Grammar*. University of Chicago Press.

Stanley Peters and Robert W. Ritchie. 1973. On the generative power of transformational grammars. *Information Sciences*, 6:49–83.

Wiebke Petersen. 2004. A mathematical analysis of Pāṇini's sivasutras. *Journal of Logic Language and Information*, 13(4):471–489.

Gordon E. Peterson and Harold L. Barney. 1952. Control methods used in the study of vowels. *Journal of the Acoustical Society of America*, 24:175–184.

Gordon E. Peterson and Frank Harary. 1961. Foundations in phonemic theory. In R. Jakobson, editor, *Structure of Language and its Mathematical Aspects*, pages 139–165. American Mathematical Society, Providence, RI.

Ion Petre. 1999. Parikh's theorem does not hold for multiplicities. *Journal of Automata, Languages and Combinatorics*, 4(1):17–30.

Joseph Picone and George R. Doddington. 1989. A phonetic vocoder. In *Proceedings of the ICASSP-89*, pages 580–583.

Kenneth L. Pike. 1967. *Language in Relation to a Unified Theory of the Structure of Human Behavior*. Mouton, The Hague.

Jean-Eric Pin. 1997. Syntactic semigroups. In G. Rozenberg and A. Salomaa, editors, *Handbook of Formal Language Theory*, volume 1, pages 679–746. Springer.

Steven Pinker. 1984. *Language Learnability and Language Development*. Harvard University Press.

Carl Pollard. 1984. *Generalized Phrase Structure Grammars, Head Grammars, and Natural Language*. PhD thesis, Stanford University.

Carl Pollard and Ivan Sag. 1987. *Information-based Syntax and Semantics. Volume 1. Fundamentals*. CSLI, Stanford, CA.

William J. Poser. 1990. Evidence for foot structure in Japanese. *Language*, 66: 78–105.

Paul M. Postal. 1964. *Constituent Structure*. Mouton, The Hague.

David M.W. Powers. 1998. Applications and explanations of Zipf's law. In D.M. W. Powers, editor, *NEMLAP3/CONLL98: New methods in language processing and Computational natural language learning*, pages 151–160. ACL.

Graham Priest. 1979. The logic of paradox. *Journal of Philosophical Logic*, 8: 219–241.

Graham Priest, Richard Routley, and J. Norman. 1989. *Paraconsistent Logic: Essays on the Inconsistent*. Philosophia-Verlag, Münich.

Alan S. Prince. 1993a. *In Defense of the Number i. Anatomy of a Linear Dynamical Model of Linguistic Generalizations*. Rutgers University Center for Cognitive Science Technical report 1., Piscataway, NJ.

Alan S. Prince and Paul Smolensky. 1993. *Optimality Theory: Constraint Interaction in Generative Grammar*. Rutgers University Center for Cognitive Science Technical Report 2., Piscataway, NJ.

Arthur Prior. 1958. Epimenides the Cretan. *Journal of Symbolic Logic*, 23:261–266.

Arthur Prior. 1961. On a family of paradoxes. *Notre Dame Journal of Formal Logic*, 2:16–32.

Geoffrey K. Pullum and Barbara C. Scholz. 2002. Empirical assessment of stimulus poverty arguments. *The Linguistic Review*, 19:9–50.

William C. Purdy. 1992. A variable-free logic for mass terms. *Notre Dame Journal of Formal Logic*, 33(3):348–358.

James Pustejovsky. 1995. *The Generative Lexicon*. MIT Press.

Hilary Putnam. 1961. Some issues in the theory of grammar. In R. Jakobson, editor, *Structure of Language and its Mathematical Aspects*, pages 25–42. American Mathematical Society, Providence, RI.

M. Ross Quillian. 1969. The teachable language comprehender. *Communications of the ACM*, 12:459–476.

Willard van Orman Quine. 1956. Quantifiers and propositional attitudes. *Journal of Philosophy*, 53:177–187.

Willard van Orman Quine. 1961. Logic as a source of syntactical insights. In R. Jakobson, editor, *Structure of Language and its Mathematical Aspects*, pages 1–5. American Mathematical Society, Providence, RI.

Panu Raatikainen. 1998. On interpreting Chaitin's incompleteness theorem. *Journal of Philosophical Logic*, 27:569–586.

Michael O. Rabin. 1963. Probabilistic automata. *Information and Control*, 6:230–245.

Lawrence R. Rabiner. 1989. A tutorial on Hidden Markov Models and selected applications in speech recognition. *Proceedings of the IEEE*, 77(2):257–286.

Lawrence R. Rabiner and Ronald W. Schafer. 1978. *Digital Processing of Speech Signals*. Prentice-Hall.

Stephen Read. 2002. The Liar paradox from John Buridan back to Thomas Bradwardine. *Vivarium*, 40:189–218.

Alonzo Reed and Brainerd Kellogg. 1878. *An Elementary English Grammar*. Clark and Maynard, New York.

Robert E. Remez. 1979. Adaptation of the category boundary between speech and nonspeech: A case against feature detectors. *Cognitive Psychology*, 11:38–57.

Greg Restall. 2007. *Modal models for Bradwardine's truth*. ms, University of Melbourne.

Nick Riemer. 2006. Reductive paraphrase and meaning: A critique of Wierzbickian semantics. *Linguistics and Philosophy*, 29:347–379.

Jorma Rissanen. 1978. Modeling by the shortest data description. *Automatica*, 14:465–471.

Abraham Robinson. 1966. *Non-standard Analysis*. North-Holland.

Emmanuel Roche and Yves Schabes, editors. 1997. *Finite-State Devices for Natural Language Processing*. MIT Press.

James Rogers. 1998. *A Descriptive Approach to Language-Theoretic Complexity*. CSLI, Stanford, CA.

John Robert Ross. 1969. Guess who? In R. Binnick, A. Davison, G. Green, and J. Morgan, editors, *Papers from the 5th Regional Meeting of the Chicago Linguistic Society*, pages 252–286. Chicago Linguistic Society.

Brian J. Ross. 1995. PAC learning of interleaved melodies. In *IJCAI Workshop on Music and Artificial Intelligence*, pages 96–100.

Pascale Rousseau and David Sankoff. 1978. Advances in variable rule methodology. In David Sankoff, editor, *Linguistic variation: models and methods*, pages 57–69. Academic Press.

Jerzy Rubach. 1997. Extrasyllabic consonants in Polish: Derivational Optimality Theory. In I. Roca, editor, *Constraints and Derivations in Phonology*, pages 551–582. Oxford University Press.

Jerzy Rubach. 2000. Glide and glottal stop insertion in Slavic languages: A DOT analysis. *Linguistic Inquiry*, 31:271–317.

Arto Salomaa. 1973. *Formal Languages*. Academic Press.

Christer Samuelsson. 1996. Relating Turing's formula and Zipf's law. In *Proceedings of the 4th Workshop on Very Large Corpora*, pages 70–78, Copenhagen.

Wendy Sandler. 1989. *Phonological Representation of the Sign: Linearity and Nonlinearity in American Sign Language*. Foris Publications, Dordrecht.

Edward Sapir. 1921. *Language*. Harcourt, Brace and World, New York.

Lawrence K. Saul and Fernando Pereira. 1997. Aggregate and mixed-order Markov models for statistical language processing. In C. Cardie and R. Weischedel, editors, *Proceedings of the Second Conference on Empirical Methods in Natural Language Processing*, pages 81–89, Somerset, NJ. ACL.

Ferdinand de Saussure. 1879. *Mémoire sur le système primitif des voyelles dans les langues indo-européennes*. Teubner, Leipzig.

Paul Schachter. 1976. The subject in philippine languages: topic, actor, actor-topic, or none of the above. In Charles Li, editor, *Subject and topic*, volume 3, pages 57–98. Academic Press.

Ronald W. Schafer and Lawrence R. Rabiner. 1975. Digital representation of speech signals. *Proceedings of the IEEE*, 63(4):662–667.

Roger C. Schank and Robert P. Abelson. 1977. *Scripts, Plans, Goals and Understanding: An Inquiry into Human Knowledge Structures*. Lawrence Erlbaum, Hillsdale, NJ.

Roger C. Schank. 1972. Conceptual dependency: A theory of natural language understanding. *Cognitive Psychology*, 3(4):552–631.

Roger C. Schank. 1973. *The Fourteen Primitive Actions and Their Inferences*. Stanford AI Lab Memo 183.

Annette Schmidt. 1985. The fate of ergativity in dying Dyirbal. *Language*, 61: 278–296.

Moses Schönfinkel. 1924. On the building blocks of mathematical logic. In J. van Heijenoort, editor, *From Frege to Gödel: A Source Book in Mathematical Logic 1879–1931*, pages 355–366. Harvard University Press.

Manfred R. Schroeder. 1977. Recognition of complex acoustic signals. In T.H. Bullock, editor, *Life Sciences Research Report*, pages 323–328. Abakon Verlag, Berlin.

Marcel Paul Schützenberger. 1960. Un probleme de la théorie des automates. *Publications Seminaire Dubreil-Pisot*, 13(3).

Marcel Paul Schützenberger. 1965. On finite monoids having only trivial subgroups. *Information and Control*, 8:190–194.

Fabrizio Sebastiani. 2002. Machine learning in automated text categorization. *ACM Computing Surveys*, 34(1):1–47.

Thomas Sebeok and Francis Ingemann. 1961. *An Eastern Cheremis Manual*. Indiana University Press, Bloomington.

Jerry Seligman and Lawrence S. Moss. 1997. Situation theory. In J. van Benthem and A. ter Meulen, editors, *Handbook of Logic and Language*, pages 239–309. North-Holland, Amsterdam.

Elisabeth O. Selkirk. 1984. On the major class features and syllable theory. In Mark Aronoff and Richard T. Oehrle, editors, *Language Sound Structure*. MIT Press.

Andrew Senior and Krishna Nathan. 1997. Writer adaptation of a HMM handwriting recognition system. In *Proceedings of ICASSP '97*, volume 2, pages 1447–1450, New York. IEEE Computer Society Press.

Jean Serra. 1982. *Image Analysis and Mathematical Morphology*. Academic Press.

Mehmet Sezgin and Bülent Sankur. 2004. Survey over image thresholding techniques and quantitative performance evaluation. *Electronic Imaging*, 13(1):146–165.

Petr Sgall, Eva Hajicova, and Jarmila Panevová. 1986. *The Meaning of the Sentence and Its Semantic and Pragmatic Aspects*. Reidel/Academia, Dordrecht/Prague.

Ben J. Shannon and Kuldip K. Paliwal. 2003. A comparative study of filter bank spacing for speech recognition. In *Proceedings of the Microelectronic Engineering Research Conference*.

Claude E. Shannon. 1948. A mathematical theory of communication. *Bell System Technical Journal*, 27:379–423, 623–656.

Claude E. Shannon and Warren W. Weaver. 1949. *The Mathematical Theory of Communication*. University of Illinois Press, Urbana.

Stuart M. Shieber. 1986. *An Introduction to Unification-Based Approaches to Grammar*. CSLI, Stanford, CA.

Stuart M. Shieber. 1992. *Constraint-Based Grammar Formalisms*. MIT Press.

Takeshi Shinohara. 1990. Inductive inference from positive data is powerful. In *Proceedings of the 1990 Workshop on Computational Learning Theory*, pages 97–101, San Mateo, CA. Morgan-Kaufmann.

Herbert A. Simon. 1955. On a class of skew distribution functions. *Biometrika*, 42:425–440.

Herbert A. Simon. 1960. Some further notes on a class of skew distribution functions. *Information and Control*, 3:80–88.

Herbert A. Simon. 1961a. Reply to Dr. Mandelbrot's post scriptum. *Information and Control*, 4:305–308.

Herbert A. Simon. 1961b. Reply to 'final note' by Benoit Mandelbrot. *Information and Control*, 4:217–223.

Charles C. Sims. 1994. *Computation with Finitely Presented Groups*. Cambridge University Press.

Daniel Sleator and David Temperley. 1993. Parsing English with a link grammar. In *Third International Workshop on Parsing Technologies*, pages 277–291.

Henry Smith. 1996. *Restrictiveness in Case Theory*. Cambridge University Press.

Ray J. Solomonoff. 1960. *A Preliminary Report on a General Theory of Inductive Inference*. Report ZTB-138, Zator Co., Cambridge, MA.

Ray J. Solomonoff. 1964. A formal theory of inductive inference. *Information and Control*, 7:1–22, 224–254.

Harold L. Somers. 1987. *Valency and case in computational linguistics*. Edinburgh University Press.

Andrew Spencer and Arnold M. Zwicky. 1998. *The Handbook of Morphology*. Blackwell.

Richard Sproat. 2000. *A Computational Theory of Writing Systems*. Cambridge University Press.

Anekal Sripad and Donald L. Snyder. 1977. A necessary and sufficient condition for quantization error to be uniform and white. *IEEE Transactions on Acoustics, Speech, and Signal Processing*, 25(5):442–448.

Frits Staal. 1962. A method of linguistic description: The order of consonants according to Pāṇini. *Language*, 38:1–10.

Frits Staal. 1967. *Word Order in Sanskrit and Universal Grammar*. Reidel, Dordrecht.

Frits Staal. 1982. *The Science of Ritual*. Bhandarkar Oriental Research Institute, Pune, India.

Frits Staal. 1989. *Rules Without Meaning. Ritual, Mantras and the Human Sciences*. Peter Lang, New York.

Mark Steedman. 2001. *The Syntactic Process*. MIT Press.

Kenneth N. Stevens and Sheila E. Blumstein. 1981. The search for invariant acoustic correlates of phonetic features. In P. Eimas and J. Miller, editors, *Perspectives on the Study of Speech*, pages 1–38. Lawrence Erlbaum, Hillsdale, NJ.

Stanley S. Stevens and John Volkman. 1940. The relation of pitch to frequency: A revised scale. *American Journal of Psychology*, 53:329–353.

Patrick Suppes. 1970. Probabilistic grammars for natural languages. *Synthese*, 22:95–116.

Patrick Suppes. 1973. Semantics of context-free fragments of natural languages. In J. Hintikka, J. Moravcsik, and P. Suppes, editors, *Approaches to Natural Language*, pages 370–394. Reidel, Dordrecht.

Patrick Suppes and Elizabeth Macken. 1978. Steps toward a variable-free semantics of attributive adjectives, possessives and intensifying adverbs. In K.E. Nelson, editor, *Children's Language*, volume 1, pages 81–115. Gardner Press, New York.

Henry Sweet. 1881. Sound notation. *Transactions of the Philological Society*, 2:177–235.

Anna Szabolcsi. 1987. Bound variables in syntax – are there any? In J. Gronendijk, M. Stokhof, and F. Veltman, editors, *Proceedings of the 6th Amsterdam Colloquium*, pages 331–351, Amsterdam. Institute for Language, Logic, and Information.

Róbert Szelepcsényi. 1987. The method of forcing for nondeterministic automata. *Bulletin of the European Association for Theoretical Computer Science*, 33:96–100.

Pasi Tapanainen and Timo Järvinen. 1997. A non-projective dependency parser. In *Proceedings of the 5th Conference on Applied Natural Language Processing*, pages 64–71.

Alfred Tarski. 1956. The concept of truth in formalized languages. In A. Tarski, editor, *Logic, Semantics, Metamathematics*, pages 152–278. Clarendon Press, Oxford.

Lucien Tesniére. 1959. *Élements de syntaxe structurale*. Klincksieck, Paris.

Wolfgang Thomas. 2003. *Applied Automata Theory*. ms, RWTH Aachen.

Richmond H. Thomason, editor. 1974. *Formal Philosophy: Selected papers of Richard Montague*. Yale University Press.

Richmond H. Thomason. 1977. Indirect discourse is not quotational. *The Monist*, 60:340–354.

Richmond H. Thomason. 1980. A note on syntactical treatments of modality. *Synthese*, 44:391–395.

Richmond H. Thomason. 1997. Nonmonotonicity in linguistics. In J. van Benthem and A. ter Meulen, editors, *Handbook of Logic and Language*, pages 777–832. North-Holland, Amsterdam.

Höskuldur Thráinsson. 1978. On the phonology of Icelandic preaspiration. *Nordic Journal of Linguistics*, 1:3–54.

Axel Thue. 1906. Über unendliche Zeichenreihen. *Kra. Vidensk. Selsk. Skrifter. I. Mat. Nat. Kl.*, 7:1–22.

Axel Thue. 1914. Probleme über Veranderungen von Zeichenreihen nach gegeben Regeln. *Skr. Vid. Kritiania, I. Mat. Naturv. Klasse*, 10.

Boris Trakhtenbrot and Yan Barzdin. 1973. *Finite Automata: Behavior and Synthesis*. North-Holland, Amsterdam.

Oivind Due Trier, Anil K. Jain, and Torfinn Taxt. 1996. Feature extraction methods for character recognition – a survey. *Pattern Recognition*, 29(4):641–662.

Nikolai Sergeevich Trubetskoi. 1939. *Grundzüge der Phonologie*. Vandenhoeck and Ruprecht, Göttingen.

Alan Turing. 1950. Computing machinery and intelligence. *Mind*, 59:433–460.

Zsolt Tuza. 1987. On the context-free production complexity of finite languages. *Discrete Applied Mathematics*, 18:293–304.

Robert M. Vágó. 1976. Theoretical implications of Hungarian vowel harmony. *Linguistic Inquiry*, 7(2):243–263.

Leslie G. Valiant. 1984. A theory of the learnable. *Communications of the ACM*, 27(11):1134–1142.

Robert Van Valin. 2005. *A Summary of Role and Reference Grammar*. ms, SUNY Buffalo.

Michiel van Lambalgen. 1989. Algorithmic information theory. *Journal of Symbolic Logic*, 54:1389–1400.

Dick C. van Leijenhorst and Theo P. van der Weide. 2005. A formal derivation of Heaps' Law. *Information Sciences*, 170(2–4):263–272.

Jan P.H. van Santen, Richard W. Sproat, Joseph P. Olive, and Julia Hirschberg, editors. 1997. *Progress in Speech Synthesis*. Springer.

Zeno Vendler. 1967. *Linguistics and Philosophy*. Cornell University Press, Ithaca, NY.

Henk J. Verkuyl. 1993. *A Theory of Aspectuality*. Cambridge University Press.

K. Vijay-Shanker, David J. Weir, and Aravind K. Joshi. 1987. Characterizing structural descriptions produced by various grammatical formalisms. In *Proceedings of ACL'87*, pages 82–93.

Alessandro Vinciarelli and Samy Bengio. 2002. Writer adaptation techniques in off-line cursive word recognition. In *Proceedings of the 8th IWFHR*, pages 287–291.

Albert Visser. 1989. Semantics and the Liar paradox. In D. Gabbay and F. Guenthner, editors, *Handbook of Philosophical Logic*, volume IV, pages 617–706. D. Reidel, Dordrecht.

Walther von Dyck. 1882. Gruppentheoretische Studien. *Mathematische Annalen*, 20:1–44.

Karl von Frisch. 1967. *Dance Language and Orientation of Bees*. Belknap Press, Cambridge, MA.

Atro Voutilainen. 1994. *Designing a Parsing Grammar*. Technical Report 22, Department of General Linguistics, University of Helsinki.

Hisashi Wakita. 1977. Normalization of vowels by vocal tract length and its application to vowel identification. *IEEE Transactions on Acoustics, Speech and Signal Processing*, 25(2):183–192.

James S. Walker. 2003. An elementary resolution of the Liar paradox. *The College Mathematics Journal*, 35:105–111.

Dacheng Wang and Sargur N. Srihari. 1989. Classification of newspaper image blocks using texture analysis. *Computer Vision, Graphics and Image Processing*, 47:327–352.

Hao Wang. 1960. Proving theorems by pattern recognition – I. *Communications of the ACM*, 3(4):220–234.

William S.-Y. Wang. 1969. Competing changes as a cause of residue. *Language*, 45:9–25.

William S.-Y. Wang and Chin-Chuan Cheng, editors. 1977. *The Lexicon in Phonological Change*. Mouton, The Hague.

Raymond L. Watrous. 1991. Current status of Peterson-Barney vowel formant data. *Journal of the Acoustical Society of America*, 89(5):2458–2459.

Bo Wei and Jerry D. Gibson. 2000. *Comparison of distance measures in discrete spectral modeling*. IEEE Digital Signal Processing Workshop, Hunt, TX.

David J. Weir. 1992. Linear context-free rewriting systems and deterministic tree-walking transducers. In *Proceedings of ACL'92*, pages 136–143.

Roulon S. Wells. 1947. Immediate constituents. *Language*, 23:321–343.

William Welmers. 1959. Tonemics, morphotonemics, and tonal morphemes. *General Linguistics*, 4:1–9.

Janet F. Werker and Richard C. Tees. 1984. Cross-language speech perception: Evidence for perceptual reorganization during the first year of life. *Infant Behavior and Development*, 7:49–63.

Kenneth Wexler and Peter W. Culicover. 1980. *Formal Principles of Language Acquisition*. MIT Press.

William Dwight Whitney. 1887. *Sanskrit Grammar*. Harvard University Press, Cambridge MA.

Edmund T. Whittaker. 1915. On the functions which are represented by the expansions of the interpolation theory. *Proceedings of the Royal Society of Edinburgh Section A*, 35:181–194.

Bernard Widrow. 1960. Statistical analysis of amplitude quantized sampled-data systems. *Trans. AIEE (II: Applications and Industry)*, 79:555–568.

Anna Wierzbicka. 1972. *Semantic Primitives*. Athenäum, Frankfurt.

Anna Wierzbicka. 1980. *The case for surface case*. Karoma, Ann Arbor.

Anna Wierzbicka. 1985. *Lexicography and conceptual analysis*. Karoma, Ann Arbor.

Anna Wierzbicka. 1992a. *Semantics, culture, and cognition: Universal human concepts in culture-specific configurations*. Oxford University Press.

Heike Wiese. 2003. *Numbers, Language, and the Human Mind*. Cambridge University Press.

John Wilkins. 1668. *Essay toward a real character, and a philosophical language*. Royal Society, London.

Karina Wilkinson. 1988. Prosodic structure and Lardil phonology. *Linguistic Inquiry*, 19:325–334.

Edwin Williams. 1976. Underlying tone in Margi and Igbo. *Linguistic Inquiry*, 7:463–84.

John C. Willis. 1922. *Age and Area*. Cambridge University Press.

Terry Winograd and Fernando Flores. 1986. *Understanding Computers and Cognition: A New Foundation for Design*. Ablex, Norwood, NJ.

Moira Yip. 2002. *Tone*. Cambridge University Press.

Anssi Yli-Jyrä. 2003. Describing syntax with star-free regular expressions. In *Proceedings of the EACL*, pages 379–386.

Anssi Yli-Jyrä. 2005. Approximating dependency grammars through intersection of star-free regular languages. *International Journal of Foundations of Computer Science*, 16(3):565–580.

Anssi Yli-Jyrä and Kimmo Koskenniemi. 2006. Compiling generalized two-level rules and grammars. In *Proceedings of the 5th International Conference on NLP (FinTAL 2006)*, pages 174–185. Springer.

Victor H. Yngve. 1961. The depth hypothesis. In R. Jakobson, editor, *Structure of Language and its Mathematical Aspects*, pages 130–138. American Mathematical Society, Providence, RI.

G. Udny Yule. 1924. A mathematical theory of evolution. *Philosophical Transactions of the Royal Society*, B213:21–87.

Tania S. Zamuner, LouAnn Gerken, and Michael Hammond. 2005. The acquisition of phonology based on input: A closer look at the relation of cross-linguistic and child language data. *Lingua*, 115(10):1403–1426.

Frits Zernike. 1934. Beugungstheorie des Schneidenverfahrens und seiner verbesserten Form, der Phasenkontrastmethode. *Physica*, 1:689–704.

George K. Zipf. 1935. *The Psycho-Biology of Language; an Introduction to Dynamic Philology*. Houghton Mifflin, Boston.

George K. Zipf. 1949. *Human Behavior and the Principle of Least Effort*. Addison-Wesley.

Eberhard Zwicker, Georg Flottorp, and Stanley S. Stevens. 1957. Critical bandwidth in loudness summation. *Journal of the Acoustical Society of America*, 29:548–557.

Arnold M. Zwicky. 1985. Clitics and particles. *Language*, 61:283–305.

Index

Only significant mentions are indexed. Entries in **bold** give the definition.

μ-law, 203
ω-word, 6
$a^n b^n$, 40
k-string, **35**
śivasūtras, 29
do-support, 165

A-law, 203
ablaut, 61
aboutness, 212
abstract phonemes, 28
abstract segment, **28**
acceptance, **16**
accommodation, 157
accusative, 90, 104
accusativus cum infinitivo, ACI, 101
Acehnese [ACE], 103
action, **134**
active, 89
adadjective, 84
adjectival phrase, AP, 85
adjective, 66, 84
adjunct, 102, 166
adverbial, 66
adverbial phrase, AdvP, 85
affix, **61, 62**
affixation, 61
affricate, 42
agglutinating languages, 62
agreement, 10, **91**
aktionsart, 107
algebra, 151
all-pole filter, 223
alphabet, **17**
ambisyllabicity, 54
amplitude, 33
antara, 30
anubandha, 28, 29
anuvṛtti, 191
Arabic, 57, 60, 63
  Cairene [ARZ], 58
Aramaic [CLD], 53
Araucanian [ARU], 58
argument, 166
artificial intelligence, AI, 106, 211, 248
ascender, 209
aspect, 63
aspirated, 28
assimilation
  across word boundary, 61
  anticipatory, 233
  nasal, 31
  perseveratory, 233
  tonal, 34
  voice, 192
association, 48
association line, 37
association relation, 42
atelic, 107
autosegmental theory, 34
axiomatic method, vii

Büchi automaton, 6
baby talk, 195
bag of words, 216
bandpass filter, 224
barber paradox, 143
bark scale, 226
bee dance, 27
Berber [TZM], 53
Berry paradox, 6, 145
biautomaton, **43**
bilanguage, **39**
bilattice, **158**
binarization, 238
bistring, 34, **35**
  b-catenation, 39
  alignment, 39
  concatenation, 39
  fully associated, **37**
  proper, **37**
  t-catenation, 39
  well-formed, **35**
blank, **35**
block coding, 184
blocking, 65
blocking off, **115**
bound form, 25
bound morpheme, **61**
boundary, **241**
boundary tone, 53
bounded counter, **113**
bounding box, 240
bracketing, 86
  improper, 54
Brahmi script, 54
broadcast quality, 203

c-categorial grammar, 170
cancellation, **82**
cascade product, **135**
case, 63, 90
Catalan numbers, 127
categorial grammar, **83**
  bidirectional, 14
  unidirectional, 14, **83**
categorical perception, 194
category system, **170**
central moment, **242**
cepstral means normalization, CMN, 238
cepstrum, 226
certificate, 9, **14**
channel symbol, 229
character recognition, 201
character size, 241
characters
  English, 10
  Han, 201
  Hangul, 201
Cheremis
  Eastern [MAL], 58
Cherokee [CER], 54
child-directed speech, CDS, 195
Chinese, 53, 232
Chinese (Mandarin), 34
Chomsky hierarchy, 20
Chomsky normal form, 16
citation form, 63
class node, 31
clipping error, 202
clitic, 57, 60
closing, **241**
clustering
  supervised, 230
  unsupervised, 229
coarticulation, 233
coda, 54
code
  Huffman, 180
  Morse, 23
  scanning, **44**
  Shannon-Fano, 180
codebook, **229**
coder, 44
colon, 56
combinatory categorial grammar, CCG, 5, 88
communications quality, 203
competence, 21, 27
composition (of weighted transductions), **123**
compositionality, 61, 149, 177
compound stress rule, 33
compounding, 60
concatenation, **17**
conceptual dependency, CD, **108**
concord, 91
connected component, 240
connected speech, 28
consonant, 42, 54

conspiracy, 11
constituents, 86, 88
constraint ranking, 67
construction, 110, **154**
construction grammar, 110, 248
containment, **26**
context-free grammar, CFG, 10, **18**
extended, 87
context-free language, CFL, **18**
context-sensitive grammar, CSG, 9
context-sensitive language, CSL, **19**, 165
contour tone, 34
control verb, 96
controversies, 2
convention, 23, 192
Stayman, 23
convolution, **224**
coreference, 92
corpus, **69**
corpus size, **69**, 213
count vector, **127**
cover, **80**
cross entropy, **185**
cybernetics, 1, 248
Cypriot syllabary, 54

data
dialectal, 21
historical, 21
dative, 90
decibel, 203
decidability, 17
deep case, 166
definition, 4
extensive, 4
intensive, 4
ostensive, 4
degree, 63
deletion, 48
delinking, 48
delta features, 236
density
Abel, **114**
Bernoulli, **117**
combinatorial, **117**
natural, **114**
dependency grammar, **92**
dependent, 85
derivation, **64**
derivational history, 10
descender, 209
descriptive adequacy, 140
descriptivity, 16, 156
deterministic FST, **41**, **124**
dhātupāṭha, 167, 248
diachrony, 20
diacritic, **28**, 65
dilation, **241**
diphthong, 42
direct product, **80**
disambiguation, 24, 152
discontinuous constituents, 88
discourse representation theory, DRT, 108, 164
discovery procedure, 26, 133, 195
discrete Fourier transform, DFT, 225
dissimilation, 193
distinctive features, **30**
distribution, **79**
ditransitive, **80**
divisor, **80**
document classification, 212
document frequency, **212**
double articulation, 53
downdrift, 237
downsampling, 227, 238
duality of patterning, 53
Dutch, 97
Dyirbal [DBL], 106

edge marker, #, 19
elsewhere, 66
empty string, λ, **17**
enclitic, 60
endocentric construction, 85
energy spectrum, **225**
English, 77, 79, 81, 98
American, 21
British, 21
general, 212
Modern, 21
Old, 21
entity, 159
entropy, **180**
epenthesis, 67
equality, 10
erosion, **241**
ETX, 184

Euclid, vii
exocentric construction, 85
explanatory adequacy, 140
extension, 152
extrametricality, 54

face recognition, 201
fast Fourier transform, FFT, 226
feature assignment, **30**
feature bundle, 30
feature decomposition, **30**
feature detection, 194, 209
feature geometry, 31
Fibonacci numbers, 13, 18, 20, 179
filter, **220**
  causal, **220**
  stable, **220**
filter bank, 224
final state, **115**
fingerprint recognition, 201
finite autosegmental automaton, 44
finite index, **41**
finite state automaton, FSA, **41**, 184
  deterministic, DFSA, **41**
  weighted, **124**
finite state transducer, FST, 40, **41**
  deterministic, **124**
  length-preserving, **124**
  probabilistic, **124**
first order language, FOL, 12
first order Markov process, **206**
floating element, **35**
font weight, 241
foot, 53, 56
form, 23
formal universal, 32
formant, 205
fragment, 15
free form, 25
free group, **12**
free monoid, **12**
free morpheme, **61**
French, 58
frequency, 33, **70**
frequency response, **220**
frequency spectrum, 221
fricative, 29
function word, 213

G.711, 217
G3, **239**
gaṇapāṭha, 248
garden path, 132
gender, 63
generalized phrase structure grammar, GPSG, 98
generalized quantifiers, 153
generation, 9
  direct, **10**
  free, **16**
generative capacity, **122**
generative grammar, 9
generative semantics, 108
generic, 153
genetic code, 27
genitive, 90, 100
geometrical probability distribution, **189**
German, 103
gloss, 17, 64
government, **91**
grammar, 9
grammaticality, ix, 15
Greek, 63
Greibach normal form, 233
group-free, 135
Gunwinggu [GUP], 63

Hamming window, 224
handprint, 241
hangman paradox, 148
head, 85
Head Feature Convention, HFC, 91
head grammar, 5, 88
Heaps' law, 76
Herbrand model, 12
Herdan's law, 72
hiatus, 11
hidden Markov model, HMM, **129**, **208**
  continuous density, 208
  diagonal, 237
  run, 130
  tied mixture, 237
high (vowel feature), 31
high tone, H, 34
highpass filter, 224
Hindi, 90
Hiragana, 54
homomorphism, **80**, 151

homorganic, 31
Hungarian, 58, 62, 64, 90, 171
hyperintensionals, 152

identification in the limit, iitl, **190**
image skeleton, **242**
immediate constituent analysis, ICA, 86
immediate dominance/linear precedence, IDLP, 98
improper parentheses, 54
impulse, **220**
impulse response, **220**
incorporating languages, 62
incorporation, 61
indexed grammar, 5
indexical, 164
infinitesimals, 5
infix, 61
inflecting languages, 62
inflection, **64**
informant judgments, 33
information extraction, 211
information gain, **185**
insertion, 48
instrumental, 90
integers, base one, **17**
intension, 152
interpretation relation, 24
intransitive, **80**
Inuktitut [ESB], 63
inverse document frequency, IDF, **214**
Irish, 103
Itakura-Saito divergence, **231**

Japanese, 103

kāraka, 166
Karhunen-Loève transformation, **243**
Kikuria [KUJ], 56
Kiribati [GLB], 56
Kleene closure, λ-free, $^+$, **17**
Kleene closure, *, **17**
knowledge representation, KR, 106, 211
Korean, 31
Kullback-Leibler (KL) divergence, **185**
Kwakiutl [KWK], 60

labeled bracketing, **87**
labial, 29
lambda move, 41
language, **18**
formal, **18**
noncounting, **133**
probabilistic, **112**
regular, **18**
weighted, **112**
language acquisition, 21
language of thought, 106
Lardil [LBZ], 56
Latin, 56, 63, 90, 91, 101
lattice, 158
left congruence, **41**
left factor, **17**
length, **17**
length distribution, **126**
length-preserving FST, **41**, **124**
lexeme, **61**, 63
lexical category, 66, 79, 80
lexical functional grammar, LFG, 89
lexical phonology and morphology, LPM, 50, 232
lexicality, 84
Liar paradox, 6, **142**, 143
linear bounded automaton, LBA, **19**
linear discriminant analysis, LDA, 244
linear machine, 216
linear predictive coding, LPC, 223
link grammar, 15
Linux, 119
Lloyd's algorithm (LGB algorithm), 230
locative, 90
log PCM, 203
logic
four-valued, 163
multivalued, 15
paraconsistent, 150
Lomongo [LOL], 34
low (vowel feature), 31
low tone, L, 34
lowpass filter, 224

Macedonian [MKJ], 58
machine learning, 211
machine learning features, 122
machine translation, 211
major class, 54
markedness, 30, 68, 192
Markov, vii, 7, 206

mathematical linguistics, 1
mathematical morphology, MM, 241
maximal projection, 85
maximum likelihood linear regression, MMLR, 238
McGurk effect, 210
Mealy FSA, 41
mean opinion scale, MOS, 203
meaning, 23
mel scale, 225
melody, **35**
membership problem, 9, 20
metarule, 192
mid tone, M, 34
MIL, **92**
minimal sufficient statistic, **191**
minimum description length, MDL, 191
minimum form, 25
model structure, **159**
model theory, 15
modification, 92
Mohawk [MOH], 88
monostratal theory, 10
Montague grammar, MG, 15, 151
mood, 63
Moore FSA, 41
mora, 53, **55**
morpheme, 3, 60
morphosyntactic feature, 63
morphotactics, 53, 61, 185
motherese, 195
multistratal theory, 10
mutual information, **185**
Myhill-Nerode equivalence, **41**

n line, 209
nasality, 32, 50
natural class, 29, **30**
natural kind, 161
nearest neighbor classifier, 230
neogrammarians, 21
neutralization, 204
no crossing constraint, NCC, 35
nominative/accusative case pattern, 90
nonconstituent coordination, NCC, 169
noncounting, 11, **133**, 197
nonintersective modifier, 143
nonterminal, **18**
normativity, 16, 156
noun, 66
Ntlaka'pamux [THP], 195
nucleus, 54
number, 63
numeral, 66

oblique, 90
odds, **121**
onomatopoeic words, 23
onset, 54
opacity, semantic, **144**
opacity, phonological, 69
opacity, syntactic, 89
opening, **241**
operation, 151
optical character recognition, OCR, 219
overestimation error, **119**
overgeneration, **15**, 33

Pāṇini, vii, 8, 28–30, 57, 66, 68, 76, 94, 100, 167, 179, 191, 248
page decomposition, 238
palindrome, 18
paradigm, **64**
paradox, 143
parataxis, 98
parenthetical expression, 119
parochial rule, 5, 193
part of speech, POS, 66, 79, 80
partial similarity, 24
particle, **62**
passive, 89
passive transformation, 89
pattern recognition, 201
pause, 25, 28
performance, 27
performative, 147
person, 63
phonemic alphabet, 26
phonetic interpretation, 28
phonetic realization, 28
phonotactics, 53, 185
phrase, **85**
pitch period, 224
place of articulation, 31
POS tag, 81
possession, 92
postulate, 26
poverty of stimulus, 200

power spectrum, 225
pratyāhāra, 29
preaspiration, 50
prefix, **18**, **61**
prefix complexity, 188
prefix-fee code, **180**
pregroup grammar, 138
prepositional phrase, PP, 85
preterminal, 81
principal component analysis, PCA, **243**
principles and parameters, 103
probabilistic finite state automaton, PFSA, **123**
probably approximately correct, pac, 190, **198**
process
  transitive, 206
proclitic, 60
production, 11, **18**
projection, 85
projection profile, 240
pronoun
  resumptive, 164
propositional attitude, 144
prosodic word, 53, 233
pulse code modulation, PCM, 203
push-down automaton, PDA, 184

quefrency, 226
query parsing, 212
question answering, 211
quinphone, 236

raising verb, 96
rank, **70**, **134**
reconstruction error, 229
rectification of names, 2
reflection, **241**
regular relation, 40
relation, 159
relational grammar, **92**
reproduction alphabet, **229**
reset semigroup, **134**
reversal, **18**, 39
rewrite rule, **18**
rhyme, 55
Richard paradox, 6
right congruence, **41**
right factor, **17**
rigid designator, 161
ritual, 108
Romanian, 171
root, **62**
ruki, 29, 233
rule, 10
  production, 11
  WFC, 11
rule to rule hypothesis, 149
run-length encoding, RLE, 239
Russian, 90

sāvarṇya, 31
Saho [SSY], 60
Sanskrit, 57, 233
Sapir-Whorf hypothesis, 49, 103
satisfaction, 10
saturation, **117**
scanning independence, 43
segment, **28**
segmentation, 210
selectional restriction, 81
self-delimiting
  code, **187**
  function, **186**
semiautomaton, 115
sentence diagram, 92
sentential form, 10
sideband, 223
sign, 23, **169**
signal, **220**
  bandlimited, **222**
  quasistationary, 224
  stationary, 224
  stationary in the wide sense, 224
signal energy, 203
signal process, **206**
  ergodic, 206
  stationary, 206
  stochastic, 206
signal to quantization noise ratio, SQNR, 203
silence, 28
sluicing, 175
sociolinguistics, 120
sonority hierarchy, **55**
sound change, 21
sound law, 28
source-filter model, 226

span, **35**
speaker adaptation
  CMN, 238
  MMLR, 238
  VTLN, 238
speech recognition, 40
split systems, 56, 103
spreading, 39
square-free word, **197**
start symbol, S, **18**
state machine, 115
stem, **62**
stratum, 10, 92
stress, 33, 53
strict category, **80**
string, **17**
stringset, **18**
stroke width, 241
structural change (of CS rule), **19**
structural decomposition, 242
structural description (of CS rule), **19**
structuring element, 241
STX, 184
subband, 225
subdirect product, **30**
substantive universal, 32
substructure, **80**
suffix, **61**
superfiniteness, **197**
suppletion, 63
suprasegmentals, 33
surface form, 28, 40
Swahili, 63
syllabic alphabet, 53
syllable, 53, 54
SYN, 184
syncategoremata, 80
synchrony, 20
syntactic congruence, **41**, **137**
syntactic monoid, 79
syntactic semigroup, 79
synthetic quality, 203
system, 220

T-scheme, 143
tagmemics, 92
tagset, 81
telic, 107
telltale, **197**
template, 60
tense, 63
terminal, **18**
text frequency, **212**
text to speech, TTS, 40
thematic role, 99
theta role, 99
tier, **34**
token, 11
toll quality, 203
tone, 33, 50
topic, 63, 212
topic detection, 212
topic model, **212**
trace, 111, 169
transducer
  weighted, 123
transfer function, 220
transformation group, 134
transformation monoid, 134
transformation semigroup, 134
transient state, **115**
transition, 41
  weighted, 123
transitive, **80**
tree adjoining grammar, TAG, 5, 88
trigger, 29
triphone, 236
truth, 149
  analytic, 155
  by T-scheme, 143
  data label, 131, 230
  in axiomatic systems, 15
  near-, 51
truth condition, 152, 167
truth value, 15, 107, 158
Tsez [DDO], 90
Turkish, 63
type, 11
type lifting, 133
type-logical grammar, 138

umlaut, 61
unaspirated, 28
underestimation error, **119**
undergeneration, **15**
underlying form, 28, 40, 63
underspecification, 135
ungrammaticality, ix, 17

unrecoverable, 105
unification grammar, 14
universal grammar, UG, 5, 68, **191**, 193
universal probability distribution, **189**
universality, 144
universe, **159**
unmarked member of opposition, **63**
unvoiced, 30
utterance, 23

valence, 94, **102**, 105
Vapnik-Chervonenkis (VC) dimension, 199
variable binding term operator, VBTO, **164**
variable rule model, **121**
variable rules, 5
vector quantization, VQ, **229**
verb, 66
verb phrase, VP, 85
vertex chain code, 242
Vietnamese, 63
Viterbi algorithm, 130
vocal tract length normalization, VTLN, 238
vocative, 90
voice, 63
voiced, 30
vowel, 31, 42, 54
vowel harmony, 50
vowel height, 31
vulnerability, **115**

Wang tiling, **11**
Warao [WBA], 58
Warlpiri [WBP], 104
weight, **112**
well-formedness condition (WFC), 11
Weri [WER], 59
word, 25, 60, **61**
fragment, **26**
frequency, 70
word problem, 20
word stress, 57
WWW, 212

Yokuts [YOK], 60

z transform, **220**
Zernike moment, **243**
Zipf's law, 70, 182